A Law of Blood-ties - The 'Right' to Access Genetic Ancestry

Alice Diver

A Law of Blood-ties - The 'Right' to Access Genetic Ancestry

Springer

Alice Diver
School of Law
University of Ulster
Londonderry
United Kingdom

ISBN 978-3-319-01070-0 ISBN 978-3-319-01071-7 (eBook)
DOI 10.1007/978-3-319-01071-7
Springer Cham Heidelberg New York Dordrecht London

Library of Congress Control Number: 2013947944

Printed on acid-free paper

Springer is part of Springer Science+Business Media (www.springer.com)

For my family

Foreword

Dr Alice Diver has produced an impressive study examining a subject that is integral to every social order and yet remains highly controversial and sensitive. Throughout human evolution, genetic kinship and natal parenthood have been viewed as the natural basis for the family. By contrast, social kinship and familial relationships, developed through such techniques as adoption, have been considered inferior and representing less-respected bonding. As Diver eloquently explains, the emphasis on developing ties akin to biological or genetic kinship has been so strong that even when adoptions have taken place, there have been equally strong efforts to remove memories of previous relationships and bonds. In the process of erasing past identities and relationships, legislative and administrative machinery of states have deployed various strategies, including permanently sealing original birth records, placing a ban on kin contact, and encouraging the renaming of the child. While generating a sense of assurance and security in the newly established bonds, these arrangements systematically exclude the involvement of the genetic parents in key decisions related to their biological offspring. In lamenting the practices of many contemporary societies and while expressing concerns over attempts to hide genetic kinship, Diver notes that '[t]he need for genetic identity cannot simply be ignored; origin deprived persons should not be expected to simply develop good "coping mechanisms" and quietly accept that they will never have a right to access their ancestry, or that such a right must always become weightless in law when set against the privacy right of the other triad members. ...[o]rigin deprivation can lead to harmful outcomes; as such, enshrining it as a normative feature of social kinship, rather than as an exceptional occurrence, amounts to a form of highly discriminatory unequal treatment' (Conclusions: Chap. 2).

In addition to sociologists, anthropologists, and family law practitioners, the study has much attraction for international and comparative lawyers, as well as human rights advocates. Diver builds an argument for the establishment and global recognition of a right that she terms as the human right to 'avoid origin deprivation'. In her investigations, she finds the current state of international law significantly limited. As Diver rightly observes, the UN Convention on the Rights of the Child (1989) fails to provide an explicit recognition to familial origins of the child.

Although the best interest of the child remains a paramount consideration for the Convention, such primacy does not take account of the need to preserve genetic heritage identity. The evolving jurisprudence of the Convention may well develop provisions contained, inter alia, in Article 8 and Article 10 to expand the notion of family to allow greater kin contact. That said, the reticence to establish identity of the biological parents is also prevalent at the domestic level, including within the constitutional, administrative, and societal framework of the United Kingdom. The UN Human Rights Committee has criticised the UK in that 'children born out of wedlock, adopted children or children born in the context of a medically assisted fertilization do not have the right to know the identity of their biological parents' (Concluding observations: United Kingdom of Great Britain and Northern Ireland, Committee on the Rights of the Child 31st Session (9 October 2002) CRC/C/15/Add.188, para 31).

As Diver notes in her study, while donor anonymity has been overturned by UK legislation since 1 April 2005, there are considerable limitations in that the parents remain under no obligation to inform the child of his or her conception through assisted reproductive techniques. Amidst these disappointments, Diver does point to variations and variables amongst traditions and values. De facto adoptions e.g. the Islamic *Kafalah*, though not completely immune from its own shortcomings—including difficulties of application—nevertheless provides a useful alternative model. At least in principle, for preserving genetic identity and the possibility of a continuing relationship with the biological parents. There are other models as well, and it is at least arguably the case that amidst modern developed societies, there is a gradual realisation towards greater acceptance of recognising the right to 'avoid origin deprivation'.

With such maturity of analysis and originality of arguments, Diver has produced an excellent study. I wholeheartedly commend this monograph, which in my view will prove to be a reference point for the future.

London, UK Javaid Rehman

Acknowledgements

Sincere thanks are due to Professor Javaid Rehman (Brunel Law School, London) for kindly agreeing to write the foreword to this text and to Sandra Wickenhäuser (Law Editor for Springer International) for providing me with the opportunity to publish this monograph. I would also extend thanks to my PhD supervisors, Mr Ciaran White (School of Law, University of Ulster) and Professor Brian Taylor (School of Sociology and Applied Social Studies, University of Ulster), for their expert guidance and advice throughout the writing of the doctoral thesis upon which this book was based. The comments of Professor Judith Masson (University of Bristol) and Professor Martin Preston-Shoot (University of Bedfordshire), who acted as my external examiners, were also invaluable, as were those of Professor Beverly Brown, who provided detailed suggestions in respect of an earlier draft of the text.

I would also like to thank Professor Emily Hipchen (University of West Georgia, and editor of *ASAC*) for permitting the use of Chap. 4 (previously published in 2012 as an article in the *ASAC* Journal of the Alliance for the Study of Adoption and Culture) and for her endless patience with my many attempts to master the MLA referencing system. I am very grateful also to my colleagues and former colleagues at the University of Ulster, particularly to Professor Mary McColgan (School of Sociology and Applied Social Studies, University of Ulster), Professor Fionnuala ní Aoláin (Transitional Justice Institution, University of Ulster; University of Minnesota), Dr Lynn Ramsey (Letterkenny Inst. of Technology), Dr Gabriele Porretto (The UN, Kosovo), Dr Liam Thornton (University College, Dublin), Professor Christine Bell (Edinburgh University), Professor Joshua Castellino (Middlesex University), Dr Eugene McNamee (Head of School), Grainne McKeever, Dr Thomas Murphy, and Dr Jacinta Miller (all School of Law, University of Ulster), for their helpful and insightful comments on earlier drafts of the various chapters. (Errors contained herein are entirely my own.) I am very grateful also to Janice McQuilkin (Law School Librarian, Magee) and Emer Carlin (School of Law/TJI Secretary) for providing me with unflagging support and assistance in terms of enabling the research and maintaining my sanity.

Thanks are also due to my tortured spouse, Gerry, and children (Ellen, John, Celie, and Eva) for putting up with me throughout and generally seeing to their own dinners.

Acknowledgments

Contents

Chapter 1
Introduction

1.1 Introduction

The decision to use certain terms was made in the interests of clarity in distinguishing between the various rights and interests of those who might be most profoundly affected by origin deprivation or created forms of legal relatedness. The concept of 'the triad' may be a highly contentious one for anyone who regards themselves as having unwillingly lost a child to adoption; it is a useful term however, for example in relation to analysing the various rights hierarchies that can arise post-adoption or gamete donation. The use of the term 'birth mother' also often provokes hurt and anger amongst that community; it will be used here as sparingly as possible, again in the interests of clarity and generally in the context of case law and policy, to highlight for example the consequences of having one's genetic connections vetoed.[1] Where any terms such as 'real', 'natural' or artificial' are used in relation to parentage or conception these will be as quotations from cases or research on the issue of origin deprivation. The term 'relinquished' is used to represent the loss of genetic connection rather than to necessarily suggest that gently provided consents to adoption are necessarily the norm. Equally, the term 'removed child' appears at times, to acknowledge that substitute child care is often grounded in an acute need to achieve child protection and prevent abuse or neglect at the hands of natal family members.

[1] As a sealed-records Quebec adoptee I would make the suggestion that the prefix 'birth' should not necessarily be omitted from rights discourses on adoption and gamete donation. Despite the argument that its usage may reduce the role of genetic mothers (and fathers) to that of narrowly defined, biological input, the term does sharply underscore the unique nature of the losses that can occur where the blood-tie is removed or relinquished in respect of severing the profound connection between child, original parent and denying their shared heritage. Arguably, the notion of the '*birthright*' is made more poignant by virtue of the 'birth' prefix, perhaps carrying a bit more weight in terms of persuading decision-makers that, in cases involving genetic ancestry and heritage, they are often tasked with protecting more than just a basic set of rights.

A. Diver, *A Law of Blood-ties - The 'Right' to Access Genetic Ancestry*,
DOI 10.1007/978-3-319-01071-7_1, © Springer International Publishing Switzerland 2014

The book's purpose is not to denigrate the practices of adoption or assistive reproduction (or the work done by social workers and the family law practitioners) but to highlight how some of the legal and social processes underpinning these created forms kinship have served to create and perpetuate discrimination and inequality of treatment, and in some cases, to violate basic human rights. It argues that a much more child-centric focus is needed at the level of domestic decision-making, to better embed the principle that the welfare of the child must be the paramount concern.[2] The book's central arguments focus mainly upon the situation of the genetically 'kinless' in terms of preventing harms, losses and rights-inequality, rather than on examining in great detail the interests or experiences of others involved in adoption or gamete donation processes. It aims primarily to establish that a 'law of blood-ties' is gradually emerging from amongst the various strands of jurisprudence and this may yet be of use in embedding a more justiciable right to genetic connection and biological identity at domestic level.

Access to ancestry should be a normative rather than exceptional feature of creating social kinship ties; permanent vetoes on accurate, identifying, birth information should only occur in exceptional circumstances, with child welfare paramountcy serving as the guiding principle. By the same token, where laws and policies have systematically protected the rights of one group over those of another then such a template merits scrutiny, and calls for reform.

Genetic connections, despite having much socio-cultural and psychological significance, remain almost weightless in terms of rights, laws and policies. No juridical right currently attaches for example to the psychological need, or sociological desire, to access biological identity, or repair natal bonds that have been legally severed during the processes of social 'kinning'.[3] The validity or otherwise of this assertion is evaluated here against the backdrop of both international and domestic law frameworks, and across a range of jurisdictions. The conflicting socio-cultural and psychological aspects of origin deprivation are examined in the opening chapters to highlight the wide range of harms that can flow from the loss of genetic kinship. The double-edged nature of the concept of relatedness, in respect of the laws and cultural norms which have grown up around it, is also discussed, with a view to contextualising the sometimes equivocal judicial discourses referred to in the later chapters on 'blood-tie jurisprudence'. Arguably, the ambivalence that can be found in some of the judgments echoes the tone of some of the 'fear or revere ancestry' tales of traditional folkloric warnings on 'kinless-ness'[4] not least in respect of the stigmatising effects of being regarded as 'other' to one's kinfolk, social or natal.

---

[2] A key assumption of this monograph is that a significant number of origin-deprived persons will probably, at some stage in their life, attempt to seek out some level of basic information on their genetic ancestry. On searching see further Lifton (1990), pp. 85–92; Carp (1998); Triseliotis (1985), pp. 19–24.

[3] Howell (2003), pp. 465–484.

[4] See for example Bremmer (1999), pp. 1–20; Kenna (2001).

The suggestion will also be made that within some open records jurisdictions (i.e. where original birth certificates are not generally subject to permanent closure[5]) other factors may act to diminish the socio-legal significance of one's original blood-ties. Judicial bars on kin contact,[6] the practice of separating genetic siblings post-adoption,[7] or the granting of a very wide degree of discretion to social parents on issues such as information release or paternity testing[8] have frequently served to weaken the bonds of natal kinship, sometimes to the point where they essentially seem 'irrelevant' or perhaps incapable of being repaired. Many of the ethical issues surrounding promises of confidentiality made to 'triad adults'[9] have also, in some jurisdictions, translated into rigid legal frameworks of absolute secrecy.[10] These can easily 'orphanise' (and perhaps permanently infantilise) origin deprived persons. Arguably, the 'right' to identity within this context effectively becomes more of a non-right,[11] rendered void by the 'weightier' over-arching privacy rights of triad adults. The welfare paramountcy principle, in focussing on the 'best interests of the child'[12] also remains open to a wide variety of interpretations in such contexts; this may be further compounded by the presence of judicial 'balancing exercises' which might afford priority to parental interests.[13]

The case law selected for analysis in the later chapters does not focus solely on situations involving social kinship triads but looks also at those wider issues that

---

[5] See for example the United Kingdom, via The Adoption Act 1976 (as amended by S 60 of the Adoption and Children Act 2002, enacted 30 December 2005) and Northern Ireland's equivalent legislation, Article 54 of The Adoption (NI) Order 1987. Note however the power of veto which vests in the inherent jurisdiction of the High Court, in respect of ordering the non-release of identifying information: see *R v Registrar-General ex p Smith* [1991] 2 QB 393 (Court of Appeal).

[6] See for example *Re K* [2002] NIFam 13; *Re H* [1981] 3 FLR 386.

[7] See *Webster (The Parents) v Norfolk County Council & Ors* (Rev 1) [2009] EWCA Civ 59.

[8] See for example *Re P (A Child)* [2008] EWCA Civ 499; *Re H & A (Children)* [2002] EWCA Civ 383.

[9] The triad refers here to the 'triangle of relationships' that exists or arises in respect of social kinship: genetic parents, social parents and adoptee or donor-gamete child. On the issue of using the term 'triad' see further http://motherhooddeleted.blogspot.com/2009/08/myth (accessed 01.02.12); http://bastardette.blogspot.com/2007/10/ethics (accessed 17.03.12); on '*Respectful Adoption Language*' see further http://www.originscanada.org (accessed 02.02.11).

[10] See Baldassi (2004–2005), pp. 212–265.

[11] On the concept of rights versus 'no-rights' see further Hohfeld (1913), p. 16.

[12] See the United Nations Convention on the Rights of the Child (1989) Article 3 (1) which states that: '*In all actions concerning children, whether undertaken by public or private social welfare institutions, courts of law, administrative authorities or legislative bodies, the best interests of the child shall be a primary consideration*' available at http://www2.ohchr.org/English/law/crc.html (accessed 21.07.011). Domestic provisions [such as Article 1(4) of Adoption and Children Act 2002 in England and Wales] frame the best interests of the child as 'paramount' rather than 'primary' however—see http://www.opsi.gov.uk/acts2002/ukpagea_20020038 (accessed 29.07.11).

[13] On the balancing exercise see further Herring (2003).

may arise where a psycho-social need for blood-tied relatedness remains unmet.[14] Identity rights, best interests-led, child welfare paramountcy and the child's 'right to know' parentage are common themes within the jurisprudence.[15] As these blood-tie cases suggest, genetically displaced children seem at times to require a greater degree of long-term, psychological welfare protection than is currently provided for within some of the basic child welfare checklists found in domestic law or policy. If decision-makers do not accept that a longing to be genetically 'kinned with' others constitutes a key psychological need, then they will continue to frame this aspect of identity as a non-rights-bearing concept, perhaps classing it as mere curiosity on the part of the adoptee.[16] The concept of legal relatedness, especially where it has arisen via biological connectedness, can also be easily dispensed with via legal process. Domestic judiciaries therefore play a key role in interpreting and applying the various provisions that might yet confer a juridical right to knowable genetic ancestry.[17] By ordering the opening of closed records or the overturning of adoptive placements, granting or forbidding kin contact, or by delegating actual decisions on such issues to parents or social workers, domestic courts are responsible for a wide variety of outcomes. Many such decisions carry profound, long-term psychological and social consequences, not just for adoptees but for all members of the social kinship triad and possibly also for other genetic relatives outside of it, such as siblings or grandparents.

Chapter 2 asks why visible ties of kinship are so highly regarded in certain contexts yet so easily denied or disregarded in others. The pragmatic, functionalist nature of kinship (avoidance of danger, ensuring clan survival) was a theme common to many of the theorists cited here; the wider notion of clanship might arise not only through shared genealogical ancestry, but via a common social history or a shared sense of cultural identity.[18] It looks also to the socio-cultural consequences of being rendered genetically kinless and examines how the disparate notions of legal and biological relatedness might be regarded in differing contexts, suggesting that a wide range of social benefits depend on whether or not one has

---

[14] The term 'social kinship' will be used here to refer to the non-genetic forms of relatedness, as created by law (e.g. adoption, Special Guardianship, marriage to a child's mother) custom (de facto 'indigenous' adoption, *kafalah*) or through assisted reproductive technologies (gamete donation or surrogacy).

[15] See for example *Odièvre v France* [2003] 1 FLR 621; *Frette v France* [2004] 38 EHRR 21 (42); *Re L* [2007] EWHC 1771 (Fam) (20 July 2007); *DeBoer v DeBoer* 509 US 1301 [1995]; *Baby Boy Richard v Kirchner* 513 US 1138 [1995].

[16] See further Nelkin and Lindee (1995): WH Freeman, on the socio-cultural significance of 'genetic essentialism'.

[17] Choudry (2003), p. 119.

[18] Strong bonds of 'clanship' may arise between individuals who otherwise lack commonal genetic ancestry. See Morgan (1871), Murdock (1949), and Pasternak (1976). See also however Carsten (2004).

been legally and visibly 'kinned' with non-stranger others.[19] It does not seek to frame the genetic mode of kinship as innately superior to that of social kinship, but focuses instead upon the question of why the concept of genetic kinship can often provoke a double-edged response of 'fear or revere'. The seminal texts of several kinship anthropologists (Radcliffe-Brown,[20] Levi-Strauss[21] and Goody,[22] for example) provide a definitional starting point, asking whether biological connections can, on their own, give rise to a sense of relatedness.[23] Generally, the research also seems to frame the notion of relatedness as a non-discrete concept, arguing that it overlaps with many of the cultural notions of kinship, and with a wide range of other purposive 'social realities', such as the desire to protect one's property, achieve psychological well-being, or avoid societal injustice. Some of the research discussed here seems to support the argument that self-perceived 'otherness' can affect not only the relationships between social and natal kinfolk, but also serve to influence the way in which a healthy sense of 'self-hood' might be achieved.

In other words, where original onomastic identity has been lost, hidden away or removed, it may be replaced by a profound need to search for authentic, identifiable ancestry, if only to avoid being deemed 'other'.[24] By the same token, access to genetic truths and realities (such as accurate narratives or knowledge of 'clan' surnames) might be seen as enabling strong kinship links and a socio-cultural sense of belonging. Lepri's work highlights for example the apparent dangers and perhaps unvoiced concerns associated with taking in unknown 'outsiders'[25] whilst Miall's study of adoptive parents found that many felt themselves at times to be regarded as socio-culturally inferior to genetic parents.[26] Conversely, the significance of the blood-tie may be completely superseded by the nurturing, protective effects of social ties. Hage for example stressed the value of feeding as a much better means of generating kinship than that of mere biology.[27] Ancestor reverence however could prevent the harms of origin deprivation by preserving or appeasing one's 'ancestral substance'.[28] Highly visible, elaborate graveside rituals[29] may preserve

[19] A number of 'kinning' devices can be found within domestic property law in some common law regions (e.g. the 'good conscience' constructive trust based on good kinship behaviours, or promissory estoppel claims over familial legacies). See for example cases such as *Re Johnson* [2008] NI Ch 11; *Little (Junior) v Maguire* [2007] NI Ch 7; *McKernan v McKernan* [2006] NI Ch 6.

[20] Radcliffe-Brown (1952).

[21] Levi-Strauss (1949, 1963).

[22] Goody (1973).

[23] See also for example Evans-Pritchard (1951) and Aginsky (1935), pp. 450–457.

[24] See for example Ortner and Whitehead (1981).

[25] Lepri (2005), pp. 703–724.

[26] Miall (1987), pp. 34–39 at p. 34.

[27] Hage (1999), p. 67.

[28] Grace (2008), pp. 257–262 at p. 257. See also Grace et al. (2008), pp. 301–314.

[29] Roesch-Rhomberg (2004), p. 83.

order amongst the wider community or protect vulnerable outsiders from the inherent dangers of their own 'otherness'. Other ceremonies may be used to reassure social parents fearful of having to defend themselves against the 'prior claims'[30] of original parents or angry kinfolk.

Many of the de facto, informal modes of customary adoption emphasize the difficulties of created relatedness, in a way that formalised western adoption models have traditionally seemed unable, or perhaps reluctant, to do. Issues of loss, fear, 'bereavement' and the need to grieve for lost biological relatives, perhaps publicly and repeatedly, are acknowledged and addressed. Intricate rituals aimed at replacing, removing or mimicking the blood-tie, achieving purification or preventing stigma, promoting psychological healing or removing aggrieved or malign 'hereditary ghosts'[31] may be carried out. A wide range of elaborate, perhaps fluid-based ceremonies exist to repair, replace or ward-off the genetic connection; these rituals may be akin to secular, legal or customary ceremonies that enable social kinship, such as birth record-sealing or the re-naming of adoptees.

Chapter 3 examines the psychological aspects of origin deprivation and broken attachments in a bid to better understand the emotional basis of our desire to 'be related' to others. Carp's contention that genetic relatedness continues to occupy a position of 'privilege' is relevant here, not least in relation to gauging whether norms of strict secrecy were actively developed to protect triad members from their innate sense of dissimilarity and quietly discourage searches for biological kinfolk.[32] A number of academics highlight the 'doomed quality'[33] that some adoptees tend to display.[34] Wegar's research on the issue of sealed birth records in the United States lends support to the argument that adoptees have long been labelled as different, and perhaps at times somehow inferior to, those children who have been raised by their natal families. As such, the dominant model of adoption research appears to have largely been one of individualized pathologies.[35] Bowlby's seminal work on the need for strong attachments in early childhood is also discussed here;[36] despite being initially regarded with a degree of ambivalence, it did serve to challenge the widely held view that abandoned or orphaned infants were incapable of grieving, due to their having 'immature egos' and that their need

[30] Hargreaves (2006), pp. 261–283 at p. 279.

[31] Levy-Shiff (2001), pp. 97–104 at p. 103; see also Berg (2003), pp. 194–207; see further Morgan (1877).

[32] Carp (1998), p. viii.

[33] Carsten (2000), pp. 687–703 at p. 691.

[34] Some writers seem to express dislike for the term 'adoptee'. See for example Trinder et al. (2004), p. 3 where it is suggested that the word tends to denote a 'category rather than a person.' If it is accepted that origin deprivation is a form of discriminatory treatment, the categorization of adopted persons as a uniquely disadvantaged group does not seem entirely inappropriate. The term also perhaps captures the degree of passivity that adopted persons might be subject to when entering the process, in terms of their not usually providing consent.

[35] Wegar (1997), pp. 97–118.

[36] Bowlby (1958), pp. 350–371.

for substitute care-giving was easily met, grounded as it was in a pragmatic need to simply receive nourishment. Reference is also made here to some of the earlier, controversial pieces of empirical research which tended to stress the 'debilitating' effects of illegitimate birth status. Such works are included to illustrate how biased social attitudes might serve to actively influence social policies and public perceptions.[37]

Other key texts referred to include: Frisk, on 'genetic ego'[38]; Erikson, on 'identity crisis'[39]; Sandler and Joffe, on the psychological benefits of physically resembling one's parents[40]; Rogers, on the attainment of 'personhood'[41] and Lifton, who described how the creation of an 'artificial self'[42] might serve as a coping mechanism. The focus however is on asking whether social kinship processes must inevitably create a permanent 'marker for difference.'[43] Theories on the 'rehabilitative' properties of created kinship, drawn from empirical research, seem to ask whether long-term stigma arising from an innate feeling of 'otherness' is an inevitable outcome, in societies where biological connection has been traditionally prized as a cultural norm. Arguably, the main ethical issues, such as promises of confidentiality to triad adults, have perhaps left law and policy makers with little option but to construct and enable frameworks of absolute secrecy, whether over birth or gamete donation.

As much of the qualitative research on international adoptions also perhaps suggests, many respondents do cite long term psychological difficulties: generally they associate their problems with having an ongoing sense of 'discontinued identity.'[44] If it is accepted that origin deprived persons run the risk of suffering from some form of 'genealogical bewilderment,'[45] then it perhaps follows that this can constitute a uniquely harmful form of social disenfranchisement. Genetic 'non-origin' may produce long term effects, not least a sense of having to cope throughout one's lifetime with powerful 'refusals of belonging'[46] As later chapters will seek to argue, origin deprived persons could be viewed in law and custom as a uniquely disadvantaged minority group, worthy of special legal protections, on the basis of the harms that they may be likely to suffer.

---

[37] Goddard (1912). The text argued for the permanent institutionalization (rather than the adoption or fosterage) of illegitimate children within the United States. The subsequently discredited research apparently arose from Goddard's 'work' with one of his female patients in the Training School for Backward and Feeble-Minded Children.

[38] Frisk (1964), p. 31.

[39] Erikson (1968).

[40] Sandler and Joffe (1969), pp. 585–595.

[41] Rogers (1961).

[42] Lifton (1990), pp. 85–92.

[43] Melosh (2002), p. 2.

[44] Ryburn (1995), pp. 41–64 at p. 42.

[45] Sants (1964), pp. 133–141.

[46] Yngvesson and Mahoney (2000), pp. 77–110 at p. 79.

As Antze and Lambeck have argued, memories remain crucial to the construction of personal identities and 'selfhood'.[47] The psychopathologies of non-attachment often comprise of quantitative studies involving adoptees and abused children.[48] The negative effects of having insecure or broken 'attachment' bonds highlight how child welfare needs may differ significantly where origin deprivation is involved. Such research frames the child's 'need to know' as a basic welfare entitlement, rather than as a personal desire to uncover 'non-essential' genealogical information. Much of the work also challenges early adoption policies and practices, which tended to advocate well-meaning, fictive options, such as telling adoptees that they had been orphaned, placing their Adoption Orders or original birth certificates in safe deposit boxes, or simply 'not telling' children that they had been adopted.[49]

Later studies on the need to access accurate information highlight more clearly the potential harms of genetic kinlessness, by stressing how problems can arise. As Wegar suggested, examination of 'the structure of adoption as a social institution' must be a central objective of any bid to redress the unique, recursive harms of origin deprivation. Blaming 'individual shortcomings'[50] of triad children or expecting them to develop strong coping mechanisms, is unacceptable. The early policies of removing the 'social flaws'[51] of illegitimate children (by placing them with new kinfolk and hiding their original background) contrast sharply with the more recent emphasis on 'Positive Adoption Language'[52] and the child-protective aims of modern practice.[53] Such policies are primarily concerned with meeting the acute needs of highly vulnerable children, rather than serving to expand family units, 'cure' infertility, or hide illegitimacy. A brief outline of the various policy changes that have shaped legislative reforms in the open records jurisdictions referred to is also included, by way of suggesting a possible policy template for reforming practice in closed records or veto-bound systems.

Chapter 4 asks whether a juridical 'right' to genetic kinship (via knowable ancestry or meaningful contact) might yet be found to exist amongst international law provisions. It expands upon some of the questions raised by the previous chapters, such as whether a stable sense of national identity might be underpinned by collective kinship, or by common connection to place.[54] Smith, for example,

---

[47] Antze and Lambeck (1996).

[48] See for example Schechter (1960), p. 21.

[49] See for example Wellisch (1952), p. 41 who observed that identification with one's relatives was often easier where there was some degree of physical resemblance. He suggested that, as a result, adoption might be better suited to orphans.

[50] Wegar (1997).

[51] See for example Kornitzer (1959), p. 102; Kellmer-Pringle (1967), p. 25.

[52] See for example Spencer (1979), p. 450 who argued that terms such as 'natural mother' seek to remove any recognition of relatedness between parent and adopted child. She advocated instead the use of 'emotionally correct' words.

[53] See Triseliotis (1973).

[54] Eriksen (2004), pp. 49–62 at p. 49.

argued that the strong bonds of '*ethnie*' (ethnic community) are often based mainly upon shared folkloric wisdoms, customs, common memories, and identification with territorial heritage.[55] The 'growing consciousness of a personal right to compose one's identity'[56] in respect of nationality rights, could perhaps yet provide significant precedent in respect of preventing the loss of birthright heritage. The importance of finding ascertainable collective ethnonyms in enabling and maintaining group identity is especially clear where a displacement of cultural histories or the loss of shared 'descent mythologies'[57] has occurred, as can happen in the worst examples of 'assimilation', where a child's ethnicity might be deliberately hidden, for example in the wake of wars or conflicts, through illegal adoptions or via baby-trafficking.[58] Although there are no explicit protections for blood-tied kinship within international law, a number of significant entitlements to enjoy a sense of 'relatedness' might be inferred from some of the main provisions: the right to a name[59] or nationality,[60] the avoidance of degrading treatment[61] and the need for respect for 'cultural integrity',[62] are some examples. Wider principles on enabling equality and preventing discriminatory treatment are also relevant in respect of preventing harm amongst vulnerable persons, and will be referred to here and in later chapters.[63] Mainly however this chapter attempts to conceptualize the child's 'right' to genetic identity as an important aspect of human rights law. To do so, it looks largely to the Committee Guidance (of the Children's Convention)

[55] See Smith (1986, 1998, 1991).

[56] See Franck (1996), pp. 359–383 at p. 359.

[57] Smith (1986, 1998, 1991).

[58] On baby-trafficking see for example Meier and Zhang (2008–2009), pp. 87–130; see also Smolin (2009–2010), p. 441.

[59] See for example Articles 16 (1) (a) and 30 (1) of The Convention on Protection of Children and Co-Operation in Respect of Inter-Country Adoption (1993) ('Hague Convention on Inter-Country Adoption') available at http://hcch.e-vision.nl/upload/outline33e.pdf accessed 13.07.11.

[60] See for example The European Convention on Nationality (1997) (ETS No 166) ('The Nationality Convention') available at http://conventions.coe.int/Treaty/en/Treaties/Word/166.doc accessed 20.07.11.

[61] See for example Article 3 of the European Convention for the Protection of Human Rights and Fundamental Freedoms (1950) ('ECHR') available at http://conventions.coe.int/Treaty/en/Treaties/Html/005.htm#FN1 accessed 01.06.11.

[62] See Article 27 of The International Covenant on Civil and Political Rights (1966) ('ICCPR') available at http://www2.ohchr.org/english/law/ccpr.htm accessed 30.07.11. On the issue of international legal and customary norms of non-discrimination and the duty upon states to protect the 'cultural integrity' of individuals and groups see also Anaya (2000), pp. 97–103.

[63] See for example Article 2 of the Universal Declaration on Human Rights (1948) which states that '*Everyone is entitled to all the rights and freedoms set forth in this Declaration, without distinction of any kind, such as race, colour, sex, language, religion, political or other opinion, national or social origin, property, birth or other status.*' See also Article 25 (2) which stresses that '*Motherhood and childhood are entitled to special care and assistance. All children, whether born in or out of wedlock, shall enjoy the same social protection.*' Available at http://www.un.org/en/documents/udhr/index.shtml accessed 10.07.11.

contained in Concluding Observations and Country Reports.[64] A key factor here is that of domestic interpretation of the best interests principle, not least in relation to how signatory states actually seek to implement meaningfully the principle of child welfare paramountcy at the level of domestic proceedings.[65] It will be argued that national courts could at times allow themselves to be more fully persuaded by the 'identity rights' provisions of such instruments as the Children's Convention, especially given its focus on life-long protection of child welfare.

Arguably, the 'rights-based approach' of the Children's Convention has at least encouraged a 'decisive shift'[66] in respect of domestic judicial attitudes towards the rights of children, not least where court decisions involve issues of psychological welfare. The chapter also includes a brief examination of how identity and family life rights (and, arguably, the best interests principle itself) may be entirely overruled by the presence of a domestic veto, even in the face of regionally enacted human rights charters aimed at embedding human rights principles.[67] It also makes the tentative suggestion that the notion of genetic identity kinship has to date been generally aligned with 'second generation', socio-economic and cultural property rights, rather than framed as a stronger, civil, political entitlement.[68] As such, any rights that might be found to attach to the concepts of genetic kinship and original identity may run the risk of being classed as largely non-juridical, or subject to the problems of overcoming wide margins of appreciation, and gradual, aspirational modes of domestic implementation.[69]

Chapter 5 examines 'blood-tie' case law from The European Court of Human Rights, for example, in connection with the rights enshrined in Articles 8 (and, to a lesser extent, Article 6) of the European Convention on Human Rights. The various

---

[64] See Article 3 (1) of The United Nations Conventions on the Rights of the Child (1989) ('The Children's Convention') which states that: '*In all actions concerning children, whether undertaken by public or private social welfare institutions, courts of law, administrative authorities or legislative bodies, the best interests of the child shall be a primary consideration.*' Available at http://www2.ohchr.org/English/law/crc.html accessed 21.07.11.

[65] See Sinclair and Franklin (2000), Sinclair (2004), pp. 106–118, and Kirby and Bryson (2002).

[66] McCarthy (2004), pp. 1–32 at p. 1.

[67] See Baldassi (2004–2005), pp. 212–265; see also *Kelly v Superintendent of Child Welfare and Williams* (1980) 23 BCLR 299 (SC); *Re Adoption of BA* (1980) 17 RFL (2d) 140 (Man Co Ct) where 'identity issues' were insufficient to open sealed birth records. See however by way of contrast *Ross v PEI* (Supreme Court, Family Division, Registrar) (1985) 56 Nfld & PEIR 248 [1985] PEIJ No 1 (PEISC) (QL).

[68] Similarly, if a positive obligation to prevent origin deprivation were found to exist in human rights law, possibly as a type of cultural property right (e.g. in preventing cultural assimilation) then arguably such a duty could, in theory at least, be framed as perhaps akin to those peremptory norms that attach to the protection of Native status and title. See further Merry (1997), p. 31; Samson (2001), pp. 226–248.

[69] See also Alston (1984), pp. 607–615. See also The International Declaration on Human Genetic Data (2003) Article 3 which states that '*Each individual has a characteristic genetic make-up. Nevertheless, a person's identity should not be reduced to genetic characteristics, since it involves complex educational, environmental and personal factors and emotional, social, spiritual and cultural bonds with others and implies a dimension of freedom.*'

discourses focus on the right to respect for family and private life from the perspective of both adult and child and on the concept of child welfare paramountcy in decision-making aimed at protecting the best interests of the child.[70] The cases reflect a fairly diverse range of issues and outcomes, including such topics as familial contact, the need for identity, anonymous birth registration, paternity, child protection and parental involvement in matters of process. As Herring has argued, a narrow, rights-based approach may serve to promote the rights of adults at the expense of protecting those of their children.[71] The cases in this chapter suggest however that child welfare has increasingly become the paramount concern in cases involving disputes over parentage and parenthood.[72] That said, a fairly wide margin of appreciation still tends to attach to issues of private family life, especially where questions have arisen over whether there is a positive obligation on the part of the state to preserve the genetic connection: the keeping of birth records, domestic regulation of surrogacy or gamete donation and parental discretion over information release are particularly difficult areas in terms of finding a degree of societal consensus.[73]

A number of Dissenting Opinions are also referred to here, given their fairly clear support for the child's need for some degree of 'identity dignity'.[74] The chapter will further argue that 'gaps' in domestic law in respect of the enforcement of certain Convention rights (and in some cases, of the best interests principle) perhaps mirror the judicial equivocation that can attach to the issue of blood-ties in general. Jurists have traditionally seemed to focus on the question of whether a juridical form of 'family life' actually or potentially exists before considering, if they do so at all, the psychological importance of establishing an accurate sense of identity or whether there is much social significance in having an ascertainable ancestral heritage. Some of the case law perhaps suggests that because newborn infants cannot demonstrate that they remember their natal relationships, they are likely to lack a meaningful level of 'family life' and cannot claim to have suffered an interference with their Article 8 rights. Because of this they will then also perhaps have no right to maintain contact with any of their genetic relatives. That said, relatives from outside of the triad, who in the past might well have been excluded entirely from the remit of family life (unmarried fathers, birth siblings,

---

[70] Especially that generated in connection with Article 8 of the European Convention on Human Rights.

[71] See Herring (2003).

[72] See for example *Neulinger and Shuruk v. Switzerland* (2010) [GC], no. 41615/07 ECHR; *Y C v United Kingdom* (2012) (App 4547/10) ECHRR; *R and H v United Kingdom* (2011) 54 EHRR 28.

[73] See for example *Kearns v France* ECHR 10 January 2008 (Application no 35991/04); *S.H. And Others v. Austria*—57813/00 [2011] ECHR 1878 (3 November 2011); *Ahrens v Germany* (App 45071/09) ECHRR (22 March 2012).

[74] *Odièvre v France* [2003] 1 FLR 621 (App no 6833/74) Dissenting Opinion at para 7, per Judges Wildhaber, Sir Nicolas Bratza, Bonello, Loucaides, Cabral Bonetto, Tulknes and Pellonpää.

grand-parents, foster parents or step-parents for example) have made progress in respect of gaining legal recognition.[75]

Some of the case law seems also to suggest that a right to family life can exist, even where there is little more than genetic kinship to serve as its basis. Such jurisprudence also often provides key insights into how judges arrive at difficult, rights-balancing decisions involving competing sets of rights and interests: it also offers some degree of guidance as to how the various laws, regulations and policies surrounding this area might eventually be improved upon, to prevent or at least minimise the wide range of harms suffered by origin deprived persons.[76] In terms of procedural matters, the involvement of birth family members in the proceedings has become increasingly significant. The need for timely assessment of parenting skills and to have legal representation for birth parents at pre-hearing meetings to discuss post-adoption contact arrangements for example, might be important factors in terms of deciding whether Convention rights have been unlawfully infringed, especially where serious issues with the potential to impact upon family life rights (such as bars on contact) have arisen.[77]

Chapter 6 looks at how domestic laws and policies on determining legal parentage may serve to veto genetic connections. Donor anonymity (as enshrined in British Columbia's recent *Pratten* decision[78]) is particularly relevant, given that issues of inequality and discrimination clearly attach to this particular form of genetic identity veto. The United Kingdom's domestic jurisprudence on parentage rights is also of interest, in relation to possibly providing useful precedent on the nature of the 'right' to access an accurate genetic identity. The Human Fertilisation and Embryology Act (2008) for example has effectively re-defined some of the socio-legal aspects of genetic kinship in an attempt to bring the law into line with the realities of modern assistive reproductive techniques and to reflect the changing nature of family life and parenthood.[79] Legal disputes over parentage and parenthood are now largely settled by reference to the issue of consent; donor-gamete children of same-sex couples may be deemed legally 'fatherless' or, in surrogacy cases, essentially 'motherless.' This chapter will argue also that donor anonymity regimes at times reflect the harsher aspects of closed records adoption practice. Similar issues arise for example in respect of the child's 'right' to know their origins and the degree of discretion which might be afforded to triad parents in terms of 'telling' their children the truth about their biological heritage. The question of

---

[75] See for example *Marckx v Belgium* (1979) (application no 6833/74) (1979); *P C and S v United Kingdom* (2002) 35 EHRR 1075; *Anayo v Germany* [2010] ECHR 2083 (21 December).

[76] See for example *Neulinger and Shuruk v Switzerland* (2010) [GC] app no 41615/07.

[77] See for example *R and H v United Kingdom* (2011) 35348/06 ECHR 844; *Y C v United Kingdom* (2012) (App 4547/10) ECHRR.

[78] *Pratten v British Columbia* (*Attorney General*) [2012] BCCA 480.

[79] See also The UK's Human Fertilisation and Embryology Bill [HL] 2007–2008 Part 2 and Schedule 6, on the proposed changes to the legal definition of parenthood, available http://www.publications.parliament.uk/pa/cm2007C accessed 15.10.2011; see also The Human Fertilisation and Embryology Act 2008.

'whether to tell' remains especially relevant in relation to this 'second generation,' technology-led form of origin deprivation.[80]

Other cases on parentage examine in some detail the concepts of the blood tie and of legal relatedness; the best interests principle is also discussed here, as is the issue of how best to balance conflicting sets of rights. Some of the cases involve DNA testing, while others appear to focus more generally on the question of whether child welfare paramountcy should include a juridical right to 'biological truth'.[81] Judicial reluctance to interfere with family autonomy is again apparent, not least in cases that involve some form of kin contact dispute. Some of the dicta involving parentage disputes appear to define the concept of family life as clearly grounded in social rather than genetic relatedness. Several of the judgments provide highly conflicting precedent, especially in relation to actually defining the best interests of the triad child in such situations. Issues such as parental consent, the need to avoid causing the child to suffer the loss of a parent, and the framing of information provision as a form of indirect contact, are all discussed here. Arguably the concept of a right to genetic connection has been lent some measure of support by cases such as *Re G* (2013)[82] where the Court of Appeal granted a sperm donor father the right to seek leave to apply for ongoing direct contact with the child, and by cases involving surrogacy which highlight the importance of having a 'legal reality which matches the day to day reality'[83] of biological parentage. An Irish case involving egg donation and surrogacy has particularly stressed the significance of the blood-tie, granting legal maternity to the egg donor mother rather than to the surrogate mother, and removing an archaic legal presumption on motherhood ('*mater simper certa est*') in the process.[84]

Chapter 7 focuses on those domestic legal matters that have the ability to affect the genetic connection in open records regions. Case law for example on issues such as the removal of parental rights, challenges to adoptive placements and decisions on post adoption contact often includes some discourse on the need for blood-ties to be severed, suspended, maintained or repaired. The principle that the welfare of the child is the paramount concern in such cases has been repeatedly endorsed across a fairly wide variety of contexts. Northern Ireland's adoption law is currently under reform and merits discussion here;[85] over the past decade, this region has seen a

---

[80] See Braude et al. (1990), pp. 1410–1412; On parental privacy see however the arguments of Walker and Broderick (1999), pp. 38–44.

[81] On 'biological truth' see Franklin (2003), pp. 65–85; On the 'legal truth' of parentage see also Eekelaar (2006), pp. 54–77.

[82] *Re G* [2013] EWHC 134 (Fam) Here, a sperm donor father (who had originally been friends with the child's lesbian parents) sought the leave of the Court to make an application for contact with the child.

[83] *A v P* [2011] EWHC 1738 (Fam) para 26.

[84] *M.R & Anor -v- An tArd Chlaraitheoir & Ors* [2013] IEHC 91.

[85] See for example *RE TMH* [2009] NIFam 11; *Re EFB* [2009] NIFam 7. See also Section 1 Children Act 1989; Article 3 Children (NI) Order 1995 on the 'make no Order' principle. Adoption practice in the region came in for considerable criticism in *Down Lisburn Health and Social*

number of changes in terms of evolving judicial attitudes towards the making of orders which 'free' the child for adoption. Until fairly recently, the courts perhaps seemed to adhere to the view that 'post-adoption contact of any kind may have a limited value.'[86] Arguably, an important policy concern was to reassure prospective adopters that they would not be subject to unwelcome intrusion by birth relatives and that 'permanence' of placement was the best means of resolving such issues. Although the courts now tend increasingly to require that prospective adoptive parents should pre-agree to the enablement of some level of post-adoption contact[87], in the absence of a court order for contact being made, no legal duty to facilitate it will actually arise.[88]

This chapter also examines the emotive issue of child 'relinquishment' (voluntary and non-consensual) with a view to gauging whether devices such as Special Guardianship Orders (in England and Wales) may be regarded as a child-centric, 'halfway house' between adoption and long term fosterage.[89] Recent case law on the issue of post-adoption contact also highlights the difficulties associated with ensuring that the long-term welfare needs of vulnerable children are fully addressed in court proceedings.[90] Siblings may be permanently separated from each other, with no order for contact being made by the court which has placed them into different homes. Similarly, genetic relatives such as grand-parents may seek to raise an at-risk child, or at least preserve some degree of contact with the child who has been taken into care or is being considered for adoption. The need for placement stability is often cited during hearings aimed at precluding or limiting contact with natal kin and, in some cases, to over-rule the recommendations of expert witnesses.[91] The issue of non-consensual, birth-parental relinquishment is also discussed here,[92] with the dicta perhaps most significant in respect of the insights that they provide into the attitudes of the affected mothers.[93] Differing domestic interpretations of Article 8 ECHR rights and of the best interests of the child

Services Trust v H [2006] UKHL 36 (see the dissenting Opinion of Baroness Hale LJ). On Freeing Orders and delays See further Kelly and McSherry (2003), Kelly and McSherry (2002), pp. 297–309; '*Adopting the Future*' Consultation Report Responses (2006) DHSSPS available http://www.dhsspsni.gov.uk/adopting-the-future-consultation-report-final-2.pdf (accessed 05.10.11).

[86] *Re A* [2001] NIFam 23 per Gillen LJ.

[87] See *SEHSST v LS* [2009] NIFam 14; and Re JJ [2009] NIFam 2.

[88] See *Re NI and NS* [2001] NIFam 7 (24 March 2001). Contact often takes the form of indirect methods such as 'letter-box' contact; it may also be restricted to a few instances per year, conditional upon natal kin not having a disruptive effect upon the placement. See *Re EFB* [2009] NIFam 7.

[89] See Bainham (2007), pp. 520–523; Talbot and Kidd (2004), p. 273.

[90] *Webster (The Parents) v Norfolk County Council & Ors* (Rev 1) [2009] EWCA Civ 59 (11 February 2009).

[91] See for example *Re P (A Child)* [2008] EWCA Civ 499.

[92] See *Down Lisburn Health and Social Services Trust v H* [2006] UKHL 36. The use of 'Freeing Orders', and the lack of kin contact pending adoptive placement was particularly criticized.

[93] See for example *Re J and S* (2001) NIFam 13 (23 May 2001); Re *CBCHSST v JKF* [2000] NIFam 76.

principle are also evident, through some of the discourses on related issues: long-term fostering, the role of expert witness evidence and the long-term, 'lifelong' consequences of being placed for adoption.

Chapter 8 offers an examination of case law that has essentially framed preservation of the blood-tie as the paramount consideration. The cases selected tend to represent a blunt inversion of the traditional, 'clean-break' kinship rights template, with genetic connection seen as the central concern.[94] Although repatriation with natal kinfolk is a common outcome, with the blood-tie being afforded a very high level of judicial protection, it is often the case that extrinsic issues have brought about the reunion rather than the blood-tie per se. Time limits attaching to parental objection to adoption and adoptive placement[95] or the 'unlawful detention'[96] of a child by the prospective adoptive parents may underpin the court's reasoning and limit the scope of the eventual outcomes. Such 'blood-tie as paramount concern' case law does not generally however shrink from addressing the lifelong impacts of origin deprivation. A number of cases refer repeatedly to the issue of broken 'attachments' between kin, both social and genetic and the effect of such losses upon relatives who may be outside the traditional triadic rights remit. Equally, the concept of the blood-tie may be framed as an essential, much-valued aspect of indigenous cultural heritage rights.[97]

It will be argued here that genetic kinship in this context has clearly been promoted to a position of primacy: origin deprivation may have become a highly exceptional occurrence rather than a normative feature of social kinship under this harshly inverted template. Many of the outcomes cannot however be described as representing a best interests-led solution in respect of the child's welfare. Although the jurisprudence stresses the need to preserve or restore blood-tied kinship, the courts' decisions tend to largely turn upon the following factors: an over-arching need to preserve cultural or racial heritages (i.e. in indigenous, customary, trans-racial or international adoption contexts) or the judicial desire to prevent a breach of adult-centric rights to parenthood. These frameworks contrast sharply with the kinship severing decisions of the preceding chapters. Here, the blood-tie has been re-defined as an essential component of kinship identity, perhaps as an item of heritage or birthright, and as a key means of achieving long-term psychological welfare. Legislation (such as the Indian Child Welfare Act 1978 in the United States) may, for example, require written consents from natal parents, or from the child's custodian, before adoption can proceed.

Hearings may occur at tribal courts (applying customary tribal law) where objection to the process of formal adoption may be raised by the child's tribal elders. Similarly, Adoption Orders may be overturned, where for example fraud or duress was found to have coerced parental consents or where peripheral issues such

[94] See for example *Re Bridget R et al (Minors)* (1995) BO93520.

[95] See for example *In Re BGC* (1992) 496 NW 2d 239 (Iowa).

[96] *N & Anor v Health Service Executive & Ors* [2006] IESC 60.

[97] See further, Crawford (2000), pp. 211–236; Jackson (2003).

as parental residence on tribal land are at issue.[98] Children will generally be returned to their original parents if the requirements of the legislative provisions have not been adhered to.[99] As Hollinger has suggested, pre-occupation with concepts such as 'tribal survival'[100] clearly influence interpretation of the best interests of the child principle in such a context. Bakeis also noted that legislation such as the 1978 Act may lead to unfair treatment of triad members. Child protection issues, delays or frequent placement changes often run counter to the child's best interests.[101] The value of maintaining genetic ties and preventing origin deprivation is clearly open to question where child protection might be an issue of acute concern.[102] In non-indigenous cases, where natal parents are seeking the overturn of an adoptive placement post-relinquishment, the problems of this alternative framework also become apparent. The placed child may have formed strong attachments to his adoptive parents, or be facing a risk of abuse or neglect if returned to the original parents or other genetic relatives. An Irish Supreme Court decision (the '*Baby Anne*' case[103]) is also looked at here, given its eventual outcome and the intense controversy that surrounded it.

Chapter 9 outlines briefly how the domestic statutory checklists of the United Kingdom are aimed at promoting and protecting the best interests of the child. It then argues that a number of guiding principles can be gleaned from the various strands of jurisprudence that currently exist in respect of the case law on blood-ties, and that these offer a useful means of preventing origin deprivation without compromising the principles of child welfare paramountcy. Children's longer term needs might not be as easily sidelined, if decision-makers were obliged to consider more fully the possible effects of genetic kinlessness. Such an approach might also serve to focus attention more closely upon the duty to protect the best interests of origin deprived children. The application of the six principles[104] to a selection of 'hard cases' drawn from previous chapters (i.e. vetoes on genetic

[98] *In Re Bridget R et al, Minors* (1995) 25 USCA 1901 *op cit* at n 22.

[99] Hollinger (1988–1989), pp. 451–501.

[100] *Ibid* p. 453.

[101] Bakeis (1996).

[102] See further Hazeltine (2002), pp. 58–84.

[103] *N & Anor v Health Service Executive & Ors* [2006] IESC 60, wherein a 'voluntarily relinquished' child had lived with her prospective adopters for several years. The case turned not on the issue of her best interests but on the Constitutional rights of her birth parents to avoid State interference with their right to family life.

[104] 1. The fundamental right to an authentic, ancestral identity is often a key component of child-welfare led best interests and the right to family and private life. 2. In terms of realising the right to identity, the best interests of the child may be closely tied to such issues as contact, 'potentiality' of relatedness or relationship, or to procedural matters such as passage of time. 3. The welfare of the child is the paramount consideration. 4. The best interests of the child may require informational disclosure given the significance of the genetic connection and that a positive, juridical obligation to preserve or repair this connection exists. 5. Loss of genetic connection ought to occur only in exceptional circumstances. 6. Preserving genetic connections should not compromise the principle of child welfare paramountcy.

ancestry, overturned adoptions) allows for evaluation of the potential outcomes. Often the eventual resolutions would not have differed significantly, other than to perhaps allow for some degree of flexibility in determining whether some level of informational disclosure or kin contact might have been allowed in the best interests of the child.

The concluding chapter makes the suggestion that many closed records jurisdictions could do worse than look to the legislative frameworks of the United Kingdom, whereby original birth certificates are generally free from non-disclosure vetoes and available on reaching the age of majority[105]; the abolition of donor anonymity in this region also represents a further step away from the various legal fictions that can surround the concept of parenthood. It calls for a fuller acknowledgement of how social kinship can generate and entrench 'difference that makes a difference'[106] and asks whether legal reforms alone can effect meaningful change in terms of achieving best interests-based, child-centric solutions. Other systems of child-protective social kinship, which do not require permanent vetoes on the genetic connection or demand exclusivity of legal parentage or parenthood, may serve as useful comparative models for gauging whether origin deprivation must be a necessary, normative feature of child welfare paramount outcomes. Equally, a change in attitude in respect of how we regard the purpose and nature of parenthood might do more to bring about meaningful changes than the law could perhaps ever hope to achieve.

References

Aginsky, B. W. (1935). The mechanics of kinship. *American Anthropologist, 37*(3), 450–457.

Alston, P. (1984). Conjuring up new human rights: A proposal for quality control? *American Journal of International Law, 78*, 607–615.

Anaya, S. J. (2000). *Indigenous peoples in international law*. New York: Oxford University Press.

Antze, P., & Lambeck, M. (1996). *Tense past: Cultural essays in trauma and memory*. London: Routledge.

Bainham, A. (2007). Permanence for children: Special guardianship or adoption? *Cambridge Law Journal, 66*(3), 520–523.

Bakeis, C. D. (1996). The Indian Child Welfare Act of 1978: violating personal rights for the sake of the tribe. *Notre Dame Journal of Law, Ethics, and Public Policy 10*, 543, 569.

Baldassi, C. L. (2004–2005). The quest to access closed adoption files in Canada: Understanding social context and legal resistance to change. *Canadian Journal of Family Law, 21*, 212–265.

Berg, A. (2003). Ancestor reverence and mental health in South Africa. *Transcultural Psychiatry, 40*(2), 194–207.

Bowlby, J. (1958). The nature of the child's tie to his mother. *The International Journal of Psycho-Analysis, 39*, 350–371.

[105] The rule in *R v Registrar-General ex p Smith* [1991] 2 QB 393 (Court of Appeal) notwithstanding.

[106] Melosh (2002), p. 2.

Braude, P., Johnson, M., & Aiken, R. (1990). The human fertilisation and embryology bill goes to report stage. *British Medical Journal, 300*, 1410–1412.

Bremmer, J. (1999). Fosterage, kinship and the circulation of children in Ancient Greece. *Hellenic Studies Review, 6*, 1–20.

Carp, E. W. (1998). *Family matters: Secrecy and disclosure in the history of adoption*. Cambridge: Harvard University Press.

Carsten, J. (2000). 'Knowing where you've come from': Ruptures and continuities of time and kinship in narratives of adoption reunions. *Journal of the Royal Anthropological Institute (NS), 6*, 687–703.

Carsten, J. (2004). *After kinship: New departures in anthropology*. Boston: Cambridge University Press.

Choudry, S. (2003). The Adoption and Children Act 2002: The Welfare Principle and The Human Rights Act 1998 – A missed opportunity? *Child and Family Law Quarterly, 15*(2), 119.

Crawford, S. J. (2000). (Re) constructing bodies: Semiotic sovereignty and the debate over Kenniwick Man. In D. A. Mihesuah (Ed.), *Repatriation reader: Who owns American Indian Remains?* Lincoln: University of Nebraska Press.

Eekelaar, J. (2006). *Family law and personal life*. Oxford: Oxford University Press.

Eriksen, T. H. (2004). Place, kinship and the case for non-ethnic nations. *Nations and Nationalism, 10*(1/2), 49–62.

Erikson, E. H. (1968). *Identity: Youth and crisis*. New York: Norton.

Evans-Pritchard, E. E. (1951). *Kinship and marriage among the Nuer*. London: Oxford University Press.

Franck, T. (1996). Clan and superclan: Loyalty, identity and community in law and practice. *American Journal of International Law, 90*, 359–383.

Franklin, S. (2003). Re-thinking nature-culture: Anthropology and the new genetics. *Anthropological Theory, 3*, 65–85.

Frisk, M. (1964). Identity problems and confused conceptions of the genetic ego of adopted children during adolescence. *Acta Paedo Psychiatrica, 31*, 6–12.

Goddard, H. (1912). *The Kallikak family: A study in the heredity of feeble-mindedness*. New York: Macmillan.

Goody, J. R. (Ed.). (1973). *The character of kinship*. Cambridge: Cambridge University Press.

Grace, V. M. (2008). The psychosocial politics of paternity in the case of male donated gametes. *Women's Studies International Forum, 31*, 257–262.

Grace, V. M., Daniels, K. R., & Gillet, W. (2008). The donor, the father, and the imaginary constitution of the family: Parents' constructions in the case of donor insemination. *Social Science & Medicine, 66*, 301–314.

Hage, P. (1999). The Arapesh atom of kinship feeding ties versus blood-ties. *Journal of Law and Family Studies, 1*, 67.

Hargreaves, K. (2006). Constructing families and kinship through donor insemination. *Sociology of Health & Illness, 28*(3), 261–283.

Hazeltine, S. L. (2002). Speedy termination of Alaska native parental rights: The 1998 changes to Alaska's child in need of aid statutes and their inherent conflict with the mandates of the Federal Indian Child Welfare Act. *Alaska Law Review, 19*, 58–84.

Herring, J. (2003). Connecting contact: Contact in a private law context. In A. Bainham et al. (Eds.), *Children and their families: Contact, rights and welfare*. London: Hart.

Hohfeld, W. (1913). Some fundamental legal conceptions as applied in judicial reasoning. *Yale Law Journal, 23*, 16.

Hollinger, J. H. (1988–1989). Beyond the best interests of the tribe: The Indian Child Welfare Act and the Adoption of Indian Children. *University of Detroit Law Review, 66*, 451–501.

Howell, S. (2003). Kinning: The creation of life trajectories in transnational adoptive families. *The Journal of the Royal Anthropological Institute, 9*, 465–484.

Jackson, H. H. (2003). *A century of dishonor: The classic exposé of the plight of the native Americans*. Mineola: Dover Publications Inc.

Kellmer-Pringle, M. L. (1967). *Adoption: Facts and fallacies*. London: Longmans.

Kelly, G., & McSherry, D. (2002). Adoption from care in Northern Ireland: Problems in the process. *Child and Family Social Work, 7*, 297–309.

Kelly, G., & McSherry, D. (2003). *Review of the freeing order processes in Northern Ireland*. Belfast: DHSS Report.

Kenna, M. (2001). *Greek Island life: Fieldwork on Anafi*. Amsterdam: Harwood Academic.

Kirby, P., & Bryson, S. (2002). *Measuring the magic? Evaluating and researching young people's participation in public decision-making*. London: Carnegie Young People Initiative.

Kornitzer, M. (1959). *Adoption*. London: Putnam.

Lepri, I. (2005). The meanings of kinship among the Ese Ejja of Northern Bolivia. *Journal of the Royal Anthropological Institute (NS), 11*, 703–724.

Levi-Strauss, C. (1949). *The elementary structures of kinship*. Boston: Beacon.

Levi-Strauss, C. (1963). *Structural anthropology*. New York: Basic Books.

Levy-Shiff, R. (2001). Psychological adjustment of adoptees in adulthood: Family environment and adoption-related correlates. *International Journal of Behavioral Development, 25*, 97–104.

Lifton, B. J. (1990). The formation of the adopted self. *Psychotherapy in Private Practice, 8*, 85–92.

McCarthy, P. (2004). Making the most of international law on the right to identity: An analysis of article 8 of the United Nations Convention on the rights of the child. *Cork Online Law Review, 3*, 1–32.

Meier, P. J., & Zhang, X. (2008–2009). Sold into adoption: The Hunan Baby trafficking scandal exposes vulnerabilities in Chinese adoptions to the United States. *Cumberland Law Review, 39*, 87–130.

Melosh, B. (2002). *Strangers and kin: The American way of adoption*. Cambridge: Harvard University Press.

Merry, S. E. (1997). Changing rights, changing culture. In R. A. Wilson (Ed.), *Human rights, culture and context- anthropological perspectives*. New York: Pluto Press.

Miall, C. E. (1987). The stigma of adoptive parent status: Perceptions of community attitudes toward adoption and the experience of informal social sanctioning. *Family Relations, 36*(1), 34–39.

Morgan, L. H. (1871). *Systems of consanguinity and affinity of the human family*. Washington DC: Smithsonian Contributions to Knowledge. Anthrop Publ. (1970).

Morgan, L. H. (1877). *Ancient society*. London: MacMillan and Co.

Murdock, G. P. (1949). *Social structure*. New York: Macmillan.

Nelkin, D., & Lindee, S. (1995). *The DNA mystique: The gene as a cultural icon*. New York: W H Freeman.

Ortner, S. B., & Whitehead, H. (Eds.). (1981). *Sexual meanings: The cultural construction of gender and sexuality*. Cambridge: Cambridge University Press.

Pasternak, B. (1976). *Introduction to kinship and social organization*. Englewood Cliffs: Prentice Hall.

Radcliffe-Brown, A. R. (1952). *Structure and function in primitive society*. London and Henley: Routledge and Kegan.

Roesch-Rhomberg, I. (2004). Korean institutionalised adoption. In F. Bowie (Ed.), *Cross-cultural approaches to adoption*. London: Routledge.

Rogers, C. (1961). *On becoming a person: A therapist's view of psychotherapy*. Boston: Houghton Mifflin.

Ryburn, M. (1995). Adopted children's identity and information needs. *Children and Society, 9*(3), 41–64.

Samson, C. (2001). Rights as the reward for simulated cultural sameness: The Innu in the Canadian colonial context. In J. K. Cowan et al. (Eds.), *Culture and rights – anthropological perspectives*. Cambridge: Cambridge University Press.

Sandler, J., & Joffe, W. G. (1969). Towards a basic psychoanalytic model: The theory of parent–infant relationship. *The International Journal of Psycho-Analysis, 41*, 585–595.

Sants, H. J. (1964). Genealogical bewilderment in children with substitute parents. *British Journal of Medical Psychology, 37*, 133–141.
Schechter, M. (1960). Observations on adopted children. *Archives of General Psychiatry, 3*, 21.
Sinclair, R. (2004). Participation in practice: Making it meaningful, effective and sustainable. *Children and Society, 18*, 106–118.
Sinclair, R., & Franklin, A. (2000). *Young people's participation: Quality protects research briefing no 3*. London: Dept of Health.
Smith, A. D. (1986). *The ethnic origins of nations*. Oxford: Blackwell.
Smith, A. D. (1991). *National identity*. London: Harmondsworth, Penguin.
Smith, A. D. (1998). *Nationalism and modernism*. London: Routledge.
Smolin, D. M. (2009–2010). Child laundering and the Hague Convention on intercountry adoption: The future and past of intercountry adoption. *University of Louisville Law Review, 48*, 441.
Spencer, M. (1979). *The terminology of adoption*. Washington: Child Welfare League of America.
Talbot, G., & Kidd, P. (2004). Special guardianship orders-issues in respect of family assessment. *Family Law Journal, 34*, 273.
Trinder, L., Feast, J., & Howe, D. (2004). *The adoption reunion handbook*. London: Wiley.
Triseliotis, J. (1973). *In search of origins: The experiences of adopted people*. London and Boston: Routledge & Kegan Paul.
Triseliotis, J. (1985). Adoption with contact. *Adoption and Fostering, 9*(4), 19–24.
Walker, I., & Broderick, P. (1999). The psychology of assisted reproduction-or psychology assisting its reproduction? *Australian Psychologist, 34*, 38–44.
Wegar, W. (1997). *Adoption, identity and kinship: The debate over sealed birth records* (pp. 97–118). New Haven: Yale University Press.
Wellisch, K. E. (1952). Children without genealogy – A problem of adoption. *Mental Health, 12*, 41.
Yngvesson, B., & Mahoney, M. A. (2000). 'As One Should, Ought And Wants To Be': Belonging and authenticity in identity narratives'. *Theory Culture and Society, 17*, 77–110.

Cases

United Kingdom
A v P [2011] EWHC 1738 (Fam)
Down Lisburn Health and Social Services Trust v H [2006] UKHL 36
Little (Junior) v Maguire [2007] NI Ch 7
McKernan v McKernan [2006] NI Ch 6
R v Registrar-General ex p Smith [1991] 2 QB 393
Re A [2001] NIFam 23
Re *CBCHSST v JKF* [2000] NIFam 76
Re EFB [2009] NIFam 7
Re G [2013] EWHC 134 (Fam)
Re H & A (Children) [2002] EWCA Civ 383
Re H [1981] 3 FLR 386
Re J and S (2001) NIFam 13 (23 May 2001)
Re JJ [2009] NIFam 2
Re Johnson [2008] NI Ch 11
Re K [2002] NIFam *Re L* [2007] EWHC 1771 (Fam)
Re NI and NS [2001] NIFam 7
Re P (A Child) [2008] EWCA Civ 499
RE TMH [2009] NIFam 11

SEHSST v LS [2009] NIFam 14
Webster (The Parents) v Norfolk County Council & Ors (Rev 1) [2009] EWCA Civ 59

European Court of Human Rights
Ahrens v Germany (App 45071/09) ECHRR (22 March 2012)
Anayo v Germany [2010] ECHR 2083 (21 December)
Frette v France [2004] 38 EHRR 21 (42)
Kearns v France ECHR 10 January 2008 (Application no 35991/04)
Marckx v Belgium (1979) (application no 6833/74) (1979)
Neulinger and Shuruk v. Switzerland (2010) [GC], no. 41615/07 ECHR
Odièvre v France [2003] 1 FLR 621 (App no 6833/74)
P C and S v United Kingdom (2002) 35 EHRR 1075
R and H v United Kingdom (2011) 54 EHRR 28
S.H. And Others v. Austria - 57813/00 [2011] ECHR 1878 (3 November 2011)
Y C v United Kingdom (2012) (App 4547/10) ECHRR

Ireland
M.R & Anor -v- An tArd Chlaraitheoir & Ors [2013] IEHC 91
N & Anor v Health Service Executive & Ors [2006] IESC 60

United States/Canada
Baby Boy Richard v Kirchner 513 US 1138 [1995]
DeBoer v DeBoer 509 US 1301 [1995]
In Re BGC (1992) 496 NW 2d 239 (Iowa)
In Re Bridget R et al, Minors (1995) 25 USCA 1901
Kelly v Superintendent of Child Welfare and Williams (1980) 23 BCLR 299 (SC)
N & Anor v Health Service Executive & Ors [2006] IESC 60
Re Adoption of BA (1980) 17 RFL (2d) 140 (Man Co Ct)
Ross v PEI (Supreme Court, Family Division, Registrar) (1985) 56 Nfld & PEIR 248 [1985] PEIJ No 1 (PEISC) (QL)

Chapter 2
The Blood-Tie as Socio-Cultural 'Item': Ancestry Feared and Revered

2.1 Introduction

> "Kinship," like totemism, the matrilineal complex, and matriarchy, is a non-subject, since it does not exist in any culture known to man.[1]

Blood-ties, especially within the sanctity of socially recognised types of family units, usually tend to be afforded legal acknowledgement and to give rise to juridical obligations. Duties of care, protection and financial maintenance remain fairly normative within most jurisdictions and underpin most aspects of familial life, usually even in the wake of relationship breakdown. As Carp famously argued in respect of American cultural and social norms, the genetic form of kinship has long 'occupied a privileged position.'[2] Bowie takes this further with the suggestion that the 'ideal' family situation possibly still rests upon the assumption that natal and social parentage will be vested in the same persons.[3] Again, the position of adoptee is, according to Kirk, 'achieved,' while the role of 'relative' tends to arise naturally as a consequence of natal connection.[4]

In a wider sense, kinship may however be based as much upon social attachments as upon genetic connection. Whether it arises from biological factors or psychological bonds, it still might require some extra element of socio-cultural endorsement to qualify for the benefits that tend to accompany a recognisable form of rights-bearing, legal 'relatedness.' A variety of judicial devices might be employed to bring this about, in a manner not dissimilar to those found in kinship-bestowing rituals, as seen in a number of social anthropological studies. Name-changing, oath-swearing, healing and protective rituals or the sharing of precious, perhaps sacred substances all serve a common cause: the prevention,

---

[1] Schneider (2004), p. 271.

[2] Carp (1998), p. xiii on the long-standing 'cultural bias against adoption'. See also Holy (1996), p. 11.

[3] Bowie (2004), p. 3.

[4] Kirk (1964).

A. Diver, *A Law of Blood-ties - The 'Right' to Access Genetic Ancestry*,
DOI 10.1007/978-3-319-01071-7_2,

denial or removal of 'otherness.' Arguably, the desire to endorse psychological ties with a stamp of visible relatedness perhaps stems from a subconscious need to identify and differentiate between

> sameness and difference . . .. in the constant effort to fend off the dangers of outsiders, while protecting and reproducing the group of those who care for each other.[5]

As such, the device of kinship, irrespective of how it has arisen, also often acts as a powerful means of social control, determining,

> whom to marry, where to live, how to raise children, which ancestors to worship, how to solve disputes, which land to cultivate, which property to inherit, to whom to turn for help in pursuing common interests and many things besides.[6]

Some of the early studies on human kinship seemed to be aimed at either denying or attributing some meaningful level of 'social significance to the natural facts of procreation.'[7] It may be argued that few aspects of kinship are cross-culturally universal beyond the 'basics of biology'[8] given that the very idea of legal relatedness itself probably arose from a fairly narrow range of genetic affiliations, such as 'self-evident lines of genetic descent.'[9] More recent studies seem however to have altered and widened the focus, looking more to the complexities of societal processes and norms of behaviour to examine more profound issues of identity, ethnicity or 'self-hood.'[10] As Stone has observed,

> Gone are the days when anthropologists assume that people the world over construct their kinship in terms of presumed biological links. What we see is the re-entry of biology, not as a universal factor (and not as the only factor) behind human kinship construction and behaviour.[11]

This 'shift in contextualization'[12] perhaps provides some insight into how the notion of kinship might be best understood or described. Despite the argument that 'genealogical relations are not relations of biological or genetic connection'[13] some theorists still seem quite taken with the significance of the blood-tie, framing it as a 'symbol of relationship'[14] or as a 'verbal label' of connectedness.[15] Genetic ancestry and the desire to either be reunited with or distanced from it, clearly remains significant, socio-culturally and psychologically.

---

[5] Lepri (2005), pp. 703–724 at p. 703.

[6] Holy (1996), p. 13. See also Beattie (1964), on social anthropology's contribution to other disciplines.

[7] Holy (1996), p. 1.

[8] Schneider (2004), p. xiii.

[9] See further Levi-Strauss (1949, 1963).

[10] Holy (1996), p. 7.

[11] Stone (2004), pp. 239–241.

[12] See Bloch and Sperber (2004), pp. 438–455.

[13] Holy (1996), p. 15.

[14] *Ibid* at p. 20. See also however the arguments of Barnes (1961), pp. 296–299 on how definitions of 'descent' and 'kinship' might differ.

[15] See Scheffler (1973), p. 751.

The question of why or how human societies are able to achieve some level of social cohesion and functionality in the absence of formalised or legal frameworks is perhaps the most relevant one to be asking. Might the need to gain some workable degree of 'consubstantiality'[16] not have arisen initially from some form of 'green beard' quasi-altruism? If so, then preferential treatment could be afforded mainly to those 'trait-recognized' others, who could pass for clan members, if not for actual genetic relatives.[17] Such an ostensibly 'selfish gene' could perhaps have encouraged subconscious recognition and protection of otherwise unknown, 'non-kin' strangers on the basis that they were at least displaying some potential for relatedness, through obvious signs of physical resemblance.[18]

The ability to establish and honour such bonds of fictive relatedness could have operated as an important mechanism for group survival, not least by enabling the perception—and the avoidance—of potentially dangerous 'otherness.' Whether such a process for avoiding difference is still relevant to the ways in which we form psychological bonds today is a matter of conjecture. In respect of attachment-forming, the blood-tie, or that which best mimics it, clearly remains significant; matching children to parents on the basis of physical characteristics was not unheard of during the 'golden age' of adoption practice,[19] while gamete donation policies seem to take into account recipient preferences in terms of race, ethnicity and physical features e.g. to match those of a spouse or partner.[20]

In anthropological terms, lost or unknown genetic ancestry seems to have represented an undesirable outcome.[21] To be seen either as utterly kinless or as an unrelated 'other' was to perhaps exist outside the safety of family circles, and to possibly also be excluded from the comforts and benefits of clanship or wider

[16] Pitt-Rivers (1973), p. 92.

[17] Hamilton (1964), pp. 1–52.

[18] Dawkins (1976); see also however Lieberman et al. (2007), pp. 727–731.

[19] The 1950s/1960s era, which saw significant numbers of white, able-bodied babies relinquished at birth, due to the twin stigmas of unwed motherhood and illegitimacy; the issue of 'whether to tell' these 'orphans' of their past, was often key. See for example Kornitzer (1952) ('family life [was] threatened' and 'the institution of marriage [was] in peril.') p. 2. See also Brown (2001), p. 538 on the levels of adjustment needed in both 'intra-racial' and trans-racial adoptions.

[20] See The Joint Statement by The British Association for Adoption And Fostering ('BAAF') and The British Society for Human Genetics ('BSHG') on '*The Use Of DNA Testing To Determine Racial Background*' on how a 'good physical match to ..size, skin colour and hair colour' between child and adoptive parents may aid bonding; how genetic testing to discover racial ancestry might delay placement and not necessarily reveal the child's ethnicity. It is conceded however that, 'knowledge of a child's racial background can assist with appropriate screening for certain hereditary conditions which are prevalent in particular ethnic groups.' (p. 1) Available at www.baaf.og.uk/info/lpp/adoption/ethnictesting.pdf (accessed 02.02.09). See also Chand and Thoburn (2005), pp. 169–178 and Thoburn (2003), pp. 236–237, together with BAAF's response (24 December 2003) available http://baaf.org.uk/res/consultations/consultresponse_bme.pdf (accessed 31.01.09).

[21] See further Radcliffe-Brown (1952), p. 53, Malinowski (1922, 1929), Aginsky (1935), pp. 450–457, and Evans-Pritchard (1951).

society. The various customs that have evolved around making the foreign or unknown blood-tie 'safe' are quite telling. They tend to echo the aims and methods of court proceedings in formalised, 'classic' adoptions which might for example re-name the child, place a ban on all kin contact or permanently seal the original birth records. They generally will also remove all aspects of parental responsibility from the original family, and perhaps delegate important decisions on the best interests of the child to the new parents, or to their social workers. Such outcomes possibly serve mainly to reassure the newly related that their position is now secure, and will not face any future challenges from those who were once connected with the child's 'past.'[22] The removal or disguising of 'otherness' may however require more than this, to copper fasten the new ties of social or legal relatedness.

2.2 'Treating' Otherness: Fluid, Magic and Myth

As the folklore would have it, even where a new family member is no longer officially regarded as an outsider there may yet be some lingering, subconscious aversion to qualities which register as unfamiliar, if not potentially dangerous. Many of the myths and morality tales on adoption and fosterage contain dire warnings against the risks of taking in the stranger.[23] Timeless stories commonly feature loss, fear and retribution following child abandonment, late-discovered genetic heritage or unwanted kin reunions.[24] The deranged adoptee frequently features as the chief protagonist in many of these 'cuckoo kinship'[25] fables, and they often will also display unwelcome tendencies for patricide, regicide or some manner of incest.[26] Factors such as ingratitude, mixed loyalties, and a dangerously 'fractured' sense of identity tend to be put forward as possible reasons for their mental instability or ill-temper.[27]

Within many forms of customary, de facto adoption, a wide variety of rituals and ceremonies exist in respect of preventing such episodes and ensuring the well-being of all involved in the triad, if not the entire community. These may be used to forge or strengthen political links, mark the commemoration of historical events or conceal unwanted matrilineal identities.[28] Some Greek legends refer for example

[22] On Special Guardianship Orders in England and Wales see Lowe (1993), p. 62; Bainham (2007), pp. 520–523; Talbot and Kidd (2004), p. 273; Mitchell (2009), pp. 8–14.

[23] See further Volf (1996) and Hallam (2000).

[24] See for example Ellis Davidson (1941), pp. 70–85; Nagy (1985).

[25] Crawford (1999), p. 131; Boswell (1988), p. 84.

[26] The raising of unwanted children by other societal outcasts or by animals was also not unheard of within such morality tales. See further Visser (1986) and Patterson (1994), and Tuite (1998), pp. 289–343.

[27] Parkes (2004), pp. 587–615 at p. 600.

[28] See Anderson (2004), p. 111.

to how 'strategic alliance' fosterage[29] provided a useful means of resolving disputes or dilemmas over rights of inheritance and title succession, whilst a number of Roman myths make reference to sacred rituals aimed at embedding the bonds of social kinship.[30] The aim of providing substitute child care might not necessarily have always been the main concern; adult-centric worries over the preservation of shared, communal histories or the maintenance of family heritage and blood-lines may have been a more likely catalyst for the transfer of parenthood and parental obligations.[31]

That said, property transmission via the notion of 'heirship' and substitutive parenting by non-genetic, social kinfolk could, from the child's perspective and in terms of material gain and social advancement, amount to a highly beneficial alternative to being raised by one's poorer, original relatives.[32] As Bowie observes,

> there are societies in which adoption is not only common, but an essential and often preferred means of raising children. . . .whether or not it is immediately visible, the practice of circulating children to be raised by non-biogenetic parents is ubiquitous.[33]

That said, many of the ceremonies connected with kinship bonding suggest that attitudes towards the blood-tie might be double-edged; it is a thing that is often as much feared as it is revered. Genetic ancestry may be worshipped and actively preserved across geographic or generational distances or completely hidden away due to being an unwanted or unknown item of birthright heritage. An 'absence' of ancestry may require elaborate socio-cultural rituals to banish, repair, replace the blood-tie; not dissimilar ceremonies may however be held to cement the new bonds of social relatedness and perhaps ward off unwelcome interference from those others that exist outside the legally sanctioned family circle.[34] Conversely, where lost or broken genetic connections are in need of repair, similarly complicated ceremonies may be carried out to achieve some degree of 'repatriation,' healing or atonement for loss.

Fluid-based rites seem particularly apt where blood-ties are concerned and generally serve to emphasize the sanctity, if not also the acute fragility, of created kinship attachments. Much of the symbolism is, as ever, perhaps aimed at providing reassurance to the new kinfolk, in terms of ensuring their exclusivity of connection and providing a sort of indemnity against future harms or challenges from

---

[29] Kenna (2001) [as cited by Parkes (2004)]. See also Ensel (2002), pp. 83–96 on Mediterranean patronage-kinship links.

[30] See Bremmer (1999), pp. 1–20; Bell (1997).

[31] Anderson (2004), p. 112. Fosterage may also be motivated by a maternal desire to achieve a more empowered matrilineage rather than by an altruistic desire to nurture a vulnerable child. See also Notermans (2004), p. 48.

[32] See Alber (2004), p. 33.

[33] Bowie (2004), p. 4.

[34] Howell (2003), pp. 465–484.

aggrieved former relatives. Examples may include the use of sanctified oils in Christian god-parenting ceremonies[35] or the ingesting of breast milk in lactation-based Islamic milk kinship customs.[36] The waters and oils of baptism also to an extent symbolize and supersede the blood-tie, forging a 'spiritual cognation superior to that of mere flesh and blood'.[37] Similarly, a variety of fluids may symbolise, replace or mimic the blood-tie; practices such as the exchange of children in return for 'anointed' animals (i.e. with the bodily fluids of the relinquished or abandoned child) amongst African tribesmen[38] or the application of red sacred clays to the foreheads of lower caste, Asian adoptees (to achieve their 'purification' and enable their new kinship ties)[39] clearly also have an important element of the mystical about them. Sharing essential substances can also enable relatedness on a more pragmatic level however, for example via a communal ingestion of scarce foodstuffs, perhaps grown on land owned by the new kin-folk.[40]

A similarly powerful tradition of ritually-created permanent affinity is found within the Islamic doctrine of 'milk kinship' ('*ridā*').[41] As Parkes observed, something very closely akin to the biological blood-tie could be established via lactation with a substitute 'mother.' Consanguinity could be conferred upon any other 'milk-siblings,' who would thereby gain a strong form of 'uterine filiation.'[42] This was dependent upon belief in the sanctity of the breast milk itself, rather than on the physical act of lactation, so that the fluid, rather than the nurturing act of feeding, served as a replacement for the biogenetic substance of blood-kinship.[43] The resultant relationship was much more profound than other forms of paternal fosterage or long-term guardianship.[44] Marriage was for example forbidden between 'milk-kin', and also between wet nurse and suckled infant, because of the irrevocable relationship of legal 'closeness' ('*Garābah*') that would result.[45]

Such was the power of milk kinship that it could also be used to bring about an end to 'blood-feuds' ('*Ak' ukatshara*')[46] and, as a fealty rite, to effect political,

---

[35] See further Foster (1953), p. 1; Jussen (2000).

[36] Parkes (2005), pp. 307–329; Clarke (2007), pp. 287–304.

[37] Parkes (2004) *ibid*, p. 587.

[38] Talle (2004), p.72.

[39] Such as '*eryaa*', a sacred reddish-coloured clay; see further Treide (2004), pp. 126–141 at p. 135.

[40] In respect of immigrants being accepted as kinfolk by more established clan members, see Watson (1983).

[41] The bonds of *Ridā* are similar but not identical to those of biological descent under *Nasab*; although they constitute legal kinship they do not give rise to any rights of property succession. Parkes (2005).

[42] Clarke (2007), p. 289.

[43] *Ibid*.

[44] Khatib-Chahidi (1992), pp. 109–133. See also Keating (2001), pp. 1–9.

[45] See Parkes (2005), p. 594: This 'blood-tie' belief followed the Aristotlian assumption that breast milk somehow derived from uterine blood that had been affected by seminal fluid. As such, colustrum was not always seen as suitable for infants. See also Altorki (1980), pp. 233–244 at p. 233; Giladi (1999).

[46] Parkes (2005), p. 591.

'cliental' alliances.[47] The fluid-derived bond, once established, could also be passed down to subsequent generations, as a valuable piece of hereditable property, much like the blood-tie is perhaps used to carry inherited wealth, peerages and legitimate Royal status. Although 'stranger' adoptions in the Western sense ('*Tabannī*') are forbidden under strict Islamic law,[48] many rights may still be enabled via milk kinship. By the same token, profound stigma could attach to any child deemed to be genetically kinless (e.g. foundlings) especially where they might have been conceived outside of marriage.[49] Strong reluctance to nurse such unwanted or orphaned children has been described by Clarke, who suggests that this may arise from an aversion to creating a fosterage relationship of obligation towards the 'fatherless' and heavily stigmatized infant.[50] Modern moral dilemmas continue to surround the concept of milk kinship; discourses on the contentious use of 'milk-banks' for premature babies and the recent controversies surrounding the use of donated eggs as a means of assisting reproduction ('*ikhtilāt al-ansāh*') tend to bear this out.[51]

In terms of feelings of fear or shame, as Wiersema found through her work in South Vietnam, some potential adoptive parents expressed reluctance over receiving a child with an unknown heritage or nature, and perhaps become preoccupied with concerns over what might happen when the child comes of age. They might equally be swayed by a preference for perpetuating only their own, knowable blood-line.[52] As Miall has further suggested, some degree of reflected societal 'stigma potential' might be felt by infertile adopters who may have some degree of self-perceived 'second-rate' status over their substitute parenthood. These feelings may be compounded by some unvoiced, subconscious bias against their child's absent 'genetic past'[53] or by the fear that they might yet grow up under the shadow of a 'spoiled identity.'[54] In any event, it still seems that, however unspoken or subtle, 'in the nuances of language and informal social exchange, adoption is marked as difference.'[55]

[47] This was often the case in pre-modern Iranian societies. See Khatib-Chahidi (1992).

[48] In respect of changing one's surname, gaining rights of succession, and the falsification of the truth'; see Clarke (2007), p. 294.

[49] Sonbol (1995), pp. 45–67. The category of '*Zinā*' is perhaps closest to the concept of illegitimacy, where the child is related only to his mother with no degree of paternal affiliation.

[50] Parkes (2005). Roman milk kinship did serve however to establish adoptive paternity unlike the narrower 'Koranic recognition' found within the classic Islamic model. See also Lynch (1986), Princeton University Press, Princeton on the creation of 'spiritual kinship' through Christian god-parenting.

[51] Clarke (2007), p. 296.

[52] See Wiersema et al (2006), p. 54.

[53] Miall (1987), pp. 34–39 at p. 34.

[54] See Goffman (1963).

[55] Melosh (2002), p. 2.

2.3 De Facto Adoption: *'to take and cause to grow'*

Many models of customary adoption do not necessarily demand complete severance of the blood-tie;[56] at times they are much more akin to a form of long-term fosterage or guardianship. They also seem generally to tend towards acknowledgement of the double-edged significance of the blood-tie; permanent severance of original ancestry or kinship is not always a pre-requisite to joining the new family, with vulnerable children sometimes gaining additional kinfolk rather than being given replacement relatives or a life veiled in secrecy. In Demian's study for example, the psychological aspects of transferred attachments seem to have been afforded some degree of significance, at least in relation to knowing which group the child might still 'belong' to post-relinquishment. Against such a template for preserving genetic connection, the clean-break, sealed records methodologies of 'classic' legal adoptions could perhaps be framed as representing a highly 'inadequate container for identity.'[57]

Informal Suau adoptions similarly tend to presume that both sets of triad parents (i.e. natal and adoptive) should be fully in support of the process. Should an adoptee choose to return to his genetic parents permanently, for example, some form of material compensation could be paid by them to the social parents 'in the form of pigs, money or both.'[58] Whereas formalized models of adoption might change 'relations into rights for the purpose of manipulating them in court,'[59] these customary models would instead transform kinship relationships into valuable items or transactable 'objects.' To some extent, this could serve to minimize the risks to the adopted child, in respect of preventing the loss of their sense of authentic 'personhood.'[60]

The dual purpose of such an informal arrangement could be to provide a solution to social crises (such as infertility, infant mortality, orphaning) or to re-allocate or share out children as precious 'resources.'[61] The psychological aspects of losing original family members are to some extent at least acknowledged here. In some areas a child would for example only be moved to the permanent foster home (usually of an unmarried or childless relative) after the birth of a younger sibling, so that the relinquishing mother would 'not stay without a child.'[62] Again, there would be scope for overturning an unsatisfactory arrangement.[63] The emphasis on parental altruism and the high esteem in which the newly formed family would be held, also

[56] Demian (2004), p. 103.

[57] *Ibid* p. 105.

[58] Disputes over the level of such compensation are heard in the village court; issues over kinship or land ownership are beyond its jurisdiction however. *Ibid* p. 107.

[59] *Ibid.*

[60] *Ibid* at p. 108.

[61] Treide (2004), p. 127.

[62] Alber (2004), p. 36. See further Alber (2003), pp. 487–506.

[63] Kraeger (1980) [as cited by Treide (2004)]; see also Carroll (1970), pp. 3–17; Brady (1976).

suggests that there was considerable honour to be found in the equitable 'sharing' of the care and parenting of longed-for children. The social re-integration of illegitimate children or of those whose parents were divorced might also be achieved.

Treating children as precious 'commodities' also features in Maasai customary practices, although here Talle differentiates between the pragmatism of inter-ethnic adoption (e.g. during famine or drought) and the altruism of creating new social kinship bonds 'within the patrilineal or extended Maasai family.' The latter model depended upon there being a shared and 'deep-rooted worldview of commonality as a form of cultural and social practice and lived experience.'[64] Shared parenthood perhaps represented an extension of those rituals surrounding other collectively 'owned' scarce and sacred essential items such as homes and livestock. Interestingly, a fairly wide range of persons might have 'shares' in the precious children, with the term '*yieyio*' ('mother') used to not only define but to include as caregivers other female relatives, such as aunts and stepmothers.[65] Successful childbearing carried much social esteem and was linked to a number of fiscal benefits such as an extra allocation of scarce livestock; deep cultural stigma would be attached to female infertility ('*enkitok olupi*').[66] Adoption of a child into a childless 'house' was a very highly regarded act, to the extent that it could even 'restore order' among the wider community.[67] Talle further noted that Maasai customary adoptions were not so much shrouded in secrecy, as they were simply not mentioned. How children might find out that they were adopted is not described, but the fact that maternal relinquishment may not always occur willingly is noted.

The use of legal fictions to create a sort of contracted-for relatedness could also occur in de facto adoptions. As Witherspoon's study of the Navajo people for example found, a genetic father's legal relatedness to his children might depend almost entirely upon his remaining in a committed relationship with the children's mother.[68] This echoes to some extent how legal parentage might be determined or

---

[64] Talle (2004), p. 65.

[65] *Ibid.*

[66] *Ibid* at p. 69. Talle further notes how, '*children and livestock go together, singly they have minimal worth...if a Maasai were to choose between livestock and children, however, most Maasai would prefer the latter because children may generate livestock through sons' raiding during the period when they are young 'warriors'...and through bride-wealth at daughters' marriages.*' See also Goodenough (1970), p. 398 on how adoption may be regarded as a 'transaction in parenthood as a form of property'—[as cited by Talle (2004), p. 97].

[67] A heifer would be exchanged as ritualistic 'compensation' for the loss of a child ('*ngishu oroto enkerai*'). Here, the use of bodily fluid 'seals' the adoption with the child urinating upon the animal. The original and adoptive mother both drink the fat obtained from a slaughtered ram to effect and endorse their new kinship bond, and (it was thought) to promote lactation in the adoptive mother. Talle (2004), p. 72.

[68] It would be likely to terminate upon marital estrangement. See Witherspoon (1975); see also *Michael H v Gerald D* 491 US 110 (1989) which ascribed legal parenthood to the child's social father (by marriage) rather than on the basis of genetic paternity. Notermans also describes how maternal grandmothers might alter a child's birth certificate to preserve matrilineal descent links, in unmarried births (2004).

sanctioned in other, non-indigenous contexts for example e.g. via matrimony or through a partner providing consent to fertility treatment.[69] Such additional, 'legalised' aspects might also feature in other regions; as Anderson discovered some models of substitute child care in Papua New Guinea ('*oala*') remained distinct from their more informal fosterage processes. Name transfer might be regarded as an essential element, for example to enable property entitlements and a sense of heritage identity 'placement' within the adoptees' new 'social landscape'.[70] Deep socio-cultural, even quite gendered taboos might also work to hinder the adoption of children from male matrilineages so that a man could not adopt his sister's child. This served to prevent the build-up of powerful matrilineal kinship descent groups which might have ultimately combined to challenge patrifiliative allocations of land, place name or heritage ('*ro*').[71]

Demian's work is perhaps also particularly significant in respect of how various terms were found to differentiate between the types of customary adoption there. '*Abiudo* '*i*' for example means to 'take a different one', whilst '*abihela* '*i*' translates into 'to take and cause to grow.'[72] Nurture is emphasised here over nature as a key component of parenting, although the child 'returning to the natal parents is always an option' with contact between birth relatives not necessarily being frowned upon by the wider community.[73] The naming of Yapese adoptees similarly often involved retaining a territorial pool of communal set names, which together held much significance in respect of maintaining for the child an authentic sense of collective heritage, via 'place, estate and ancestors.'[74]

Halbmeyer further noted that the transplant model of adoption tended to be 'entirely absent' from his study, with infants still regarded as 'children of their parents, even if adopted by others.'[75] Within such frameworks, parental responsibility could still be forfeited however, if for example child neglect had occurred. If birth parents had stopped feeding a child, they would lose the right to engage in any subsequent decision-making over that child's future welfare. The child's need for familial attachments also seems to be acknowledged here, where the 'movement of children to other subgroups. . .[was] tolerated at least as long as a close relative or parent moves with them.'[76] Where relinquishment was completely non-consensual

[69] See for example the United Kingdom's Human Fertilisation and Embryology Act 2008 s 35–41 available http://www.legislation.gov.uk/ukpga/2008/22/pdfs/ukpga_20080022_en.pdf (accessed 12.12.12).

[70] Anderson (2004), p. 113.

[71] *Ibid.*

[72] Demian (2004), p. 102.

[73] *Ibid* p. 103.

[74] Treide (2004). Official registration of such de facto adoptions was rare, with presumptions of mutual trust, gratitude and generosity between triad members generally forming the basis for this model of substitutive child care.

[75] Halbmeyer (2004), p. 147.

[76] *Ibid.*

however (for example, following missionary intercession) this was likely to be perceived as a form of 'loss, robbery or predatory intervention.'[77]

Alber's suggestion is a logical one; the wider, basic notion of kinship is perhaps best understood as comprising not merely of 'biologically based social relations, but as social relations of belonging, often expressed by metaphors coming from the biological arena.'[78] He found Northern Benin customs of fosterage to be fairly widespread, culturally desirable phenomena which often simply rendered genetic parenthood 'invisible.'[79] Social parents might easily assume the basic functions and duties of genetic parenthood, but children would still generally retain their original clan name, even where the exact identity of their actual birth parents was perhaps not revealed to them. Their placement would also occur at a fairly young age to try and minimize the risk of psychological trauma; the process was aimed mainly at enabling their self-confidence and providing them with an education and some sense of '*sekuru*' (respect, gratitude or shame) in the presence of their elders, which their biological parents had been deemed incapable of providing.[80] The arrangement served as a sort of societal apprenticeship, with both child and new parents deriving a considerable degree of mutual, material benefit from the arrangement.[81] In this sense at least, it was felt that 'parental change d[id] not damage the child.'[82]

This assertion seems slightly at odds however with some of the 'very emotional stories' of how some customary adoptees had eventually learned the true identities of their biological parents, in later life. Alber also outlined how some birth mothers would secretly make contact with their relinquished child, either to help care for them or to provide them with small gifts. Birth fathers, conversely, seemed to embrace more fully the socio-cultural 'norms of distance' that were meant to apply, sending their children away if they were approached by them. Notermans further noted that there was a wide range of 'conflicts that may occur in foster situations'[83] where for example mothers might be tempted to actively discriminate against those foster children who had come into their care from different descent lineages. The blood-tie was also highly relevant in terms of eligibility to foster a child;

> though kinship networks are large and people in a polygynous society have numerous (half) sisters and (half) brothers, only full siblings are considered good and reliable foster parents.[84]

---

[77] *Ibid.*

[78] Alber (2004), p. 33.

[79] *Ibid* p. 34.

[80] *Ibid* p. 43.

[81] *Ibid.* Migration to urban areas and increased access to formal schooling appeared to be leading to a gradual decrease in the practice however.

[82] *Ibid* p. 44.

[83] Notermans (2004), p. 49.

[84] *Ibid* p. 56.

Lepri's processual forms of kinship ('*queanahue*') in rural Northern Bolivia seem similarly unfixed, remaining open to possible challenges from original kin, in sharp contrast to the uncomplicated, blood-tied form of relatedness that would simply arise at birth.[85] Within the Ese Ejja's definitions, social kinfolk were viewed as 'non-others ('*uapapojiama*') and thus in a forbidden category in terms of sexual relationships, as either 'parallel relatives' or 'cognatic affines.'[86] As ever, double-edged aspects of otherness remained; those who had become kinned with through marriage, might remain as '*pia-jaaji*' (enemy, or 'other') unless they managed to establish strong links of residence with their new relatives ('*enije-jaaji*').[87] Fears of encountering such 'otherness' retained significance, despite Anderson's assertion that informal adoptions tend to 'add relations' and be largely directed at 'creating a proper belonging for the child.'[88]

As in classic models of closed adoption, frameworks of secrecy seem to be used here to embed a sort of 'transactionary contract' and hide unwanted heritages or unhappy personal histories, such as infertility or neglect.[89] The 'flow of knowledge' to children about their genetic ancestry could still be subject to parental vetoes; identity 'rights' in this context perhaps become 'handles by which anthropological theorists transformed the parent–child relationship into a transactable object.'[90] By the same token, some form of equitable 'trusteeship' perhaps attaches to these transactional aspects of created, social kinship, lending them an increased level of sanctity or social esteem. In any event, these customary models for social kinship provide a useful contrast to the more formalised processes of legal adoption. They do not necessarily demand that complete origin deprivation occurs; they suggest that a variety of forms of parenting may exist and serve to benefit the vulnerable child; they also acknowledge that the blood-tie is highly significant and might need to be replaced or supplemented with other substances to avoid a placement-disruptive sense of 'otherness.' Most significantly they often emphasise that children are a precious commodity and that parenthood brings not just an increased sense of social esteem but a wide range of sacred obligations of care and protection. Perhaps most significantly, they highlight that an absence of ancestry or a loss of birth right heritage can be harmful and might require some form of healing process to be carried out.

85 Lepri (2005), p. 703.

86 *Ibid* at p. 708.

87 Such deep mistrust of 'strangers' (which included non-cognatic affines) apparently arose from the long history of violent conflict both within and beyond tribal networks; *ibid* p. 710.

88 Anderson (2004), p. 112. The social powerlessness of unmarried birth mothers is also referred to here p. 118.

89 Demian (2004), p. 100.

90 *Ibid* p. 97.

2.4 Hereditary Ghosts and Beggar Spirits: Healing, 'Sorcery' and Ritualised Reverence

In terms of preserving genetic ancestries, communal veneration of relatives tends to rank highly as a social custom. Preventing needless 'orphanisation' of the living and avoiding the anger of deceased, neglected or lost ancestors, seems to have been of central concern. Pre-occupation with invisible yet potentially malign kinfolk might also underpin the act of name-giving. The power of the blood-tie was often also such that it could survive death, with the spirits of biological ancestors having the ability to 'pass as "ghosts" into women during pregnancy'.[91] As Brosnan noted in respect of the Shinto religion, absolute sanctity might attach to the duty of preserving family ties, 'so that there may be always someone to worship at the ancestral tablets.'[92]

Li et al. have also observed that the practice of Chinese uxorilocal marriage stressed the profound importance of being kinned with, especially where the hereditary transmission of family wealth was involved.[93] Formal adoptions in imperial Japan also encompassed an impressive range of motives, including the 'adoption of extramarital lovers ... to reduce inheritances taxes'.[94] The relationships here were based upon adult 'contracts' to engage in symbiotically beneficial kinship arrangements, in contrast to child-protective substitute parenting.[95] Enabling such illicit 'non-kinfolk' inheritance of family property also demonstrates not only the 'breadth of the concept' of kinship, but the significance of how 'extralegal norms' can come to be associated with the various fictions of legal relatedness.

The 'house' systems ('*ie seido*') of Asian kinship were similarly generally comprised of ancestors, but could be expanded to include as-yet unborn, descendant genetic kinfolk. Interestingly, house membership did not necessarily depend upon the presence of the blood-tie; slaves could also be adopted into a house and gain the status of social kin-folk.[96] As Yanagisako has further observed in relation to Japanese Americans, such a means of creating kinship could also move beyond

[91] Treide (2004), pp. 126–141.

[92] Brosnan (1922), p. 332 at p. 334 [as cited by Bryant (1990), p. 299 at p. 302].

[93] Here the husband of a woman lacking male siblings might be adopted by her parents to enable family property succession to occur. See further Li et al. (2001), p. 12.

[94] Bryant (1990), p. 304. Adoption among the Samurai (warrior classes) was however subject to greater judicial regulation given the financial implications of appointing successors.

[95] *Ibid* p. 328. Such processes involved consensual, adult adoptees ('*yoshiengumi*') yet still outnumbered the number of child protective 'stranger' adoptions involving abandoned infants ('*miseinen yoshiengumi*'). Mediation played an important part in ensuring that subsequent disputes over property or inheritance might be resolved without recourse to judicial intervention and the risk of public opprobrium. See further Befu (1963), p. 1328; Shimizu (1987), p. 585.

[96] *Ibid*. Eldest sons remained within their patrilineal line ('*honke*') while younger sons were free to establish discrete, branched lineages of their own ('*bunke*').

consanguineous links of blood affinity to include the spouses of one's siblings.[97] Asian agnatic adoptions allowed for a 'ritual heir to succeed a sonless couple' to preserve clan lineages, although the blood-tied kin-folk would retain their positions of superiority. Wives would generally be subsumed into their husband's lineage, so that they might be included in key ancestral rituals, post mortem, rather than being rendered kinless. Posthumous adoption could also ensure the continued observance of ancestral worship rites and preserve social status amongst the living.[98]

This form of social relatedness could also arise from the exchange of mortuary offerings ('*koden*').[99] Addison outlined how amends for lost ties of kinship were often made through the use of elaborate and highly visible public ceremony involving banqueting, or the burning of paper money. The maintenance of a designated temple for those whose descendants might have subsequently become 'extinct' was also commonplace. Repossessing previously relinquished place names within kinship groups was also possible.[100] The ancient feudal system of the *Dozoku* stressed the need and duty to actively sustain and make visible the links with one's clan history. Chinese 'clan temples' similarly honoured and nurtured ancestral links via the preservation of sacred family tables,[101] as do many South Korean graveside kinship rituals ('*chesa*').[102] Evans-Pritchard also described how a form of 'ghost-marriage' could also occur amongst the Nuer tribes-people, where the obligations of kinship might require a younger brother to marry the widow of his deceased sibling. Ceremonies would be held to prevent the spirit of the recently 'unkinned' dead husband returning to 'haunt' his new family. Any children born subsequently would also be seen as having been fathered by the deceased kin's spirit.[103] As with some western models of assisted reproduction, avoiding disputes or ill-feeling on legal parentage and warding off unwanted interference perhaps seems to be a fundamental concern.

More pragmatically, many Polynesian 'kindred groups' and a number of early Semitic and Indo-European tribes used their models of 'ancestor worship' as a means of strengthening their extended clan networks and ensuring survival in times of crisis.[104] Welfare-led Confucian doctrines of filial piety underpinned sacred duties of veneration in relation to genetic forebears,[105] whilst Suau mortuary

[97] Yanagisako (1978), pp. 15–29; see further Yanagisako (1985).

[98] A brother's son (*samch'on*) took precedence over the child of a cousin in respect of succession however. See Roesch-Rhomberg (2004), p. 83.

[99] Yanagisako (1985), p. 18.

[100] Addison (1925); see further Traylor (1988); Baker (1979).

[101] Addison (1925).

[102] The ritual honouring of one's in-laws during marriage ceremonies ('*p'yebaek*') where both bride and groom must bow deeply in gratitude to the kinfolk of the groom is also sacred. Roesch-Rhomberg (2004), p. 3.

[103] Evans-Pritchard (1951).

[104] See further Foster (1995) and Moore (1998).

[105] See also Traylor (1988).

exchanges were framed as essential acts, allowing mourners to be released from the grieving process.[106] Broken links of kinship might give rise to a number of such dangers, including that of witchcraft accusations.[107] Berg describes how healing rituals would be performed by the Xhosa-speaking peoples of South Africa, indoors and within a system of concentric, protective circles ('*Mandalas*') which served to shield the 'patient' from the harmful influence of outside world, in a manner perhaps not very far removed from some of forms of Western psychotherapy e.g. those inspired by Jungian wisdom. Other tales stressed the malevolence of such 'beggar spirits,' describing them as disturbed souls who had either died without any relatives, or who were not now being afforded sufficiently respectful 'rites of tendence' by their genetic descendants. They posed a significant danger to the rest of the community because they were believed to be 'subject to no law'.[108]

The link between honouring ancestry and achieving good mental health is also stressed by Berg, who notes that the concepts of kinship and relatedness are usually 'given expression in life cycle rituals'.[109] Two interlinked notions attach to such a belief system; classic ancestor reverence and the wider notion of '*Ubuntu*' (which translates as compassion or innate humanity). Within these examples, the need to establish social kinship seems again to be underpinned by a fear of giving offence to those unpredictable, kinless others who might be likely to seek revenge on whoever was the cause of their state of exclusion.[110] Repatriation via healing spells ('*intlombe*') could restore genetic links weakened or broken by adverse circumstances however. Such ceremonial reunions in a sense brought estranged kin back together by strengthening the ties between those vulnerable, 'disconnected' former members of the group and the rest of the 'kinned with' community, helping to maintain a sort of social order.[111] In terms of preventing the stigma of illegitimacy, a variety of public or private rituals also existed, such as those Yapese processes which would allow for young and disadvantaged villagers ('*milingai*') to be made eligible for adoption by a higher-caste, land-owning family, once 'purifying enchantments' are carried out.[112] Therapeutic ceremonies aimed at replacing or repairing blood-ties may also see the spirits of dead enemies subsequently 'incorporated' into a household, for example following episodes of

[106] Demian (2004), p. 108. See also Damon and Wagner (1989) and Battaglia (1992), pp. 3–18.

[107] Berg (2003), pp. 194–207 at p. 197. See also Bion (1967), pp. 271–280 on the 'containment' provided for infants by mothers.

[108] *Ibid.*

[109] Ibid p. 196.

[110] Fears over origin deprivation and kinship loss might attach to a variety of activities. Thus the right to engage in ancestor veneration may be lost to one's fellow clan members if an individual moves away from ancestral lands. A form of quasi-adoption may however enable proxy rituals to still be carried out through fictive kin living in neighbouring or distant lands. See Treide (2004), p. 134.

[111] Berg (2003). See also Grimassi (2001), on the more superstitious aspects of ancestor worship, linked to accusations of '*stregheria*'.

[112] Treide (2004), p. 135.

warfare, again to avoid the creation of potentially dangerous, kinless others.[113] Conversely, sorcery by female kin might be seen as dangerous, with some matrilineal heritages opting for concealment;

> Some lineages are said to use *iaboua* more often than others and if one person uses the magic, the whole matrilineage shares the responsibility. . .nobody likes to draw attention to their matrilineal identity.[114]

With many of the healing, repairing or cleansing rituals described here, there seems to be an acceptance that the various processes and purposes of social kinship can sometimes place all members of the triad 'firmly outside the order of nature' and perhaps create a permanent 'marker for difference.'[115] Arguably, they also address the tendency towards the sort of 'doomed quality'[116] that can attach to social kinship. Subjects that might well provoke acute discomfort or permanent silences in more formalised settings (parental infertility, illegitimacy, ingratitude, aggrieved former relatives, mental health problems) seem to find an easier expression in many of the rituals referred to here. The diverse range of symbols, gestures and ceremonial acts seem to reflect how pain, loss and suffering can sometimes be the main catalysts for social parenting.

As with the 'classic' adoption model, children may have been orphaned or bereaved, or suffered severe abuse or abandonment. Adoptive parents may have been infertile, while birth parents may have had to submit to non-consensual relinquishment of their children through poverty, illegitimate birth or social pressure. That 'healing' rites are sometimes also needed to appease wounded ancestors, placate a disgruntled community or perhaps erase or minimise stigma at least also indicates a useful degree of awareness of how the less noble aspects of human nature might operate in such circumstances. Similarly, the 'second-choice' nature of some forms of social parenthood is perhaps acknowledged through the use of sacred items (consecrated oils, sacred colourful clay) or via the careful usage of 'next-best' substances such as breast milk or scarce, self-grown foodstuffs.

Perhaps where children are viewed and valued as a scarce 'resource,' disputes over exclusivity of parentage or parenting rights are inappropriate. Against a backdrop of war or famine, pragmatic approaches to kinship and child protection are based more upon a con-substantive need to survive rather than on an adult-centric desire to conceal infertility or child illegitimacy, by permanently extinguishing all links with their original genetic relatives.[117] In any event, the message of these de facto adoptions seems to be twofold; the blood-tie clearly matters but can be satisfactorily replaced or have its ill-effects held at bay by the efforts of nurture, the use of other healing substances or the introduction of some

[113] Halbmeyer (2004), p. 146.

[114] Anderson (2004), p. 121.

[115] Wegar (1997), pp. 97–118 at p. 98; see also Wegar (2000), pp. 363–370.

[116] Carsten (2000), p. 691.

[117] See Wright (1994).

cleansing 'magic.'[118] Because it matters, the blood-tie may also be seen as quite dangerous if ignored; it may well be necessary to have it revived or repaired via atoning gestures involving other losses or sacrifices (money, food, time) and by taking great care to preserve at least some aspect of it (tribal or place names, knowledge of kinfolk). Tying these two strands together as ever is the need to avoid being regarded as the kinless, potentially dangerous 'other.'

2.5 Conclusion

As Carsten has argued, where crafted forms of relatedness are based upon quasi-contractual exchanges of rights and responsibilities, children are the parties most affected by such 'contracts.' This is so in spite of the fact that they were most likely drawn up and agreed to by the triad adults or by those given charge of protecting social kinship bonds. Thus,

> from the point of view of the child, adoptive kinship is stripped of the elements of choice or preference which anthropologists generally attribute to friendship or 'fictive kinship'.[119]

Simply framing social kinship as identical to biological relatedness, or trying to avoid its mention once it has occurred, clearly requires some degree of 'mutual pretence awareness'[120] by all concerned. Where there is a perceived need to fix the 'flaws' of relinquished children (for example by having them 'fit' well into their new forever family) then institutionalised forms of secrecy, family silences and the permanence wrought by sealed records systems perhaps seems entirely necessary and appropriate.[121] Much like the English Poor Laws, harsh norms of secrecy can quickly excise unwanted genetic heritages but also serve to punish unmarried relinquishing mothers, perhaps as a form of rehabilitative 'self-mortification'[122] aimed at returning them to 'respectable' society. Alternatively a cultural norm of simply 'not telling' might be used as a protective shield against the child (or wider society) learning unwelcome genetic truths.[123] Elaborately contrived back stories could similarly help to re-define the narrative of a child's adoption, with, as occurred in the past, some adoptees simply being told that they had been orphaned, rather than relinquished.[124] In respect of gamete donation, the issue of whether to tell gives rise to similar dilemmas in terms of a 'right' to know.

---

[118] See further Carsten (1995), pp. 223–241.

[119] Carsten (2000), p. 693.

[120] Glaser and Strauss (1965).

[121] Kornitzer (1952), p. 48. See also Kornitzer (1959).

[122] Benet (1976), p. 60; see also Lockridge (1947).

[123] See also Rodger (1996).

[124] Melina (2010), p. 120.

It is maybe not entirely surprising that some kinship theorists seem, like a number of domestic jurists, to almost prefer avoiding discussion of the more difficult issues surrounding the need for natal relatedness, such as whether genetic identity rights are, or ought to be, fully justiciable, and whether the potential for harm might require that the concept of non-ancestry should be included in discourses involving definitions of best practice and the best interests of the child. It is perhaps slightly ironic that a 'reduction in interest'[125] in the study of kinship seemed to occur throughout the 1970s just as major legislative reforms on access to birth records began to be introduced within a number of jurisdictions.[126] Whether the urge to search or achieve reunion with natal relatives derives from some socio-cultural pressure to do so or from some inherent need to find related others who are likely to resemble us in some way, is perhaps extraneous to the issue of whether there should be greater legal recognition of genetic links. The fact that generally preventable forms of harm can flow from permanent types of origin deprivation is the central issue, especially given the lack of coherent judicial or legislative approach to the notion of a right to genetic identity and blood-tied kinship.

A sort of birthplace 'lottery' seems to exist: in many jurisdictions the ability to access ancestry (via information or through some form of kin contact) might be viewed as a basic right or strictly limited (if not completely prevented) by parental vetoes. This in itself suggests that, irrespective of why so many people are driven to search for information or reunion, such activities still seem to be viewed with some measure of suspicion by the wider public. If it is accepted however that genetic kinlessness has the potential to be damaging to one's welfare or well-being, then the question of whether current law and policy frameworks are perpetuating injustice and inequality of treatment against a significant section of the population, needs to be asked. The need for genetic identity cannot simply be ignored; origin deprived persons should not be expected to simply develop good 'coping mechanisms'[127] and quietly accept that they will never have a right to access their ancestry, or that such a right must always become weightless in law when set against the privacy rights of the other triad members. As the next chapter will argue, origin deprivation can lead to harmful outcomes; as such, enshrining it as a normative feature of social kinship, rather than as an exceptional occurrence, amounts to a form of highly discriminatory unequal treatment.

---

[125] Schneider (2004), p. viii.

[126] See for example s.51 of The Adoption Act 1976 (England and Wales) (as amended) which generally allows adoptees to access their original birth certificate on gaining the age of majority.

[127] See Friedlander (2003), pp. 745–752 at p. 751 on the need for coping skills. See also Schneider (2004), p. 107 on the need for 'biogenetic substance' links between birth relatives, and how it may be 'symbolically interchangeable' with feelings of affection, or the desire for lasting familial cohesion. Schneider later admitted however that his narrow anthropological focus (upon white, middle-class, co-habiting American households) would not as easily apply to the wide range of non-nuclear family relationships which tend to exist within modern, ethnically diverse societies.

References

Addison, J. T. (1925). *Chinese ancestor worship: A study of its meaning and its relations with Christianity*. Cambridge: CLC, SPCK.

Aginsky, B. W. (1935). The mechanics of kinship. *American Anthropologist, 37*(3), 450–457.

Alber, E. (2003). Denying biological parenthood: Fosterage in Northern Benin. *Ethos, 68*(4), 487–506.

Alber, E. (2004). "The Real Parents Are the Foster Parents": Social parenthood among the Baatombu In Northern Benin. In F. Bowie (Ed.), *Cross-cultural approaches to adoption* (p. 33). London: Routledge.

Altorki, S. A. (1980). Milk kinship in Arab Society: An unexplored problem in the ethnography of marriage. *Ethnology, 19*(2), 233–244.

Anderson, A. (2004). Adoption and belonging in Wogeo, Papua New Guinea. In F. Bowie (Ed.), *Cross-cultural approaches to adoption*. London: Routledge.

Bainham, A. (2007). Permanence for children: Special guardianship or adoption? *Cambridge Law Journal, 66*(3), 520–523.

Baker, H. D. R. (1979). *Chinese family and kinship*. New York: Columbia University Press.

Barnes, J. A. (1961). Physical and social kinship. *Philosophy of Science, 28*(3), 296–299.

Battaglia, D. (1992). The body in the gift: Memory and forgetting in Sabarl mortuary exchange. *American Ethnologist, 19*, 3–18.

Beattie, J. (1964). *Other cultures: Aims, methods and achievements in social anthropology*. London: Cohen and West.

Befu, H. (1963). Patrilineal descent and personal kindred in Japan. *American Anthropologist, 65*, 1328.

Bell, C. M. (1997). *Ritual: Perspectives and dimensions*. Oxford: Oxford University Press.

Benet, M. K. (1976). *The politics of adoption*. New York: The Free Press.

Berg, A. (2003). Ancestor reverence and mental health in South Africa. *Transcultural Psychiatry, 40*(2), 194–207.

Bion, W. R. (1967). Notes on memory and desire. *Psychoanalytic Forum, 2*(3), 271–280.

Bloch, M., & Sperber, D. (2004). Kinship and evolved psychological dispositions: The mother's brother controversy reconsidered. In R. Parkin & L. Stone (Eds.), *Kinship and family: An anthropological reader* (pp. 438–455). London: Blackwell.

Boswell, J. (1988). *The kindness of strangers: The abandonment of children in Western Europe from late Antiquity to the Renaissance*. London: Allen Lane.

Bowie, F. (Ed.). (2004). *Cross-cultural approaches to adoption*. London: Routledge.

Brady, I. (1976). *Transactions in kinship, adoption and fosterage in Oceania*. Honolulu: University of Hawaii Press.

Bremmer, J. (1999). Fosterage kinship and the circulation of children in ancient Greece. *Hellenic Studies Review, 6*, 1–20.

Brosnan, J. F. (1922). The law of adoption. *Columbia Law Review, 22*, 332.

Brown, L. E. (2001). The role of perceived similarity to parents in adopted children's adjustment. *Dissertation Abstracts International: Section B: The Sciences and Engineering, 62*(1-B), 538.

Bryant, T. (1990). Sons and lovers: Adoption in Japan. *American Journal of Comparative Law, 38*, 299.

Carp, E. W. (1998). *Family matters: Secrecy and disclosure in the history of adoption*. Cambridge (US): Harvard University Press.

Carroll, V. (1970). Introduction: What does "Adoption" mean? In V. Carroll (Ed.), *Adoption in Eastern Oceania* (pp. 3–17). Honolulu: University of Hawaii Press.

Carsten, J. (1995). The substance of kinship and the heat of the hearth: Feeding, personhood and relatedness among the Malays. *American Ethnologist, 22*(2), 223–241.

Carsten, J. (2000). *Cultures of relatedness: New approaches to the study of kinship*. Cambridge: Cambridge University Press.

Chand, A., & Thoburn, J. (2005). Research review: Child and family support services with minority ethnic families: What can we learn from research? *Child and Family Social Work, 10*(2), 169–178.
Clarke, M. (2007). The modernity of milk kinship. *Social Anthropology, 15*(3), 287–304.
Crawford, S. (1999). *Childhood in Anglo-Saxon England*. Stroud: Sutton Publishing.
Damon, F. H., & Wagner, R. (1989). *Death rituals and life in the societies of the Kula ring*. DeKalb: Northern Illinois University Press.
Dawkins, R. (1976). *The selfish gene*. Oxford: Oxford University Press.
Demian, M. (2004). Transactions in rights, transactions in children: A view of adoption from Papua New Guinea. In F. Bowie (Ed.), *Cross-cultural approaches to adoption*. London: Routledge.
Ellis Davidson, H. R. (1941). Fostering by giants in Old Norse Sagas. *Med Aev, 10*, 70–85.
Ensel, R. (2002). Co-lactation and fictive kinship as rites of incorporation and reversal in morocco. *Journal of North African Studies, 7*, 83–96.
Evans-Pritchard, E. (1951). *Kinship and marriage among the nuer*. London: Clarendon.
Foster, G. (1953). Cofradia and Compadrazgo in Spain and Spanish America. *Southwestern Journal of Anthropology, 9*, 1.
Foster, R. J. (1995). *Social reproduction and history in Melanesia: Mortuary ritual, gift exchange and custom in the Tanga Islands*. Cambridge: Cambridge University Press.
Friedlander, M. L. (2003). Adoption: Misunderstood, mythologized, marginalized. *The Counseling Psychologist, 31*, 745–752.
Giladi, A. (1999). *Infants, parents and wet-nurses: Medieval Islamic views on breastfeeding and their social implications*. Boston: Brill.
Glaser, B. G., & Strauss, A. L. (1965). *Awareness of dying*. New York: Aldine.
Goffman, E. (1963). *Stigma: Notes on the management of spoiled identity*. New Jersey: Prentice Hall.
Goodenough, W. H. (1970). Transactions in parenthood. In V. Carroll (Ed.), *Adoption in Eastern Oceania*. Honolulu: University of Hawaii Press.
Grimassi, R. (2001). *Hereditary Witchcraft*. Woodbury: Llewellyn Publications.
Halbmeyer, E. (2004). "The One Who Feeds Has The Rights": Adoption And fostering of kin, affines and enemies among the Yukpa and other Carib-speaking Indians of Lowland South America. In F. Bowie (Ed.), *Cross-cultural approaches to adoption*. London: Routledge.
Hallam, E. (2000). *Cultural encounters: Representing otherness*. London: Street and Maxwell.
Hamilton, W. D. (1964). The genetical evolution of social behaviour. *Journal of Theoretical Biology, 7*, 1–52.
Holy, L. (1996). *Anthropological perspectives on kinship*. London: Pluto Press.
Howell, S. (2003). Kinning: The creation of life trajectories in transnational adoptive families. *The Journal of the Royal Anthropological Institute, 9*, 465–484.
Jussen, B. (2000). *Spiritual kinship as social practice: Godparenthood and adoption in the early middle ages*. Delaware: University of Delaware Press.
Keating, J. (2001). Struggle for identity: Issues underlying the enactment of the 1926 adoption of Children Act. *University of Sussex Journal of Contemporary History, 3*, 1–9.
Kenna, M. (2001). *Greek island life: Fieldwork on Anafi*. Amsterdam: Harwood.
Khatib-Chahidi, J. (1992). Milk kinship in Shi'ite Islamic Iran. In V. Maher (Ed.), *The anthropology of breast-feeding: Natural law or social construct?* (pp. 109–133). Oxford: Berg.
Kirk, H. D. (1964). *Shared fate: A theory of adoption and mental health*. New York: Free Press of Glencoe.
Kornitzer, M. (1952). *Child adoption in the modern world*. New York: Philosophical Library.
Kornitzer, M. (1959). *Adoption*. London: Putnam.
Kraeger, P. (1980). *Traditional adoption practices in Africa, Asia, Europe and Latin America. International Planned Parenthood Federation*. London: International Office.
Lepri, I. (2005). The meanings of kinship among the Esse Ejja of Northern Bolivia. *Journal of the Royal Anthropological Institute (NS), 11*, 703–724.

Levi-Strauss, C. (1949). *The elementary structures of Kinship*. Boston: Beacon.
Levi-Strauss, C. (1963). *Structural anthropology*. New York: Basic Books.
Li, S., Feldman, M. W., & Li, N. (2001). A comparative study of determinants of uxorilocal marriage in two counties of China. *Social Biology, 48*, 125–150.
Lieberman, D., Tooby, J., & Cosmides, L. (2007). The architecture of human kin detection. *Nature, 445*, 727–731.
Lockridge, F. (1947). *Adopting a child*. New York: Greenburg.
Lowe, N. (1993). Adoption placement orders-freeing by another name. *Journal of Child Law, 5*, 62.
Lynch, J. H. (1986). *Godparents and kinship in early Medieval Europe*. Princeton: Princeton University Press.
Malinowski, B. (1922). *Argonauts of the Western Pacific*. New York: Dutton & Co.
Malinowski, B. (1929). *The sexual life of savages in North Western Melanesia*. London: Routledge and Keegan Paul.
Melina, L. R. (2010). *Raising adopted children: Practical reassuring advice for every adoptive parent*. New York: Harper Collins.
Melosh, B. (2002). *Strangers and kin: The American way of adoption*. Cambridge: Harvard University Press.
Miall, C. E. (1987). The stigma of adoptive parent status: Perceptions of community attitudes toward adoption and the experience of informal social sanctioning. *Family Relations, 36*(1), 34–39.
Mitchell, J. (2009). *Adoption and special guardianship: A permanency handbook*. Bristol: Jordans.
Moore, A. (1998). *Cultural anthropology: The field study of human beings*. Lanham: Rowman and Littlefield.
Nagy, J. F. (1985). *The wisdom of the outlaw: The boyhood deeds of Finn in Gaelic narrative tradition*. Berkeley: University of California Press.
Notermans, C. (2004). Fosterage and the politics of marriage and kinship in East Cameroon. In F. Bowie (Ed.), *Cross-cultural approaches to adoption*. London: Routledge.
Parkes, P. (2004). Fosterage kinship and legend: When milk was thicker than blood? *Comparative Studies in Society and History, 46*(3), 587–615.
Parkes, P. (2005). Milk kinship in Islam: Substance, structure, history. *Social Anthropology, 13*(3), 307–329.
Patterson, N. W. (1994). *Cattle Lords and Clansmen: The social structure of Early Ireland*. New York: Garland.
Pitt-Rivers, J. (1973). The kith and the kin. In J. R. Goody (Ed.), *The character of kinship*. Cambridge: Cambridge University Press.
Radcliffe-Brown, A. R. (1952). *Structure and function in primitive society*. London: Routledge.
Rodger, J. (1996). *Family life and social control*. London: Macmillan.
Roesch-Rhomberg, I. (2004). Korean institutionalised adoption. In F. Bowie (Ed.), *Cross-cultural approaches to adoption*. London: Routledge.
Scheffler, H. W. (1973). Kinship, descent and alliance. In J. J. Honigman (Ed.), *Handbook of social and cultural anthropology*. Chicago: Rand Mc Nally & Co.
Schneider, D. M. (2004). What is kinship all about? In R. Parkin & L. Stone (Eds.), *Kinship and family: An anthropological reader*. London: Blackwell.
Shimizu, A. (1987). Ie and Dozoku: Family and descent in Japan. *Current Anthropology, 28*, 585.
Sonbol, A. (1995). Adoption in Islamic Society: A historical survey. In E. W. Fernea (Ed.), *Children in the Muslim Middle East* (pp. 45–67). Austin: University of Texas Press.
Stone, L. (2004). Contemporary directions in kinship. In R. Parkin & L. Stone (Eds.), *Kinship and family: An anthropological reader* (pp. 239–241). London: Blackwell.
Talbot, C., & Kidd, P. (2004). Special guardianship orders-issues in respect of family assessment. *Family Law Journal, 34*, 273.
Talle, A. (2004). Adoption practices among the pastoral Maasai of East Africa. In F. Bowie (Ed.), *Cross-cultural approaches to adoption*. London: Routledge.

The British Association for Adoption and Fostering/The British Society for Human Genetics ('BSHG') Joint statement: 'The Use of DNA Testing to Determine Racial Background.' Available at www.baaf.og.uk/info/lpp/adoption/ethnictesting.pdf. Accessed 2 Feb 2009.

Thoburn, J. (2003). The colour of difference: Journeys in transracial adoption. *Child and Family Social Work, 8*(3), 236–237.

Traylor, K. L. (1988). *Chinese filial piety*. Bloomington: Eastern Press.

Treide, D. (2004). Adoptions in Micronesia: Past and present. In F. Bowie (Ed.), *Cross-cultural approaches to adoption* (pp. 126–141). London: Routledge.

Tuite, K. (1998). Achilles and the Caucasus. *Journal of Indo-European Studies, 26*, 289–343.

Visser, M. (1986). Medea: Daughter, sister, wife, mother; natal family versus conjugal family in Greek and Roman Myths about women. In M. Cropp et al. (Eds.), *Greek tragedy and its legacy: Essays presented to DJ Conagher*. Calgary: Calgary University Press.

Volf, M. (1996). *Exclusion and embrace: A theological exploration of identity, otherness, and reconciliation*. Nashville: Abingdon.

Watson, J. B. (1983). *Tairora culture: Contingency and pragmatism*. Seattle: University of Washington Press.

Wegar, K. (1997). *Adoption, identity and kinship: The debate over sealed birth records*. New Haven: Yale University Press.

Wegar, K. (2000). Adoption, family ideology and social stigma: Bias in community attitudes, adoption research and practice. *Family Relations, 49*, 363–370.

Wiersema, N. J., et al. (2006). Consequences of infertility in developing countries: Results of a questionnaire and interview survey in the south of Vietnam. *Journal of Transnational Medicine, 4*, 54.

Witherspoon, G. (1975). *Navajo kinship and marriage*. Chicago: University of Chicago Press.

Wright, S. (1994). Culture in anthropology and organizational studies. In S. Wright (Ed.), *Anthropology of organizations*. London: Routledge.

Yanagisako, S. J. (1978). Variance in American kinship: Implications for cultural analysis. *American Ethnologist, 5*, 15–29.

Yanagisako, S. J. (1985). *Transforming the past: Tradition and kinship amongst Japanese Americans*. Stanford: Stanford University Press.

Chapter 3
The Blood-Tie: 'Properly Locked Drawers' and a 'Doomed Quality'

3.1 Introduction

> In the context of adoption, birth does not imply certainty or endurance or solidarity. It is thus emptied of most of the symbolic meaning it has in the dominant discourse of kinship.[1]

As the preceding chapter sought to argue, the preservation and pursuit of 'ancestral substance'[2] and 'authentic narrative'[3] tend to be highly regarded activities across a wide variety of cultures. Common to most societies is a shared desire for genealogical facts, family mythologies and remembrance of 'clan' histories.[4] Equally, if the concept of kinship has its 'foundations in human cognition' it seems likely to also be underpinned by a number of 'psychological universals.'[5] This lends support to the argument that having or acquiring an accurate set of collective memories may serve as a key catalyst for gaining an essential sense of individual 'selfhood.'[6] As Stiffler has observed in respect of family reunions, there is often much fascination over resemblance, shared coincidence and instances of apparent synchronicity.[7] The discovery of significant similarities between separated natal kinfolk (such as commonalities of name, educational attainment or religious belief) arguably lends some measure of support to some of the theories on behavioural genetics, and to the idea of a 'DNA blueprint' somehow influencing important life events. Interest in the concept of how genetically templated behavioural traits might exist amongst

[1] Carsten (2000), pp. 687–703 at p. 691.

[2] Grace (2008), pp. 257–262 at p. 257.

[3] On the importance of authenticity see further Ingvesson and Mahoney (2000), pp. 77–110; on the need for accurate narrative, see Dunbar and Grotevant (2004), pp. 135–161.

[4] See also Tulving (1972), pp. 382–403 on how life experiences tend to be remembered in episodic memory whilst generic propositions (gleaned from information supplied by others or from actual experience) are usually stored in semantic memory.

[5] Jones (2004), pp. 211–215 at p. 211. See also Jones (2003), pp. 320–350.

[6] See further Antze and Lambeck (1996).

[7] Stiffler (1993).

A. Diver, *A Law of Blood-ties - The 'Right' to Access Genetic Ancestry*,
DOI 10.1007/978-3-319-01071-7_3,

separated blood relatives perhaps also suggests that a positive form of 'genetic essentialism'[8] might be influencing scientific research and social policy in this area.[9]

This chapter will mainly examine the various theories on the psychological effects of genetic relinquishment; it looks in particular at how to prevent or minimise 'refusals of belonging'[10] where original attachments have been removed and the blood-tie has been lost or severed.[11] It also traces how various research pathways and policy developments have led to an increased social awareness of how non-ancestry might impact adversely upon long-term psychological well-being and give rise to unresolved issues of 'diffuse grief.'[12] The notion that the child's welfare-led best interests should be a paramount consideration in such situations also seems grounded upon the work discussed here, despite (or perhaps because of) the controversial nature of some of theories on the psychology of blood-ties.

3.2 Strangest Situations

Some of the earliest theories on the psychological aspects of adoption and illegitimacy are not particularly edifying. The idea for example that a degree of 'feeble-mindedness' (not to mention moral laxity and mental instability) could be transmitted to the illegitimate child persisted in a number of jurisdictions until well into the last century. In some parts of the United States and Canada tests for 'mental hygiene' were routinely carried out, to preclude from the placement selection process any infants deemed 'unfit' for adoption. The system also tended to exclude children with physical or mild learning disabilities and those who seemed to be of 'mixed race' parentage. Perhaps unsurprisingly, children conceived in incest or through 'unacceptable' maternal promiscuity were often also blacklisted for placement.[13] The eugenicist Stoneman issued a chilling warning that the 'mysteries' generated by illegitimate conception could pose a grave danger to the entire process of adoptive care. He advocated for a considerable degree of caution on the part of any overly optimistic, sentimentally motivated agency workers, who might find themselves charged with 'disposing of such children' to unsuspecting, altruistic adoptive parents.[14]

[8] See Cahn (2000–2001), pp. 1–27 at p. 7.

[9] Nelkin and Lindee (1995). See also Nelkin and Lindee (1998), pp. 662–663.

[10] Ingvesson and Mahoney (2000).

[11] See further Lifton (1990), pp. 85–92.

[12] Ryburn (1995), pp. 41–64 at p. 42.

[13] See Kenworthy (1921), p. 499; Zenderland (1998).

[14] Stoneman (1926), p. 10.

Such thinking to an extent tied in with the controversial, and later completely discredited, findings of Goddard who declared that an entire branch of one family involved in his research had been physically and mentally affected by the recessive, negative tendencies and latent genetic traits (or, more accurately, the promiscuity) of one of their slow-witted, unmarried female ancestors. 'Inherited' traits such as delinquency, insanity, mental disability and poverty apparently abounded along her bloodline, in sharp contrast to the other side of their family, who were happily descended from 'legitimate' ancestors (who were, presumably, better-behaved in respect of their sexual activities). These legitimate descendants had become much more socially advantaged, accomplished and intelligent than their illegitimate cousins.[15] Goddard thus advocated for the complete segregation of such 'morally lax' members of society via mass institutionalisation.[16] Slingerland similarly cautioned against the dangers of placing such 'low grade', mentally defective children with intellectually normal adoptive families. This would, it was argued, help avoid a series of irritating 'calls for continuous explanations' from the wider community which adoptive parents might 'find embarrassing' to deal with. He also suggested that it might be best to avoid altogether any adoptive placements that might give rise to racial confusion for example, by taking care not to mix the 'very swarthy with the decidedly blond.'[17]

Kirschner's thoughts on violently disturbed or homicidal adoptees were similarly heartening, framing adoption as 'a critical factor in the psychopathology of many adopted killers'[18] and arguing that 'the psychodynamic underlying the syndrome in adoptees is unique'.[19] The homicide cases referred to share a number of relatively mundane characteristics, not least the adoptive parents' verbalised pre-occupation with 'bad seed' pre-destination, a 'rigid façade of denial' and high levels of fear and secrecy over the adoptive process itself.[20] Similarly, it was noted that the psychotic adoptees involved in this empirical study tended to display an

> inability to integrate parental and self-images (which) results in an impairment of the development of the superego, failure to develop a coherent sense of self and a predisposition to dissociation.[21]

---

[15] Goddard (1912). The apocryphal morality tale of the apparently degenerate, intellectually challenged wing of the pseudonymed 'Kallikak Family' argued for permanent institutionalization (rather than adoption or fosterage) of such illegitimate children within the United States. The research was supposed to have been inspired by the case of one of the female patients at the comfortingly-named Goddard's Training School for Backward and Feeble-Minded Children. See however Gould (1996), who suggested that a number of the photographs referred to in Goddard's work had been obviously retouched by the author to render the faces of the 'poor side' of the family even more sinister-looking.

[16] This later turned into a proposal for enforced sterilization for disabled persons, to prevent the transmission of their 'genetically coded' physical characteristics and apparent moral degeneracy.

[17] Slingerland (1919), p. 125.

[18] Kirschner (1992), pp. 323–333 at p. 323.

[19] *Ibid* p. 325. See further Herrnstein and Wilson (1985); Mednick et al. (1984), pp. 891–894.

[20] *Ibid* p. 328.

[21] *Ibid* p. 326; see also Kendler and Greunberg (1984), pp. 555–564.

In other words, the threat of rejection, or a perceived risk of imminent abandonment, could apparently precipitate the dangers of 'an extreme dissociative' state.[22] Such factors feature regularly within a number of high profile cases of adoptive patricide, and seem to also echo some of the darker morality tales on social kinship, involving adoptees' repressed fantasies or intense emotional responses.[23] Less disturbing theories on the prevalence of seemingly predetermined and unavoidable 'maladjustment' amongst many fostered or adopted children were also put forward however.[24] As Krugman for example suggested, some adopted adults appeared to simply demonstrate a significant pre-occupation with imaginary notions based largely upon possibly harmful 'biological fantasies.'[25] Not all were prone to inevitable psychosis however, despite apparently having a sort of innate vulnerability to being a little oddly behaved.[26]

Perhaps unsurprisingly, there was little consensus on the key question of whether to tell adoptees of their unusual 'condition' or hidden background;[27] the 'widespread and deep-seated . . .disapproval regarding illegitimacy'[28] prevalent at the time would probably have been of central concern to policy-makers and to those tasked with advising adoptive parents on the difficulties surrounding 'telling.' Kornitzer's advice perhaps typifies the conflicted thinking of the time; despite calling for an atmosphere of familial 'openness', she also seemed to initially advocate that

> those adopters who tell little or nothing to anybody are the wise ones. . .documents should be kept in a safe place, best of all in a safe deposit, otherwise in a properly locked drawer to which there is no casual access.[29]

Bowlby's work on 'attachment theory' grasped a number of nettles; drawing upon the presumption that 'family romance'[30] imaginings might be fairly prevalent amongst many adoptees, he argued against much of the current wisdom which suggested that many of their emotional problems in childhood tended to arise from inherent psychological conflicts rather than from the actual trauma of sustaining a fractured family life.[31] He looked to groups of children who had been separated

---

[22] *Ibid* p. 331.

[23] *Ibid* p. 326. See also Derdeyn (1979), pp. 215–225; Weiss (1985), pp. 763–774.

[24] See for example Knight (1941), pp. 65–74; Stonesifer (1942), p. 161; Holman (1953), pp. 654–688; Kadushin (1966), pp. 30–39.

[25] Krugman (1964), pp. 349–358 at p. 336. She refers primarily to the early work of Schechter (1960), p. 21; Peller (1963), pp. 145–153; see also Wieder (1977), pp. 1–22.

[26] See for example Schechter (1960), p. 21.

[27] Krugman (1964). See further Tarachow (1937), pp. 401–412; Cowan (1938), pp. 454–458.

[28] Kellmer-Pringle (1967), p. 27.

[29] Kornitzer (1952), p. 102. See also Kornitzer (1959).

[30] Freud (1957). See further Horner and Rosenberg (1991), pp. 131–148.

[31] Working with Robertson, Bowlby then focused upon a group of 'affectionless' children who had been separated from their parents through hospitalisation or institutionalization (Robertson 1953). See further Bretherton (1992), pp. 759–775 at p. 761. See also Klein (1984) (originally published 1932) on the role of child aggression in fantasy, the 'depressive position', 'Object Relations Theory' and the significance of the 'paranoid-schizoid' position.

from their natal parents at an early stage in their development,[32] providing a seminal report for the World Health Organisation ('WHO') which challenged the simple but widespread belief that the mother–infant bond was founded merely upon the vulnerable infant's pragmatic need to receive basic sustenance from whichever substitute care-giver was available or willing to provide it at that point in time.

This wider, ethological approach to infant attachment also underpinned his arguments on the child's fundamental need to engage in social bonding. The 'primary drive' to identify with a central parent figure at such an early stage of development arose from the innate need to achieve some enduring sense of 'belonging'. His 'instinct theory' focused upon the levels of separation anxiety that would arise normally in infancy upon maternal loss and challenged the contention that bereaved infants were largely incapable of mourning due to their having an inadequately developed ego. The accepted wisdom that the presence of a surrogate parent might automatically lower or even cancel out the profound effects of separation anxiety was similarly questioned.[33] He further argued that, in respect of infants affected by these broken early attachments, a significantly impaired ability to form meaningful relationships might well occur in later life, especially where a long series of substitute care givers would also perhaps be involved in their upbringing.[34] The sense of 'belonging' was essential to the development of a 'normal' personality, which would eventually be capable of forming healthy, permanent attachments in adulthood.[35]

Such research also established that fears in young children generally tended to arise from two sources: the perception of imminent danger (the instinctual variety and the type that is socially acquired at a later stage of development) and the absence of the main 'attachment figure'. Either of these could lead to very strong, adverse emotional responses.[36] A deep-seated need to be able to flee from the unknown and from otherness, to regain the protection of the 'attachment figure' perhaps also underpins the psychological basis of the need for visible relatedness. The 'retreat from the strange and novel' tended to be accompanied by the urge to find 'familiarity preserving, stress-reducing behaviours.'[37] Arguably, a desire for reunion with one's original kinfolk in adulthood could perhaps be viewed as having a similar stimulus.[38]

---

[32] See Bowlby (1958), pp. 350–371. See also Bowlby (1951); Bowlby (1980); Ainsworth and Bowlby (1991), pp. 333–341; Bowlby (1988); Bowlby (1979).

[33] See for example Freud (1960), pp. 53–62.

[34] Bretherton (1992), p. 768.

[35] See also Pendry (1998); Schaffer and Emerson (1962).

[36] Bowlby (1973).

[37] Bretherton (1992), pp. 759–775 at p. 782.

[38] See also Marris (1991), pp. 77–90, on how 'good societies' protect attachments, keep disruptive events to a minimum, and help families to achieve cohesion.

Ainsworth and Wittig's work on the 'Strange Situation' is perhaps particularly relevant to the early experiences of relinquished persons who have been estranged from their genetic relatives.[39] Such research was not initially directed at infant adoptees despite the clear focus upon infant responses to parental loss, fears of abandonment, and the question of whether babies were capable of forming strong attachments at this early stage of their development. Arguably, the process of adoption itself represents an extended, extreme form of this experiment, apart of course from the key, final stage of speedy reunion between parent and child. It is perhaps exactly because of this uncomfortable parallel with the relinquishment process, that adopted infants seem to have been excluded from the initial research, having already been through a similarly upsetting experience, minus the maternal reunion.

Sants' work famously focused attention not on child fears of maternal abandonment but on the state of genetic confusion that the process might create or require. Examining the notion of 'genealogical bewilderment,'[40] he highlighted this 'condition' as one of the chief causes of identity-based problems amongst adolescent adoptees. The phrase has been much used by a number of theorists, across a fairly wide range of disciplines, to sum up the psychological consequences of having a disputed, unknown or unclear parentage.[41] The norms of origin deprivation are not necessarily adequately represented by the term, not least because of the gentleness of the concept of 'bewilderment.' The expression still seems to emphasize an individualized response to the 'predicament' of ancestor loss and identity confusion rather than highlighting how systems of secrecy and veto contribute to and sanction the losses associated with child relinquishment. In fairness, some adoptees might well be simply bemused at their lack of information, or temporarily prone to a sort of child-like puzzlement which might perhaps be easily cured or coped with. Others may not see their losses in quite such straightforward terms.

The need to search or achieve reunion has at times been pathologized, if not adversely medicalized, as an over-reactive, individual response to an unusual but not necessarily harmful circumstance, or perhaps ascribed to an abnormal emotional state arising from the adoptee's 'weakened' psyche. Kirschner's theories for example (on the existence of 'Adopted Child Syndrome'[42]) seem to tap into

---

[39] See Ainsworth et al. (1978), on how this test categorises infant attachments on the basis of their responses following separation and reunion.

[40] Sants (1964), pp. 133–141; see also Wellisch (1952), p. 41. Wellisch had theorized that identification with relatives is made easier where there is some degree of physical resemblance: adoption might therefore be best suited to children who had actually been orphaned. See also however Carp (1998), p. 130 on the methodological failings of these theorists in respect of their empirical sampling methods. See also however Sorosky et al. (1976), pp. 900–904.

[41] See Jones (1997), pp. 64–68 at p. 65.

[42] Kirschner (1992), pp. 323–333.

this approach, although they were not universally accepted nor have they been particularly applauded. Freidlander for example has called them 'mere myth,' contending that many of the symptoms described could have resulted from simple coincidences or perhaps from some highly conjectured thinking on the part of adoptees themselves or their misguided therapists.[43] She concedes however that many adopted people do show some signs of suffering from a reactive attachment or post-traumatic stress disorder, but suggests that such conditions are more likely to have arisen from some form of pre-adoption trauma such as abuse or neglect at the hands of one's birth family rather than simply by virtue of the loss of one's original relatives. As such, she argues that therapists need to be very careful that they are not engaging in an 'overplay [of] the adoption history' where children have been exhibiting psychological difficulties.[44]

The fact remains however that adoptees do appear to be over-represented in respect of a number of mental health studies.[45] This may be down to some aspects of the adoption process not exactly being best suited to the promotion of an ideal childhood: early infant trauma, ambiguous interpersonal communication with a succession of carers or adolescent engagement in identity fantasy as a 'psychological survival' mechanism, may be relevant factors.[46] Severe trauma can also perhaps result from receiving contradictory, remembered information, especially in respect of bereavement in infancy.[47] Kellmer-Pringle's observations on adoptees, though outdated in terminology, are perhaps not entirely inappropriate. She suggested that 'the principle need of the adopted child…is rather similar to that of handicapped children'.[48]

Post-Bowlby era research has tended to look more widely to the longer term life experiences of adoptees, for example through work involving 'separated twin studies' on biogenetic predisposition and behavioural genetics.[49] These studies provide further insight into the psychological need for mutual recognition which tends to motivate infant bonding and kinship. Domestic policies on familial contact also seem now to take into account the fact that a healthy sense of self-concept[50] will often need considerable time to either develop or heal, post-separation. This is especially so where there might have been sudden or profound severance of the

---

[43] Freidlander (2003), pp. 745–752 at p. 747. Friedlander further argues that birth family loss need not always be 'universally experienced as tragic' warning that therapists should not be tempted to promote too enthusiastically the need for kin reunions, or to view such events as necessarily being 'essential for psychological wholeness'.

[44] *Ibid* p. 751.

[45] Levy-Shiff (2001), pp. 97–104.

[46] Greunewald (1984), pp. 170–190.

[47] Bowlby (1982).

[48] Kellmer-Pringle (1967), p. 25. She advocates honesty as the best policy however; pp. 26–27.

[49] See for example Loehlin and Nichols (1976); Kempthorne (1997), pp. 2–3 and pp. 109–112.

[50] See further Freud (1949).

child's bio-genetic relationships in infancy or early childhood.[51] Frisk's theories on the existence of the 'genetic ego' and the need to protect it also stressed how an adoptee's psychological welfare was likely to depend upon the achievement of a fully rounded information-based identity.[52]

Erikson's work on the 'identity crisis', which might arise from a poorly developed personality, also emphasized the importance of psycho-social development during the infant's first year of life, including the ability to place one's trust in others. Consistency of caregiver was important, as was the ability to access positive identity formation and gain a sense of 'ego integrity'.[53] A healthy 'goal-corrected partnership' dyad, which would normally evolve between parent and child as the child grew to maturity, tended to involve information exchange, identity creation and kinship bonding processes.[54]

As Hurlock had argued earlier, 'until an individual establishes a stable concept of self, he will be uncertain about his ability, his status....and how he compares with others.'[55] Non-availability of information on one's 'original family background' could impact upon a person's ability to develop a healthy sense of personal identity.[56]

Rogers had also suggested that the attainment of 'personhood' frequently depends upon a useful level of self-discovery. Physical resemblance, particularly where assignation to a particular 'side' of the family tree occurs, might be highly significant in relation to achieving a sense of connection or relatedness to others.[57] As Sandler and Joffe further argued, the ability to physically identify with one's parents is perhaps crucial in relation to achieving a useful degree of emotional development.[58] Clear links seemed to exist between infant trauma, strain on the nascent ego, subsequent inner conflict, and the potential for psychotic pathologies.[59]

Smith and Brodzinsky's findings on 'negative affect' (over relinquishment) simlarly linked adoptees with low self-esteem and depression.[60] Their findings supported the earlier work of Cohen on the developmental challenges faced by

---

[51] See for example Arsenian (1943), pp. 225–229; Bowlby (1940), pp. 1–25; Goldfarb (1945), pp. 247–255; Heinicke (1956), pp. 105–176; Sroufe (1983). Good levels of mental health may owe as much to positive familial experiences ('nurture') as to innate genetic pre-disposition ('nature'). See also DeFries et al. (2000); Ridley (2003).

[52] Frisk (1964), pp. 6–12; see also Juffer et al. (2004), pp. 697–706 on international adoptions.

[53] Erikson (1968); see also Erikson (1970, 1980).

[54] Bowlby (1973).

[55] Hurlock (1955), p. 44; see also Ziller (1969), pp. 287–300.

[56] See Goffman (1963); see also Melosh (2002), on the shielding of adopted children from the knowledge of their own illegitimacy; Cole and Donley (1990), pp. 273–294 on the 'labelling' of children and adopters as 'suitable'.

[57] Rogers (1961); see also Rogers (1955), pp. 267–278; Nelson (1992), pp. 172–177.

[58] Sandler and Joffe (1969), pp. 585–595 at p. 585.

[59] Joffe and Sandler (1967), pp. 56–66 at p. 60.

[60] Smith and Brodzinsky (2002), p. 213.

many adoptees in relation to their 'life situation;' they may lack the 'inner realities' and 'constructive processes' needed to form healthy representations of 'self and other.'[61] Dunbar and Grotevant's categorization of four adoptive identity typologies (Unexamined, Limited, Unsettled, Integrated) further supports the view that some degree of 'narrative development' is an essential element of emotional maturation, and that the 'identity conflicts' experienced by adoptees are generally greater than those faced by non-relinquished children.[62]

The general consensus now seems to be that vulnerable individuals may be prone to experiencing greater problems of 'adjustment' and may be at risk of going on to develop a 'confused sense of self'.[63] An early, sympathetic 'truth-telling' is an essential element of the adoption process,[64] given the need to establish a clear sense of one's 'autobiographical self' which now underpins much of the discourse on the issue of the 'right' to information disclosure, whether in respect of adoption or gamete donation.[65] Key factors in addressing self-identity issues include the ability to 'resolve feelings towards parents and siblings with whom one shares no biological connection' and to cope with the 'dual identity' that exists where one has 'two sets' of parents. Similarly, lack of 'biological mutuality' may be worse where older children are being placed for adoption, given that they might well have been victims of abuse, and capable of remembering their biological relatives.[66] As Freundlich summarises,

> the clinical literature suggests a number of themes that may arise in the psychosocial development of adopted persons, including loss, abandonment and rejection, "differentness" and social stigmas.[67]

3.3 A False Self

Modern media coverage of reunions or search narratives can echo the histrionics found in many popular works of fiction[68] and strengthen the cultural perception that social kinship must inevitably lead to an 'inferior family form' of relatedness and

---

[61] Cohen (1996), p. 287. See also Park (2006), p. 738.

[62] Dunbar and Grotevant (2004), pp. 135–161.

[63] R Levy-Shiff (2001), p. 97. She cites, for example, the work of Brodzinsky and Steiger (1991), pp. 484–489; Wierzbicki (1993), pp. 447–454; Grotevant (1997b), pp. 139–161; Haimes (1987), pp. 126–135; Howe (1997), pp. 401–411.

[64] Jaffe and Fanshel (1970). On 'Late Discovery Adoptees' see further http://www.latediscovery.org/ (accessed 09.01.2011).

[65] Nelson (1993), pp. 7–14; see also Nelson and Fivush (2004), pp. 486–511.

[66] Levy-Shiff (2001), p. 102.

[67] Freundlich (2001), p. 61.

[68] *Ibid* p. 168.

to profound mental distress on the part of the adoptee.[69] Additionally, the self-stigmatizing actions of many adoptees (such as reluctance to request birth information through fear of causing offence to their adoptive parents) may further inhibit identity-formation processes. Where origin-deprived persons are however likely to 'seek to recover a sense of agency over their own past. . .this often involves a kind of retrospective bereavement process' over the loss of their genetic ties. This 'diffuse grief process' tends to follow the first stage of the 'adoption and loss' process, namely, that of acknowledging the 'difference perception'.[70]

Verrier has argued that maternal separation at birth destroys all vestiges of the pre-natal 'primal relationship' that exists between mothers and their unborn children. Because an infant is still unable to distinguish between its own identity and that of its mother, profound feelings of abandonment might be compounded by a significant 'loss of self.' The consequences might include premature ego development, resulting in an inability to place one's trust in other people for fear that they may similarly 'disappear' at some stage.[71] The trauma of the initial separation may in turn provoke differing behavioural responses: aggressively anti-social behaviour or an introverted, highly submissive demeanour, both of which reflect the presence of an outwardly 'false self.'[72]

Powell and Afifi's study looked at older adoptees and the means by which they have had to 'experience and manage uncertainty and loss.'[73] Drawing upon Berger and Calabrese's 'Uncertainty Reduction Theory'[74] (which posited that individuals look for information to reduce their uncertainty) they argue that such a coping mechanism seems to empower adoptees, possibly by allowing them to perhaps foretell or explicate the behaviour of other individuals. The ambiguous nature of adoptees' losses and their ability to engage in the process of 'grief management' were of central importance. Similarly, the indistinct nature of adoptee grief may well arise from the fact that the relatives 'lost' to them are, despite their absence, often still very much alive. This can lead to a state of 'unresolved grief' or 'non-closure;' losses may be amplified or left unaddressed, with 'mourning' perhaps looked upon by the rest of society as artificial in nature.[75] Key factors which usually dictate the level of such uncertainty stress include: the nature and amount of information or kin contact available; the degree of openness permitted by the adoptive

[69] Wegar (1997), p. 76; see also Bartholet (1993), on how the 'search movement' has added 'new sources of stigma' (as cited by Wegar 1997, p. 76).

[70] Carsten (2004), p. 689.

[71] Verrier (1993). See further Verrier (2003).

[72] On the need to create an alternative self, see also Winnicott (1960), pp. 140–152. A number of NGO's engaged in lobbying for open records have argued that terms such as 'voluntary relinquishment' represent an attempt to smooth over the realities of abandonment traumas. See for example www.bastards.org/activism, www.bastard-nation.de, www.transracialabductees.org, http://adoptionroadkill.blogspot.com, www.donorsiblingregistry.com, and www.donorsffspringhealth.com. Accessed 18.11.2011.

[73] Powell and Afifi (2005), pp. 129–151 at p. 129.

[74] Berger and Calabrese (1975), pp. 99–112.

[75] See Boss and Greenberg (1984), pp. 535–547 (as cited by Powell and Afifi 2005, p. 130)

family; and the factual circumstances of the adoption itself. Response to the uncertainties may in turn be shaped by adoptee awareness of possible outcomes. This means that a range of reactions may be possible. Contentment at simply 'not knowing' the details of one's origins might perhaps serve to lower stress levels, as might a placid acceptance of the situation, resulting in an absence of desire to ever search for information.

The wish to avoid a potentially negative outcome (such as rejection by birth family members) may also motivate some adult adoptees to opt instead for the relative comfort and familiarity of chronically lived-with, low-level uncertainty stress.[76] Residual feelings of having 'gaps in identity' and difficulties in forming lasting attachments, seem however to feature quite frequently in the experiences of the adult adoptees who took part in this research. Significantly, a number of reunited adoptees (especially in respect of international, trans-racial adoptions) did state that the acts of 'solidifying their identity and connecting with their birth mothers relieved the loss because they finally knew their origin and heritage.'[77] The 'recursive nature' of uncertainty in this context demands fuller analysis, as does the fact that such insecurities can tend to foster other types of anxiety. There may be a need for adoptees to develop the ability to self-manage their grief, possibly by deciding upon the option of not seeking out information on genetic ancestry or a reunion with birth kin. The therapeutic role played by adoptive families and health care professionals may be an essential one, in terms of permitting or encouraging adoptees to freely express their feelings of loss.[78]

This pragmatic suggestion does to an extent place the task of remedying the loss firmly upon the adoptee and upon their immediate private circle, rather than pointing out how legislative and policy reforms could perhaps help prevent the harms of total origin deprivation from occurring in the first place. That said, many psychologists also seem to take the view that adoption is not necessarily a pre-ordained 'formula for traumatic loss'[79] requiring some level of 'fantasy for psychological survival'.[80] Freidlander for example argues that adoption should be best regarded as a past event, rather than as an ongoing state of being. She stresses the very 'subtle but meaningful difference' between the two phrases, 'I was adopted' and 'I am adopted,'[81] pointing out that a wide variety of popular cultural myths have adversely reinforced the image of birth family loss as one that is invariably and 'universally experienced as tragic', with too many therapists seeking to promote reunion with biogenetic kin as unavoidably 'essential for psychological wholeness.'[82] The inability of many adoptees' to basically 'accept the adoption loss' and 'move past it' may be down to

[76] Powell and Afifi (2005), p. 131.

[77] *Ibid* p. 143.

[78] *Ibid* p. 148.

[79] Friedlander (2003), pp. 745–752 at p. 745.

[80] Greunewald (1984), pp. 170–190.

[81] Freidlander (2003), p. 746.

[82] *Ibid* p. 747.

their choosing to actively 'live their lives as protagonists in a never-ending tragedy.'[83] Promoting strong familial attachments within the substitute family is a better strategy for the development of 'good coping skills' on the part of the adoptee,[84] even though adoptees perhaps represent a 'too vulnerable population.'[85]

The 'negative messages' that can attach to loss of 'genetic continuity' may however clearly lead to social disenfranchisement.[86] As Ingvesson and Mahoney have noted in respect of international adoptions, where 'dislocation' from original history has occurred, it may serve to 'deny the possibility of a seamless narrative of origin'.[87] Traditional psychological welfare issues can arise in this context but in doing so they provide a new perspective on the subject of origin deprivation.[88] Some international adoptees for example describe their difficulties in achieving a sense of belonging, tending at times towards feelings of 'non-existence'.[89] Achieving some level of 'rootedness' depends, they argue, upon a coherent degree of blood connection or the ability to demonstrate links to an ascertainable 'national soil'. Like some 'diasporic populations' (i.e. refugees and migrants) they may be affected by their own sense of having 'broken' identity narratives.[90]

Referring to Swedish adoption relinquishment law and policy, they stress the near-impossibility of terminating the rights of birth parents within this jurisdiction. The '*blodsband*' (blood-tie) is regarded as a central element in achieving personal 'authenticity…collective belonging and social continuity.'[91] Where adoptees are denied access to their biological or social history they may 'remain immigrants for life'.[92] Patel has further outlined the need for 'adoptees to speak for themselves about their own lives'. Where gaps exist in life histories, origin deprived persons may fill these in with 'data based on assumptions', or seek to 'negotiate themselves' some manner of substitutive racial identity.[93] Such narrative identity confusion can clearly form 'a barrier to the attainment of a realistic self-concept.'[94]

[83] *Ibid* p. 749.

[84] *Ibid* p. 751.

[85] *Ibid*; she includes parents within this 'vulnerable population' although she does not make clear whether she is referring equally to all triad adults, so as to include birth parents as well as adoptive parents.

[86] Ryburn (1995), p. 42. See also Lifton (1994), p. 37 on the 'non-events' that can underpin and follow in the wake of adoption ('non-birth', 'non-conception') and the need for clear 'narrativity' in establishing a sense of connectedness to one's lineage.

[87] Ingvesson and Mahoney (2000), pp. 77–110 at p. 77.

[88] Weitzman and Albers (2005), pp. 1395–1419. See also Weitzman (2003), p. 303.

[89] Weitzman and Albers (2005); see also Cohen (1996), p. 287.

[90] Ingvesson and Mahoney (2000), p. 78. See also Featherstone (1995), p. 142, on the exclusive nature of belonging to a 'national genealogical tree'.

[91] *Ibid* p. 85. See also Boholm (1983): Acta Universitatis Gothenburgensis (cited in Ingvesson and Mahoney 2000, p. 85).

[92] *Ibid* p. 89.

[93] Patel (2005), p. 327 at p. 332.

[94] Ryburn (1995), p. 45. See also Sachdev (1991), pp. 249–264.

In sum, the general consensus seems to be that potentially adverse consequences may attach to genetic kinlessness.[95] Kinship denials may compound the initial losses, reinforce inherent socio-cultural stigma and perhaps mask discriminatory and unequal treatment, such as where birth certificates are permanently altered or sealed or medical records are unavailable to one section of the population solely on the basis of birth place or birth date.[96] Veneers of concealment needed to achieve new family 'permanence' within classic, clean-break adoptions may well have contributed to mental health difficulties or behavioural problems in adoptees.[97] As Lifton famously argued, many adoptees will at some stage of their life develop an 'artificial self' so as to cope with the psychological 'splits' of adoption. These splits can occur upon initial separation from the genetic parents, on learning of one's adopted status, or much later, through having to cope with a lifelong legacy of quiet gratitude and secrecy that can surround the social kinship process.[98] The idea of the adoptee expressing grief may itself also be subject to a sort of quiet self-sanctioning; as Ryburn noted,

> the act of separation from a still living original family does not have the finality of death. As a loss that may be largely unacknowledged it may become subtly pervasive and difficult therefore to grieve.[99]

Lindblad et al's work with groups of inter-country adoptees sampled age-matched immigrants, siblings and non-adopted persons. Similarities between the two groups were found to be greater than their differences; most of the aspects of difference remained highly significant however and could not simply be overlooked.[100] Lee's identification of four cultural socialization strategies for trans-racial adoptees included: assimilation, 'enculturation', racial inculcation, and child choice.[101] 'Enculturation' appears to be emerging as the preferred model, over that of racial assimilation,[102] with the preservation of authentic heritage identity gradually becoming a more pressing concern. This is especially so where 'not telling' is clearly not an option, such as where the adoptees' racial characteristics differ greatly from those of their social kinfolk.[103] Ignoring the biogenetic origins of one's adopted child in the hope that they might remain

[95] See Graham and Stevenson (1985), pp. 33–41.

[96] See Grotevant (1997a), pp. 3–27 on the 'pre-crisis' stage of denial in relation to imminent 'identity challenges'. See also Grotevant (2003), p. 753.

[97] Krugman (1964), pp. 349–358. See also Eiduson and Livermore (1953), pp. 795–802; Humphrey and Ounsted (1963), pp. 599–608; Goodman et al. (1963), pp. 451–456.

[98] Lifton (1990), pp. 85–92. See also Sobol and Cardiff (1983), pp. 477–483.

[99] Ryburn (1995), p. 49.

[100] Lindblad et al. (2003), pp. 190–202. See also Lee (2003), pp. 711–744.

[101] Lee (2003).

[102] Grotevant (1997b), p. 759.

[103] *Ibid.* See also the increased risk of psychiatric difficulties which attaches to the presence of socio-cultural difference; this may manifest as substance abuse, lower levels of academic attainment or greater reliance on State benefits.

undiscovered or insignificant is no longer a desirable option in relation to parenting. The importance of 'an information-rich environment' is now an accepted element in respect of embedding the bonds of best-interests-led social kinship, not least in respect of ensuring that a child's 'sense of adoptive identity is based on fact rather than fantasy or assumption'.[104]

As Brodzinsky's work on 'Adoption Communication Openness' ('ACO') found, 'attunement' (a high degree of information exchange and empathy) between kinfolk is central to ensuring a useful degree of healthy identity development.[105] The act of searching should not therefore be regarded by the social family (or wider society) as a sort of emphatic rejection or negative indictment of their parenting skills.[106] Classifying the need for ancestry as simply a 'socially induced' one, does not address the question of why a section of the population must be denied genetic ancestry.[107] Arguably, the concepts of justice, law, equality, discrimination and human rights are themselves social constructs, but no less important for that, in terms of promoting and protecting individual and communal welfare. Whether the need for accurate ancestry arises from socio-cultural pressure or innate psychological need is not the key issue in respect of reforming discriminatory laws and policies.

3.4 Secrecy and Severance in Law and Policy

Where a need to effect 'complete removal from the taint'[108] exists, a dual 'social remedy' may be achieved, redressing a legacy of disadvantage (poor parentage or, in the past, illegitimacy) that many children would otherwise be exposed to.[109] Within England and Wales (where originally birth certificates are generally available to adult adoptees) a long-standing, deep-seated socio-cultural bias against poor parentage and illegitimacy is perhaps apparent within the reports of both the Hopkinson (1921) and Tomlin (1924–1925) Committees.[110] As Keating noted,

[104] Grotevant (1997b), p. 758.

[105] Brodzinsky (2005), pp. 145–166 at p. 149.

[106] Humphrey and Humphrey (1986), pp. 133–140.

[107] O'Donovan (1988), p. 27 at p. 29. She does concede that providing accurate information on medical history could at least help in the avoidance of accidental incest between genetic relatives.

[108] Holman (1988), p. 11.

[109] Such a strategy perhaps adds a dimension of applied social control to the principle of child welfare 'paramountcy'; under the 1886 Guardianship of Infants Act, if substitute State care was deemed necessary, the issue of 'contact' with the biological kinfolk was to be regarded as moot, given that the original parents had 'voluntarily' relinquished all rights to parent.

[110] See The Report of The Committee on Child Adoption ('The Hopkinson Committee') (1921); The Child Adoption Committee ('The Tomlin Committee') First Report, (1924–1925) Vol ix. See also The Horsburgh Report (1937) para 19 which suggested some degree of concern for the worries that birth parents might face post-relinquishment.

within both documents the issue of property succession by non-genetic descendants appeared to be the most pressing matter for consideration. Many Adoption Societies had also argued that secrecy was the best option and therefore had to be legalized via the closed records, despite the opposing views of other philanthropic agencies of the time. The agencies 'were single-minded in ensuring the rights of the adopting families to remove all trace of a previous identity from their new offspring'.[111]

The Adoption Act 1926 enabled the 'total legal transplant'[112] of all rights and duties from birth parents to adopters, with adopted children gaining inheritance and intestate succession rights shortly thereafter,[113] and allowed for key aspects of the blood-tied relationship to enrich those social kinship connections that were already being enabled through informal adoptions and long-term fostering or guardianship.[114] The Children and Young Persons Acts of 1932 and 1933 continued to promote permanent 'removal' as the best option, especially where the child's past included cruelty, neglect, or 'illegitimacy'. As Heywood noted, 'neglectful' and 'substitute' homes became completely exclusive of one another; issues such as kin contact, or the possible rehabilitation of original kinship links were rarely considered.[115]

Arguably, for the weaker members of the adoption triad (unmarried mothers and illegitimate children) the system seemed at this time to be geared almost towards reinforcing personal shame and creating stigma. War-time evacuations of urban children served eventually to raise middle-class awareness of the psychological impacts of separating mothers and children.[116] Perhaps ironically, the high number of unmarried mothers after the Second World War created a sort of 'golden age' for stranger adoption, and for institutionalised secrecy. The best means of safeguarding the psychological welfare of adoptees was to 'ground them' as firmly as possible within their new family environment, and to do whatever was necessary to excise their links with their biogenetic relatives.

The Ingleby Report (1960)[117] suggested however that the Government should actively work towards preventing the irreversible breakdown of birth families. Many aspects of child protection policy seemed to have been drawn into an ongoing series of pendulum swings between high-profile child-care scandals, tragedies and government inquiries.[118] Judicial interpretation of the principle of 'best interests of the child' was perhaps not always sufficiently robust in addressing the complex

---

[111] Keating (2001), pp. 1–9, at p. 3.

[112] Allen (2003), p. 14.

[113] See The Adoption of Children Act 1926 ss 2(3), 5 (2); see also the Legitimacy Acts 1926 and 1976. See further Cretney et al. (2003), pp. 791–796.

[114] Josling and Levy (1984).

[115] Heywood (2001).

[116] Birlingham and Freud (1942).

[117] Report of the Committee on Children and Young Persons (1960). See also Seebohm (1968), The Local Authority Social Services Act (1970) which, in respect of England and Wales, was aimed at creating a unified, non-specialised Social Services department.

[118] See Parton (2004), pp. 80–94.

issues surrounding child abuse. Child protection often meant a complete and permanent loss of genetic kinship, which tended to be accompanied by episodes of 'drift' between foster homes or a cruelly long 'wait' in substitute care. Neither of these options offered a terribly satisfactory solution to the problems of protecting the 'kinless', vulnerable child.[119] Shortly after this, Goldstein, Freud and Solnit's observations on 'psychological parenting' suggested that 'contact parents' could help protect their child's long-term psychological welfare.[120] The Houghton Report (1972) recommended counselling sessions (which would later become compulsory) for searching adoptees.[121] Works such as Shawyer's 'Death by Adoption'[122] dramatically highlighted the importance of the blood-tie, raising awareness of the harms that can attach to being origin deprived. The Barclay Report (1982) further suggested that courts and parents ought not to be simply pushed into automatic acceptance of adoption as the preferred solution to acute dilemmas over child protection.[123] Early policies on the use of 'Freeing Orders'[124] however seemed to still be based upon the presumption that a child would be incapable of relating to more than one set of parents at a time.[125]

Triseliotis' fundamental suggestion has broadly underpinned and characterized much modern adoption practice in this region ever since: the opinions of adoptees ought to be taken into account, rather than simply having their basic welfare needs referred to at some point during the proceedings. He argued that many adoptees might, as they grow older, want some degree of contact with their natal kin. His examination of the 'deep psychological need' to search seemed to persuade public opinion in favour of opening previously sealed birth records.[126] Media-led fears that access to original birth certificates would see genetic parents harried by

[119] Rowe and Lambert (1973); see also Lyon et al. (2003).

[120] Goldstein et al. (1973).

[121] See The Houghton Report (1972) Cm 2288; see also The Adoption Act 1976 (as amended) and s. 60 of the Adoption and Children Act 2002 (enacted December 2005); see also Northern Ireland's equivalent legislation The Adoption (NI) Order 1987 Article 54. The Houghton Report's preference for the term '*long term welfare*' was replaced by the phrase '*throughout childhood*' in the final draft. There was debate as to whether child welfare was to be the framed as the 'first' or the 'paramount' consideration in family law cases involving the 1976 legislation. A right of veto vests in the inherent jurisdiction of the High Court to order that identifying information should not be released by the Adoption Agency under *R v Registrar-General ex p Smith* [1991] 2 QB 393.

[122] Shawyer (1979).

[123] Barclay (1982).

[124] Freeing Orders dispense with the need for birth parent consent to the adoption and effectively transfer parental responsibility to Social Services Trusts. Their use is currently under review in Northern Ireland. They were 'judged to be a failure' in respect of England and Wales in 1993 and replaced by Placement Orders (under the Adoption and Children Act 2002). See further Lowe and Douglas (2007), pp. 837–839. The continuing use of Freeing Orders in Northern Ireland came in for criticism by the House of Lords, via the dissenting opinion of Baroness Hale, in *Down Lisburn HSST v H* (NI) [2006] UKHL 36, at para 6.

[125] Goldstein et al. (1979, 1996).

[126] Triseliotis (1973); see also Triseliotis (1985), pp. 19–24.

long-lost children proved to be unfounded.[127] In *R v Registrar-General ex p Smith* [1991][128] it was further established that the right to receive one's birth information is far from absolute, and clearly subject to judicial discretion. As Neil suggests however, much of the research in this area,

> cannot fully capture the complexity of these processes but does contribute to a much needed detailed examination of one part of this system: the adoptive parent perspective on communicative openness.[129]

In respect of maintaining secrecy, a sort of 'role handicap' may also attach to infertile adoptive parents, especially if the original pain of their infertility has not been completely assuaged by the adoption process. Two options arguably exist: to either reject or acknowledge the reality of the differences that exist in respect of their family situation. Describing this scenario as 'shared fate', Kirk advocated that adoptive parents should, for the sake of their children's mental health, accept the fragile reality that is social kinship, and the dangers that can often attach to it.[130] The dual challenges faced by many adoptees and their parents clearly highlight the need for flexible, adaptive parenting especially given that the burden borne by the children might increase as they become adults.[131] As Levy-Shiff has argued,

> The experience of loss associated with adoption entails not only loss of biological parents and origin, but also loss of stability in the relationship with the adoptive parents, loss of self, loss of genealogical continuity, and loss of status associated with being different.[132]

Thus for many people, the ability to trace ancestry and build up a detailed family tree is regarded as a worthy and worthwhile pursuit; mentioning adoption in this context can provoke mixed responses, not least in terms of accessing information that has been framed as 'belonging' to others who might not want to be 'found' by the descendant that they have relinquished. Relatively mundane events may trigger a profound sense of 'displacement', where, for example, one is asked to produce an original birth certificate or provide a medical history that involves genetic information.[133] This sense of difference may be especially acute at emotionally significant times in the familial life cycle: anniversaries, holidays or birthdays. The 'enjoyment' of such events may be double-edged or serve as triggers for trauma: they could be regarded with trepidation or, as a coping mechanism, simply treated as 'non-events'.[134] As Neil has also observed, there is a clear need for social workers

[127] Haimes and Timms (1985), p. 74: see also Haimes (1988), pp. 46–61.

[128] [1991] 2 QB 393 (Court of Appeal for England and Wales). Here a psychotic prisoner attacked his cell-mate because he believed him to be his birth mother. See also the vetoing power of the High Court under s 60(2) of the Adoption and Children Act 2002.

[129] Neil (2009), pp. 5–23 at p. 21.

[130] Kirk (1964); see also Kirk (1981).

[131] Robertson (2006), p. 727

[132] Levy-Shiff (2001), p. 98.

[133] *Ibid.*

[134] See Ryburn (1995), p. 43.

to remain open-minded in respect of issues such as kin contact;[135] often a 'dynamic and transactional'[136] attitude is needed in relation to preventing origin deprivation where possible. Arguably, adoptive parents in a number of juris dictions seem less vehemently opposed to the idea of some form of post-adoption contact.[137] As Ryburn stresses,

> the open exchange of information which in many instances may most easily be achieved through contact so that it is readily available and not offered on a once and for all basis is likely to be a vital factor in adopted children's achievement of a positive sense of self-identity.[138]

3.5 New Technologies, Traditional Identity Dilemmas

The adverse effects of origin deprivation on 'early object' (interpersonal) relations have been analyzed for the best part of a century, as have the effects of opening previously sealed birth records, and engaging in searches and reunion. Such research clearly holds much significance for anyone affected by origin deprivation, regardless of how it may have arisen. The issues surrounding anonymous gamete donation arguably mirror fairly closely those of clean-break, closed records adoption practice, especially in respect of telling and limiting information release.[139] As Grace has recently observed in relation to donor insemination, modern reproductive technologies have tended at times towards creating 'the conditions for ignoring or erasing the existence of the donor as gamete provider'.[140]

Terminologies have changed however, with 'donor-assisted conception' replacing such phrases as 'artificial reproduction' and 'artificial insemination'. These terms might potentially be seen as either protecting gamete donors' anonymity or perhaps removing them entirely from the equation. They could also possibly relegate the status of the child who has been so conceived to that of a sort of 'non-natural' being. Arguably, this echoes the way in which 'natural' 'birth' or 'biological' parents are increasingly referred to as either 'first' or 'original' parents in adoption, yet may still be described as 'real parents' in many works of popular or cultural fiction. Much debate also surrounds the question of whether donor-gamete children should be told of their origins. In respect of Australian policies, Walker and Broderick have made the argument that an unnecessarily 'therapeutic

[135] Neil (2009), p. 5.

[136] Neil and Howe (2004).

[137] See Turkington and Taylor (2009), pp. 21–38.

[138] Ryburn (1995), p. 62.

[139] The issue of matching physical characteristics is also telling: the selective placement of relinquished children used to mean that they often tended to be 'matched' with their prospective adopters via shared physical features such as eye and hair colouring. See further Hardy-Brown et al. (1980), p. 143. See also Modell (1994).

[140] Grace (2008).

injunction' seems to have been laid down by many psychologists in respect of 'telling' children about their genetic background.[141] They seem to suggest that the parallels with adoption are merely 'putative' and that the notion of 'genealogical bewilderment' is largely a socio-cultural construct. They ostensibly reject the suggestion that any degree of 'community stigma' might attach to donor gamete conception or to the social status achieved through this method of family formation.[142] They similarly refuse to accept that any degree of negative secrecy might surround the process, regarding anonymity as a positive means of achieving 'confidentiality' and as basically an essential measure aimed at protecting the privacy rights of adults. They argue also that 'different labels convey different psychological, moral and political implications and construct different meanings of ostensibly the same event.'[143]

They caution psychologists against assuming the role of a self-appointed advocate for any child 'as yet not born' lest they unnecessarily or unwittingly 'psychologize' the situation and perhaps 'manufacture victims' in the process.[144] They question whether donor gamete conceived individuals would have reason to 'be any more bewildered than the rest of the community,'[145] citing anecdotal evidence of how some sperm donors often regard their contribution to the procreative process as being merely akin to 'giving blood'. They admonish the Australian government over its official advice to infertile couples that the process constitutes a form of 'pre-natal adoption' and for suggesting that a distinction exists between the biological and psychological modes of parenting.[146]

Such a stance seems to ignore much of the psychological research on adoption and overlooks the socio-cultural importance that attaches to ascertainable bio-genetic identity. It also avoids issues such as the harmful effects of receiving genetic misinformation, the potential consequences of absent or inaccurate medical records, and the implications of being severed from one's ancestral heritage. It similarly avoids key issues such as those raised by Kirk's 'social role' theory, which placed much of the responsibility for adoptee maladjustment firmly at the door of those who denied that any degree of 'otherness' or difference exists in respect of social kinship.[147]

Similarly, if it is accepted that a fairly deep sense of respect and gratitude is probably felt by many adoptees towards their social families, and that feelings of guilt may attach to the decision to search or seek reunion, then it may be argued that such emotions might be magnified for donor-gamete conceived persons. Their relationship with their parent(s) may be less straightforward, possibly approximating

[141] Walker and Broderick (1999), pp. 38–44 at p. 39.

[142] *Ibid* p. 40.

[143] *Ibid* p. 39.

[144] *Ibid* p. 41. See also Dawes (1994); Dineen (1996), p. 157 (as cited by Walker and Broderick 1999).

[145] Walker and Broderick (1999), p. 40.

[146] See The Western Australian Human Reproductive Technology Council (1995).

[147] Kirk (1964).

more to that of the conventional, genetically connected relationship, but perhaps still having an element of missing ancestry or ambiguity. Couples affected by male sub-fertility were for example at one time encouraged to engage in intercourse post-treatment so that there might be a remote possibility that any child conceived would be biologically connected to the infertile or low fertility male partner.[148] The need to suggest such an approach brings to mind some of the 'blood-tie replacement' practices described in Chap. 2, where reassurance of triad adults might have taken precedence over the psychological welfare needs of the child. Highlighting the 'authorship' of descent that paternity bestows upon children, Grace stresses the importance of being placed 'within a history of ancestors' where one's genitors are clearly identifiable.[149] Bifurcated forms of parentage that can arise post-assisted conception can have significant psychosocial implications for all members of the donor-gamete triad.

The powerful symbolism that still attaches to the blood-tie perhaps makes it unlikely that the 'liberal gloss of happy donated-gamete families where all parties are mature, honest and open' will occur automatically.[150] It is perhaps more likely that some children could, upon being told of the circumstances of their conception, tend instead towards assigning an adoptee-like 'unconscious and phantasmatic role' to their unknown genitor.[151] The parallels with the family romance pre-occupations of some adoptees are obvious. As Hargreaves recently noted,

> the dominant discourse of the primacy of biological connection is a powerful cultural theme, evident in talk about the meaning of parent–child relationships, about genealogy, or prior claims by biological relatives.[152]

Although adoption law frameworks may seem to represent 'a useful precedent' for law and policy-makers tasked with protecting the interests of donor conceived individuals, its usefulness is likely to be limited to the way in which it highlights the key dilemmas within the process.[153] Where non-disclosure of genetic identity primarily benefits infertile or relinquishing adults, it can create much detriment for genetically kinless children, aimed as it is at ensuring that the

> non-genetic parent be perceived as equally connected to the child, to ensure that the child grows as strong a bond with that parent as with the genetic parent, to maintain the appearance of a 'normal' family, to avoid distressing the child with the truth of his/her origin, and to allow the non-genetic's parent infertility – a condition that normally carries a negative stigma with it- to remain unknown to others.[154]

---

[148] Grace et al. (2008), pp. 301–314. See also Hargreaves (2006), pp. 261–283.

[149] Grace (2008), p. 258.

[150] *Ibid* p. 259.

[151] *Ibid* p. 261.

[152] Hargreaves (2006), pp. 261–283 at p. 279.

[153] Haimes (1988), p. 46.

[154] McGee et al. (2001), pp. 2033–2038 at p. 2034. See also Blyth (2002), p. 797. Available at www.bmj.com/cgi/content/full/324/7340/797/DC1. Accessed 24.11.10; Blyth (2008), pp. 157–174. Available http://webdoc.sub.gwdg.de/univerlag/2008/gsm3um wege.pdf. Accessed 01.09.010

Closed adoption policies do not often provide solution-rich templates for the new reproductive technologies to follow. They do however, along with much of the jurisprudence on genetic kinlessness, merit discussion if only for the warnings which they seem to carry about human nature. As Freeman has noted in relation to the socio-legal dilemmas raised by the use of assisted reproduction techniques,

> There has been no systematic exploration of the questions it raises which has put children, their interests and rights into the forefront. Too many of the conclusions and too much of the legislation have been hurried, often responding to scandal or panic.[155]

Our societal preoccupation with identifying the sources of our 'ancestral substance'[156] is not a recent phenomenon. Arguably, advances in reproductive technologies may mean that the need to persuade parents towards embracing meaningfully 'communicative openness'[157] is likely to remain an essential aspect of protecting the psychological welfare of children deprived (whether fully or partially) of genetic ancestry. It is perhaps not surprising that the apparently simple option of not telling children of their adoptive origins was sometimes suggested as the best means of achieving a successful placement based on the creation of strong attachments.[158] Providing information might lead to a sort of 'biological pressure to fulfill [a] genetic blueprint'[159] and create a sort of additional burden for origin-deprived children who might already be seeking a way to cope with their unique social situation.

Complete denial or sealing off of a child's origins absolves parents (natal and adoptive) of a number of parental obligations however; successfully maintaining secrecy saves genetic parents from having to provide information about themselves, the child's ancestry or the circumstances surrounding conception or relinquishment. Adoptive parents might similarly be spared the need to provide the levels of emotional support needed to cope with telling the adoptee and enabling searching or contact. Although a significant level of 'permanence'[160] might be achieved via such a harsh, 'clean-break' model for social kinship, such a strategy relies upon several gambles coming together: physical characteristics matching very well, silence from other triad members and wider family or society, and a strict adherence to upholding the various legal fictions that tend to underpin social kinship (e.g. altered or sealed birth certificate). As the various studies referred to in this chapter indicate, family life and human nature are often such that legal fictions may be quite easily and dramatically outmatched by the eventual emergence of 'biological realities.'[161]

---

[155] Freeman (1996), pp. 273–297 at p. 297.

[156] Grace (2008), p. 257.

[157] Brodzinsky (2005).

[158] See for example Wellisch (1952), p. 41 who observed that identification with one's relatives is often easier where there is some degree of physical resemblance; as such adoption might be better suited to orphaned children.

[159] Maddi (1996), p. 106.

[160] See Holman (1988).

[161] *Kroon v The Netherlands* (1994) 19 EHRR 263 para 11.

Whether such deliberate 'orphanisation' is aimed primarily at reassuring adoptive or gamete recipient parents, is a key question for those who must interpret and implement the best interests of the child principle in such contexts. A significant section of the population currently exist as 'perennial children' given that they are denied a legal right to challenge sealed records or parental vetoes on information disclosure.[162] The 'power of the law as a force for completeness, coherence and closing off "the past" '[163] should not be underestimated. If a clearer system of rights and responsibilities could yet be found to exist between biologically 'related' people, the legal significance of the blood-tie would not have to depend almost entirely upon one's position within the social kinship triad, or upon one's place or time of birth.[164] In respect of understanding the potential impacts of non-ancestry, the 'doomed quality'[165] of some adoptees should perhaps not be overlooked.

3.6 Conclusion

As Rushbrooke has argued, the need to search for genetic relatives is now essentially 'mainstream' in nature; 'deep-seated problems' seem likely therefore to perhaps affect 'a significant proportion of adoptees' throughout the world.[166] A focus on 'open, honest, non-defensive and emotionally attuned family dialogue'[167] to promote meaningful, communicative openness between children and their social parents is gradually challenging the 'pro-customer paradigm'[168] of adult privacy. It seems now to be generally acknowledged that many triad children may be vulnerable to socio-cultural exclusion and to psychological harm due to the latent sense of 'otherness' that can attach to having an absent, uncertain or unknown bio-genetic heritage. The importance of achieving a sense of 'fluent narrative'[169] is evident from the considerable body of research that exists in respect of the psychological aspects of social kinship.

The notion of genetic heritage comprises not just of the more obvious biological and socio-cultural components, but also tends to overlap with more abstract concepts such as selfhood, the 'right' to identity, and to not be deprived of one's

[162] Freundlich (2001), pp. 61–68.

[163] Ingvesson and Mahoney (2000), p. 81.

[164] See also Strathern (1995).

[165] Carsten (2000).

[166] Rushbrooke (2001), pp. 26–34 at p. 32. See also Kirton et al. (2000), pp. 6–18.

[167] See Brodzinsky (2005), on the need for 'communicative openness' in adoption.

[168] See Rushbrooke (2007). Available http://www.publications.parliament.uk/pa/jt200607/jtselect/jtembryos/169/169we79.htm (accessed 01.02.11) wherein he argues that the UK Government has a retrospective duty towards pre-1991 donor-conceived persons, to both collate their information and permit access to it, on the basis of the principle of equality of treatment and the fundamental right to genetic identity.

[169] Kirkman (2003), pp. 2229–2242 at p. 2229.

key 'attachments.' Questions over the best means of developing a strong sense of personhood, have in turn served to inform both the study of kinship, the ongoing refinement of adoption and gamete donation practice and, to an extent, the interpretation of the best interests principle.[170] Looking to how social kinship functions as an institution is important given the recursive nature of the various harms that may accompany the permanent loss of genetic identity. As Wegar has suggested, much 'adoptionology' seems to perhaps still focus largely upon individualized, adoptee pathologies rather than on the flaws and failings of the various systems for creating social kinship.[171] This has led to the perception that,

> dilemmas of adoption result from individual shortcomings (which ultimately have a biological or instinctual basis) rather than from the structure of adoption as a social institution.[172]

Carp similarly found that much archaic adoption legislation (in the United States) served to promote the 'notion that the adoptive relationship was inherently flawed'.[173] Simply placing the blame upon the 'individual shortcomings'[174] of adopted children and then expecting that they will easily manage 'their' difficulties through the use of inherent or taught coping skills, seems to echo some of the early policies on overcoming or managing child illegitimacy, insofar as they also sought to quietly erase the 'social flaws'[175] of unwanted children. That there is now a clear need for 'Positive Adoption Language'[176] seems to at least acknowledge that searches aimed at recovering lost 'origin'[177] will perhaps be undertaken by a good number of adoptees at some stage in their life.

That there may be a need to search for genetic ancestry does not detract from the fact that social kinship often has the potential to heal many of the emotional and physical traumas suffered by abused or abandoned children and to perhaps reverse those 'developmental deficits' which they are likely to have suffered at the hands of abusive carers.[178] A child's 'need to know' in such a context would not necessarily be afforded much priority over the necessity for substitute parenting aimed at enabling the healing properties of social kinship to take effect.[179] Equally, where

---

[170] *Ibid.*

[171] Wegar (1997), pp. 97–118.

[172] *Ibid* at p. xii.

[173] Carp (1998), p. 18.

[174] Wegar (1997).

[175] See for example Kornitzer (1959), p. 102; Kellmer-Pringle (1967), p. 25.

[176] See for example Spencer (1979), p. 450 who argued that terms such as 'natural mother' seek to remove any recognition of relatedness between parent and adopted child. She advocated instead the use of 'emotionally correct' words.

[177] See Triseliotis (1973).

[178] Johnson (2002), p. 39.

[179] On the benefits of substitute parenting see for example Rutter (1998), pp. 465–476 on rates of recovery from early privations (psychological and nutritional) in respect of Romanian infants adopted in the United Kingdom. See also Rutter et al. (2001), pp. 97–103 which compared the

domestic systems of social kinship operate largely as a means of protecting older, possibly abused children, the need to 'search' perhaps becomes less of an issue, with the question of kin contact possibly instead becoming the significant factor.[180]

The research discussed in this chapter seems to suggest that any 'generalisations about adoption must be made cautiously'.[181] The phrase 'origin deprivation' does however seem quite apt where children have been normatively denied access to genetic kinship. Arguably, it seems reminiscent of a term that might be used to discuss identity rights violations, for example involving civil and political freedoms, which can arise where displaced persons have been cut off from their roots. Individualised arguments over the need to protect 'psychological welfare' and socio-cultural entitlements, though valid, are perhaps unlikely to provoke the same level of legal urgency in respect of finding a domestic remedy for closed records. Given the harms that can flow from origin deprivation, a strongly remedial element of justiciability ought to attach to the loss of genetic identity and natal kinship.

If it is accepted that origin deprivation often tends to arise from matters over which triad children will have had little or no control (conception, relinquishment, birthplace) then greater judicial direction on this point perhaps becomes key, not least in terms of establishing how the best interests of the child principle might be better complied with. The chapters which follow look to the 'law' on blood-ties, as it continues to emerge from international law provisions and guidance, Strasbourg discourses, and domestic interpretations of 'family life' rights and the best interests of the child in cases involving access to ancestry.

References

Ainsworth, M. D. S., Blehar, M. C., Waters, E., & Wall, S. (1978). *Patterns of attachment: A psychological study of the strange situation*. Hillsdale: Erlbaum.

Ainsworth, M. D. S., & Bowlby, J. (1991). An ethological approach to personality development. *American Psychologist, 46*, 333–341.

Allen, N. (2003). *Making sense of the new adoption law*. London: Russell House.

Antze, P., & Lambeck, M. (1996). *Tense past: Cultural essays in trauma and memory*. London: Routledge.

Arsenian, J. M. (1943). Young children in an insecure situation. *Journal of Abnormal and Social Psychology, 38*, 225–229.

Barclay, P. (1982). *Social workers: Their role and tasks*. London: Bedford Square Press.

Bartholet, E. (1993). *Family bonds: Adoption and the politics of parenting*. Boston: Houghton Mifflin.

behaviour patterns of formerly institutionalised Romanian adoptees with those of domestic adoptees.

[180] Triseliotis (1985), pp. 19–24; see also Maluccio et al. (1986), on the primacy of the 'natural family'.

[181] Grotevant (1997b).

Berger, C. R., & Calabrese, R. J. (1975). Some explorations in initial interaction and beyond: Toward a developmental theory of interpersonal communication. *Human Communication Research, 1*, 99–112.

Birlingham, D., & Freud, A. (1942). *Young children in war time*. London: Allen & Unwin.

Blyth, E. D. (2002). How it feels to be a child of donor insemination. *British Medical Journal, 324*, 797.

Blyth, E. D. (2008). Donor Insemination and the Dilemma of the "Unknown Father". In G. Bockenheimer-Lucius, P. Thorn, & C. Wendehorst (Eds.), *Umwege zum eigenen Kind Ethische und rechtiche Herausforderungen an die Reproduktionmedizin 30 Jahre nach Louise Brown* (pp. 157–174). Göttingen: Universitätsverlag Göttingen.

Boholm, A. (1983). *Swedish kinship: An exploration into cultural processes of belonging and continuity*. Gothenburg: Acta Universitatis Gothenburgensis.

Boss, P., & Greenberg, J. (1984). Family boundary ambiguity: A new variable in family stress theory. *Family Process, 23*, 535–547.

Bowlby, J. (1940). The influence of early environment in the development of neurosis and neurotic character. *The International Journal of Psycho-Analysis, XXI*, 1–25.

Bowlby, J. (1951). *Maternal care and mental health*. New York: Schocken.

Bowlby, J. (1958). The nature of the child's tie to his mother. *The International Journal of Psycho-Analysis, 39*, 350–371.

Bowlby, J. (1973). *Separation: Anxiety and anger, attachment and loss* (Vol. 2). London: The Hogarth Press.

Bowlby, J. (1979). *The making and breaking of affectional bonds*. London: Tavistock.

Bowlby, J. (1980). *Attachment and loss: Loss, sadness and depression* (Vol. 3). London: Hogarth.

Bowlby, J. (1982). *Loss: Sadness and depression (attachment and loss)*. New York: Basic Books.

Bowlby, J. (1988). *A secure base: Clinical applications of attachment theory*. London: Routledge.

Bretherton, I. (1992). The origins of attachment theory: John Bowlby and Mary Ainsworth. *Developmental Psychology, 28*, 759–775.

Brodzinsky, D. M. (2005). Reconceptualizing openness in adoption: Implications for theory research and practice. In D. Brodzinsky & J. Palacios (Eds.), *Psychological issues in adoption: Research and practice*. New York: Greenwood.

Brodzinsky, D. M., & Steiger, C. (1991). Prevalence of adoptees among special education populations. *Journal of Learning Disabilities, 24*, 484–489.

Cahn, N. (2000–2001). Children's interests and information disclosure: Who provided the egg and sperm? Or, mommy, where (and whom) do I come from? *The Georgetown Journal of Gender and Law, 2*, 1–27.

Carp, W. E. (1998). *Family matters: Secrecy and disclosure in the history of adoption*. Cambridge: Harvard University Press.

Carsten, J. (2000). 'Knowing Where You've Come From': Ruptures and continuities of time and kinship in narratives of adoption reunions. *Journal of the Royal Anthropological Institute (NS), 6*, 687–703.

Carsten, J. (2004). *After kinship: New departures in anthropology*. Boston: Cambridge University Press.

Cohen, S. (1996). Trauma and the developmental process: Excerpts from an analysis of an adopted child. *The Psychoanalytic Study of the Child, 51*, 287.

Cole, E. S., & Donley, K. S. (1990). History, values and placement policy issues in adoption. In D. Brodzinksy & M. Schechter (Eds.), *The psychology of adoption*. New York: Oxford University Press.

Cowan, E. A. (1938). Some emotional problems besetting the lives of foster children. *Mental Hygiene, 22*, 454–458.

Cretney, S., Masson, J., & Bailey-Harris, R. (2003). *Principles of family law*. London: Thomson.

Dawes, R. (1994). *House of cards: Psychology and psychotherapy built on myth*. New York: Free Press.

DeFries, J. C., McGuffin, P., Mc Clearn, G. E., & Plomin, R. (2000). *Behavioral genetics*. New York: W H Freeman & Co.
Derdeyn, A. P. (1979). Adoption and the ownership of children. *Child Psychiatry and Human Development, 9*, 215–225.
Dineen, T. (1996). *Manufacturing victims: What the psychology industry is doing to people*. Montreal: Robert Davies.
Dunbar, N., & Grotevant, H. D. (2004). Adoption narratives: The construction of adoptive identity during adolescence. In M. W. Pratt & B. H. Fiese (Eds.), *Family stories and the life course: Across time and generations*. Mahwah: Erlbaum.
Eiduson, B. T., & Livermore, J. B. (1953). Complications in therapy with adopted children. *The American Journal of Orthopsychiatry, 23*, 795–802.
Erikson, E. H. (1968). *Identity: Youth and crisis*. New York: Norton.
Erikson, E. H. (1970). Identity crisis in perspective. In E. H. Erickson (Ed.), *Life history and the historical moment*. New York: Norton.
Erikson, E. H. (1980). *Identity and the life cycle*. New York: Norton.
Featherstone, M. (1995). *Undoing culture: Globalization, postmodernism and identity*. London: Sage.
Freeman, M. (1996). The new birth right? Identity and the child of the reproductive revolution. *The International Journal of Children's Rights, 4*, 273–297.
Freidlander, M. L. (2003). Adoption: Misunderstood, mythologized, marginalized. *The Counseling Psychologist, 31*, 745–752.
Freud, S. (1949). *The ego and the Id*. London: Hogarth Press.
Freud, S. (1957). *Origins of psychoanalysis*. New York: Basic Books.
Freud, A. (1960). Discussion of Dr John Bowlby's Paper. *The Psychoanalytic Study of the Child, 15*, 53–62.
Freundlich, M. (2001). The impact of adoption upon members of the triad. *Adoption and Ethics, 3*, 61–68.
Friedlander, M. L. (2003). Adoption: Misunderstood, mythologized, marginalized. *The Counseling Psychologist, 31*, 745–752.
Frisk, M. (1964). Identity problems and confused conceptions of the genetic ego of adopted children during adolescence. *Acta Paedo Psychiatrica, 31*, 6–12.
Goddard, H. H. (1912). *The Kallikak Family: A study in the heredity of feeble-mindedness*. New York: MacMillan.
Goffman, E. (1963). *Stigma: Notes on the management of spoiled identity*. Englewood Cliffs: Prentice Hall.
Goldfarb, W. (1945). Psychological privation in infancy and subsequent adjustment. *The American Journal of Orthopsychiatry, 15*, 247–255.
Goldstein, J., Freud, A., Burlingham, D., & Solnit, A. J. (1979). *Before the best interests of the child*. New York: Free Press.
Goldstein, J., Freud, A., Goldstein, S., & Solnit, A. J. (1996). *Best interests of the child: The least detrimental alternative*. New York: Free Press.
Goldstein, J., Freud, A., & Solnit, A. J. (1973). *Beyond the best interests of the child*. New York: Free Press.
Goodman, J. D., Silberstein, R. M., & Mandell, W. (1963). Adopted children brought to child psychiatric clinic. *Archives of General Psychiatry, 5*, 451–456.
Gould, S. J. (1996). *The mismeasure of man*. New York: Norton.
Grace, V. M. (2008). The psycho-social politics of paternity in the case of male-donated gametes. *Women's Studies International Forum, 31*, 257–262.
Grace, V. M., Daniels, K. R., & Gillet, W. (2008). The donor, the father, and the imaginary constitution of the family: Parents' constructions in the case of donor insemination. *Social Science & Medicine, 66*, 301–314.
Graham, P., & Stevenson, J. (1985). A twin study of genetic influences on behavioral deviance. *Journal of the American Academy of Child Psychiatry, 24*, 33–41.

Greunewald, D. (1984). On the nature of multiple personality: Comparisons with hypnosis. *International Journal of Clinical and Experimental Hypnosis, 32*, 170–190.
Grotevant, H. D. (1997a). Coming to terms with adoption: The construction of identity from adolescence into adulthood. *Adoption Quarterly, 1*(1), 3–27.
Grotevant, H. D. (1997b). Family processes, identity development and behavioral outcome for adopted adolescents. *Journal of Adolescent Research, 12*, 139–161.
Grotevant, H. D. (2003). Counselling psychology meets the complex world of adoption. *The Counselling Psychologist, 31*, 753.
Haimes, E. (1987). "Now I Know Who I really am": Identity change and redefinitions of the self in adoption. In T. Honess & K. Yardley (Eds.), *Self and identity: Psycho-social processes*. Chichester: Wiley.
Haimes, E. (1988). Secrecy: What can reproduction learn from adoption? *International Journal of Law Policy and the Family, 2*, 46–61.
Haimes, E., & Timms, N. (1985). *Adoption, identity and social policy: The search for distant relatives*. London: DHSS Gower.
Hardy-Brown, K., Plomin, R., Greenhalgh, J., & Jax, K. (1980). Selective placement of adopted children: Prevalence and effects. *Journal of Child Psychology and Psychiatry, 21*(2), 143.
Hargreaves, K. (2006). Constructing families and kinship through donor insemination. *Sociology of Health & Illness, 28*(3), 261–283.
Heinicke, C. M. (1956). Some effects of separating two-year olds from their parents: A comparative study. *Human Relations, 9*, 105–176.
Herrnstein, R. J., & Wilson, J. Q. (1985). *Crime and human nature*. New York: Simon and Shuster.
Heywood, C. (2001). *A history of childhood*. Cambridge: Polity Press.
Holman, P. (1953). Some factors in the aetiology of maladjusted children. *Journal of Mental Science, 99*, 654–688.
Holman, B. (1988). *Putting families first*. London: MacMillan.
Horner, T. M., & Rosenberg, E. B. (1991). The family romance: A developmental-historical perspective. *Psychoanalytical Psychology, 8*, 131–148.
Howe, D. (1997). Parent-reported problems in 211 adopted children: Some risk and protective factors. *Journal of Child Psychiatry and Psychology, 38*, 401–411.
Humphrey, M., & Humphrey, H. (1986). A fresh look at genealogical bewilderment. *The British Journal of Medical Psychology, 59*(2), 133–140.
Humphrey, M., & Ounsted, C. (1963). Adoptive families referred for psychiatric advice. *The British Journal of Psychiatry, 109*, 599–608.
Hurlock, E. B. (1955). *Adolescent development*. New York: McGraw-Hill Book Co.
Ingvesson, B., & Mahoney, M. A. (2000). 'As One Should, Ought And Wants To Be': Belonging and authenticity in identity narratives'. *Theory Culture and Society, 17*, 77–110.
Jaffe, B., & Fanshel, D. (1970). *How they fared in adoption: A follow-up study*. New York: Columbia University Press.
Joffe, W. G., & Sandler, J. (1967). Some conceptual problems involved in the consideration of disorders of narcissism. *Journal of Child Psychotherapy, 1*, 56–66.
Johnson, D. E. (2002). Adoption and the effect on children's development. *Early Human Development, 68*(1), 39.
Jones, A. (1997). Issues relevant to therapy with adoptees. *Psychotherapy, 34*(1), 64–68.
Jones, D. (2003). The generative psychology of kinship: Generating variation from universal building blocks with optimality theory. *Evolution and Human Behaviour, 24*, 320–350.
Jones, D. (2004). The Universal Psychology of Kinship: Evidence from language. *Trends in Cognitive Sciences, 8*(5), 211–215.
Josling, J. F., & Levy, A. (1984). *Adoption of children*. London: Oyez.
Juffer, F., Stams, G. J. J. M., & van Ilzendoorn, M. H. (2004). Adopted children's problem behaviour is significantly related to their ego resiliency ego control and sociometric status. *Journal of Child Psychology and Psychiatry, 45*(4), 697–706.
Kadushin, A. (1966). Adoptive parenthood: A hazardous adventure? *Social Work, 11*, 30–39.

Keating, J. (2001). Struggle for identity: Issues underlying the enactment of the 1926 Adoption of Children Act. *University of Sussex Journal of Contemporary History, 3*, 1–9.

Kellmer-Pringle, M. L. (1967). *Adoption: Facts and fallacies*. London: Longmans.

Kempthorne, O. (1997). Heritability: Uses and abuses. *Genetica, 99*, 2–3.

Kendler, K. S., & Greunberg, A. M. (1984). An independent analysis of the Danish adoption study of schizophrenia. *Archives of General Psychiatry, 41*, 555–564.

Kenworthy, M. E. (1921). The mental hygiene aspects of illegitimacy. *Mental Hygiene, 5*, 499.

Kirk, D. H. (1964). *Shared fate: A theory of adoption and mental health*. New York: The Free Press.

Kirk, D. H. (1981). *Adoptive kinship*. Toronto: Butterworths.

Kirkman, M. (2003). Parents' contributions to the narrative identity of offspring of donor-assisted conception. *Social Science & Medicine, 57*, 2229–2242.

Kirschner, D. (1992). Understanding adoptees who kill: Dissociation patricide and the psychodynamics of adoption. *International Journal of Offender Therapy and Comparative Criminology, 36*, 323–333.

Kirton, D., Feast, J., & Howe, D. (2000). Searching, reunion and transracial adoption. *Adoption and Fostering, 24*(3), 6–18.

Klein, M. (1984). The psycho-analysis of children. In R. Money-Kyrle (Ed.), *The writings of Melanie Klein* (Vol. 2). New York: Free Press.

Knight, R. P. (1941). Some problems in selecting and rearing adopted children. *Bulletin of the Menninger Clinic, 5*, 65–74.

Kornitzer, M. (1952). *Child adoption in the modern world*. New York: Philosophical Library.

Kornitzer, M. (1959). *Adoption*. London: Putnam.

Krugman, D. C. (1964). Reality in adoption. *Child Welfare, 43*, 349–358.

Lee, R. M. (2003). The trans-racial adoption paradox: History, research and counseling implications of cultural socialization. *The Counseling Psychologist, 31*(6), 711–744.

Levy-Shiff, R. (2001). Psychological adjustment of adoptees in adulthood: Family environment and adoption-related correlates. *International Journal of Behavioral Development, 25*, 97–104.

Lifton, B. J. (1990). The formation of the adopted self. *Psychotherapy in Private Practice, 8*, 85–92.

Lifton, B. J. (1994). *Journey of the adopted self: A quest for wholeness*. New York: Basic Books.

Lindblad, F., Hjern, A., & Vinnerljung, B. (2003). Intercountry adopted children as young adults: A Swedish cohort study. *The American Journal of Orthopsychiatry, 73*, 190–202.

Loehlin, J. C., & Nichols, R. C. (1976). *Heredity, environment and personality: A study of 850 sets of twins*. Austin: University of Texas Press.

Lowe, N., & Douglas, G. (2007). *Bromley's family law*. Oxford: Oxford University Press.

Lyon, C. M., Cobley, C., Petrie, S., & Reid, C. (2003). *Child abuse*. London: Jordans.

Maddi, S. R. (1996). *Personality theories: A comparative analysis*. Pacific Grove: Brooks & Cole.

Maluccio, A. N., Fein, E., & Olmstead, K. A. (1986). *Permanency planning for children: Concepts and methods*. London: Routledge.

Marris, P. (1991). The social construction of uncertainty. In C. M. Parkes, J. Stephenson-Hinde, & P. Marris (Eds.), *Attachment across the life cycle*. London: Routledge.

McGee, G., Brakman, S. V., & Gurmankin, A. (2001). Gamete donation and anonymity: Disclosure to children conceived with donor gametes should not be optional. *Human Reproduction, 16*(10), 2033–2038.

Mednick, S. A., Gabrielli, W. F., & Hutchings, B. (1984). Genetic influences in criminal convictions: Evidence from an adoption cohort. *Science, 224*, 891–894.

Melosh, B. (2002). *Strangers and kin: The American way of adoption*. Cambridge: Harvard University Press.

Modell, J. S. (1994). *Kinship with strangers: Adoption and interpretation of kinship in American culture*. Berkeley: University of California Press.

Neil, E. (2009). Post-adoption contact and openness in adoptive parents minds': Consequences for children's development. *British Journal of Social Work, 39*, 5–23.

Neil, E., & Howe, D. (Eds.). (2004). *Contact in adoption and permanent foster care: Research, theory and practice*. London: BAAF.

Nelkin, D., & Lindee, S. (1995). *The DNA mystique: The gene as a cultural icon*. New York: W H Freeman.

Nelkin, D., & Lindee, S. (1998). Media portrayals of genetics. *The American Journal of Human Genetics, 63*(2), 662–663.

Nelson, K. (1992). Emergence of autobiographical memory at age 4. *Human Development, 35*, 172–177.

Nelson, K. (1993). The psychological and social origins of autobiographical memory. *Psychological Science, 4*(1), 7–14.

Nelson, K., & Fivush, R. (2004). The emergence of autobiographical memory: A social cultural developmental theory. *Psychological Review, 1*(2), 486–511.

O'Donovan, K. (1988). A right to know one's parentage? *International Journal of Law and the Family, 2*, 27.

Park, K. E. (2006). Adoptees knowledge about and contact with birth parents and their adjustment in adolescence and young adulthood. *Dissertation Abstracts International Section A: Humanities and Social Sciences, 67*(2-A), 738.

Parton, N. (2004). From Maria Colwell to Victoria Climbie: Reflections on a generation of public inquiries into child Abuse. *Child Abuse Review, 13*(2), 80–94.

Patel, T. (2005). The usefulness of oral life (hi)story to understand and empower: The case of trans-racial adoption. *Qualitative Social Work, 4*, 327.

Peller, L. E. (1963). About 'Telling the Child' of his adoption. *Bulletin of the Philadelphia Association for Psychoanalysis, 11*, 145–153.

Pendry, P. (1998). *Ethological attachment theory: A great idea in personality?* North Western University (www.personalityresearch.org/papers/pendry.html). Accessed 16 May 2010.

Powell, K., & Afifi, T. D. (2005). Uncertainty management and adoptees' ambiguous loss of their birth parents. *Journal of Social and Personal Relationships, 22*, 129–151.

Ridley, M. (2003). *Nature via nurture: Genes, experience and what makes us human*. New York: Harper Collins.

Robertson, J. (1953). *A two year-old goes to hospital (film)*. London: Tavistock Child Development Research Unit.

Robertson, K. C. (2006). Attachment and care-giving behavioral systems in intercountry adoption: A literature review. *Children and Youth Services Review, 28*(7), 727.

Rogers, C. (1955). Persons or science: A philosophical question. *American Psychologist, 10*, 267–278.

Rogers, C. (1961). *On becoming a person: A therapist's view of psychotherapy*. Boston: Houghton Mifflin.

Rowe, J., & Lambert, L. (1973). *Children who wait*. London: ABAA.

Rushbrooke, R. (2001). The proportion of adoptees who have received their birth records in England and Wales. *National Statistics, 104*, 26–34.

Rushbrooke, R. (2007). '*Proposals To Bring Donor-Conceived People's Birth Certification Into Line With That Of All Other UK Citizens*' Memorandum (Ev 114) to the Joint Committee on the Human Tissue and Embryos (Draft) Bill June 2007.

Rutter, M. (1998). Developmental catch-up, and deficit, following adoption after severe global early privation. *Journal of Child Psychology and Psychiatry, 39*(4), 465–476.

Rutter, M., Kreppner, J. M., & O'Connor, T. G. (2001). Specificity and heterogeneity in children's responses to profound institutional privation. *The British Journal of Psychiatry, 179*, 97–103.

Ryburn, M. (1995). Adopted children's identity and information needs'. *Children and Society, 9* (3), 41–64.

Sachdev, P. (1991). The triangle of fears: Fallacies and facts. In E. Hibbs (Ed.), *Adoption: International perspectives* (pp. 249–264). Madison: International Universities Press.

Sandler, J., & Joffe, W. G. (1969). Towards a basic psychoanalytic model: The theory of parent–infant relationship. *The International Journal of Psycho-Analysis, 41*, 585–595.

Sants, H. J. (1964). Genealogical bewilderment in children with substitute parents. *The British Journal of Medical Psychology, 37*, 133–141.

Schaffer, H., & Emerson, P. (1962). The development of social attachments in infancy. *Monographs for the Society for Research in Child Development* 29(3) (serial no. 94).

Schechter, M. (1960). Observations on adopted children. *Archives of General Psychiatry, 3*, 21.

Seebohm, F. (1968). *Report of the committee on local authority and allied personal social services*. London: HMSO CMND 3703.

Shawyer, J. (1979). *Death by adoption*. Auckland New Zealand: Cicada Press.

Slingerland, W. H. (1919). *Child-placing in families: A manual for students and social workers*. New York: Russell Sage Foundation.

Smith, D. W., & Brodzinsky, D. M. (2002). Coping with birthparent loss in adopted children. *Journal of Child Psychology and Psychiatry, and Allied Disciplines, 43*(2), 213.

Sobol, M. P., & Cardiff, J. (1983). A socio-psychological investigation of adult adoptees' search for birth parents'. *Family Relations, 32*, 477–483.

Sorosky, A. D., Baran, A., & Pannor, R. (1976). The effects of the sealed record in adoption. *The American Journal of Psychiatry, 133*, 900–904.

Spencer, M. (1979). *The terminology of adoption*. Washington: Child Welfare League of America.

Sroufe, L. A. (1983). Infant-caregiver attachment and patterns of adaptation in preschool: the roots of maladaptation and competence. In M. Perlmutter (Ed.), *Minnesota symposium on child psychology* (pp. 41–81), vol. 16.

Stiffler, L. H. (1993). Adoptees and birth parents connected by design: Surprising synchronicities in histories of union/loss/reunion. *Journal of Prenatal and Perinatal Psychology and Health, 7* (4), 267–287.

Stoneman, A. H. (1926). Adoption of illegitimate children: The peril of ignorance. *Child Welfare League of America Bulletin, 5*, 8.

Stonesifer, E. (1942). The behavior difficulties of adopted and own children. *Smith College Studies in Social Work, 13*, 161.

Strathern, M. (1995). Displacing knowledge: technology and the consequences for kinship. In F. D. Ginsburg & R. Rapp (Eds.), *Conceiving the new world order: The global politics of reproduction*. Berkeley: University of California Press.

Tarachow, S. (1937). The disclosure of foster-parentage to a boy: Behaviour disorders and other psychological problems resulting. *The American Journal of Psychiatry, 94*, 401–412.

Triseliotis, J. (1973). *In search of origins: The experiences of adopted people*. London and Boston: Routledge & Kegan Paul.

Triseliotis, J. (1985). Adoption with contact. *Adoption and Fostering, 9*(4), 19–24.

Tulving, E. (1972). Episodic and semantic memory. In E. Tulving & W. Donnaldson (Eds.), *Organisation of memory* (pp. 382–403). San Diego: Academic.

Turkington, S., & Taylor, B. J. (2009). Post-adoption face to face contact with birth parents: prospective adopters' views. *Child Care in Practice, 15*(1), 21–38.

Verrier, N. (1993). *The primal wound: Understanding the adopted child*. California: Verrier.

Verrier, N. (2003). *Coming home to self: The adopted child grows up*. Baltimore: Gateway Press.

Walker, I., & Broderick, P. (1999). The psychology of assisted reproduction – Or psychology assisting its reproduction? *Australian Psychologist, 34*(1), 38–44.

Wegar, K. (1997). *Adoption, identity and kinship: The debate over sealed birth records*. New Haven: Yale University Press.

Weiss, A. (1985). Symptomatology of adopted and non-adopted adolescents in a psychiatric hospital. *Adolescence, 20*, 763–774.

Weitzman, C. (2003). Developmental assessment of the internationally adopted child: Challenges and rewards. *Clinical Child Psychology and Psychiatry, 8*(3), 303.

Weitzman, C., & Albers, L. (2005). Long term developmental, behavioural and attachment outcomes after international adoption. *Pediatric Clinics of North America, 52*(5), 1395–1419.

Wellisch, K. E. (1952). Children without genealogy – A problem of adoption. *Mental Health, 12*, 41.

Western Australian Human Reproductive Technology Council. (1995). *Questions and answers about the donation of human reproductive material: Donor insemination and sperm donation, donated eggs and donated embryos*. Perth: Health Dept. of Western Australian.

Wieder, H. (1977). On being told of adoption. *Psychoanalytic Quarterly, 46*, 1–22.

Wierzbicki, M. (1993). Psychological adjustment of adoptees: A meta-analysis. *Journal of Clinical Child Psychology, 22*, 447–454.

Winnicott, D. (1960). Ego distortion in terms of true and false self. In *The maturational process and the facilitating environment: Studies in the theory of emotional development* (pp. 140–152). New York: International UP.

Zenderland, L. (1998). *Measuring minds: Henry Herbert Goddard and the origins of American intelligence testing*. New York: Cambridge University Press.

Ziller, R. C. (1969). The alienation syndrome: A triadic pattern of self-other orientation. *Sociometry, 32*(3), 287–300.

Chapter 4
Conceptualizing the "Right" to Avoid Origin Deprivation: International Law and Domestic Implementation

> A proclamation of a right is not the fulfilment of a right any more than an airplane schedule is a flight. A proclamation may or may not be an initial step toward the fulfilment of the right listed. It is frequently the substitute of the promise in the place of the fulfilment.—Shue (1980)

4.1 Introduction

As Donnelly (2003) has suggested, at the level of domestic implementation "identity is entitled to protection only where it is an autonomous expression of the rights and values of those who carry it" (214). Although there are no explicit rights to avoid origin deprivation ("genetic non-ancestry") within international law, a number of related entitlements to access certain aspects of genetic identity do appear to exist. These include such ostensibly justiciable concepts as the right to possess a name[1] or nationality,[2] and the duty upon States to respect "cultural integrity."[3] Broader, normative principles such as the best interests of the child,[4] equality of treatment and non-discrimination[5] also merit discussion here, given that they are often either

[1] See for example Articles 16 (1) (a) and 30 (1) of *Hague*.

[2] See for example *European*.

[3] See Article 27 of *International*; and Anaya (97–103) on the international legal and customary norms of non-discrimination and the protection of cultural integrity.

[4] See Article 3 (1) of *Convention*, which states that: "In all actions concerning children, whether undertaken by public or private social welfare institutions, courts of law, administrative authorities or legislative bodies, the best interests of the child shall be a primary consideration."

[5] See for example the anti-discrimination provisions of the Universal Declaration on Human Rights (1948) ('UDHR') namely Article 7; Article 2 ('*Everyone is entitled to all the rights and freedoms set forth in this Declaration, without distinction of any kind, such as race, colour, sex, language, religion, political or other opinion, national or social origin, property, birth or other status.*'); Article 25 (2) ('*Motherhood and childhood are entitled to special care and assistance. All children, whether born in or out of wedlock, shall enjoy the same social protection.*' Available

DOI 10.1007/978-3-319-01071-7_4,

embedded in domestic legislation or have provided useful guidance for domestic proceedings.[6] This article will attempt to conceptualise the hypothetical right to avoid origin deprivation, as a basic human right. The existence of such a right, or rather the need for such a right to be more fully articulated and more consistently implemented by domestic legal systems, is underpinned by the considerable amount of research highlighting the wide range of harms that can attach to genetic kinlessness. The article will ask whether such a right might yet be found to carry sufficient weight to "embed" itself as usefully juridical concept within the sphere of domestic child law, especially where private law issues such as family formation, information release, kin contact or welfare paramountcy arise. Arguably, domestic interpretations of many socio-economic entitlements, such as the right to respect for family life, are often almost entirely resource-dependant, and therefore tend to be viewed more as privileges than as fully justiciable human rights. The "right" to receive information or to be afforded socio-cultural heritage protections, are key examples (Article 19, 1995; Marks 2006).[7]

This article also makes particular reference to one closed records jurisdiction (Quebec) where the rights of many adoptees have been almost completely "outweighed" by the use of non-disclosure vetoes, on the basis of protecting the juridically heavier interests of natal parents. This is so despite the enactment of two regional charters aimed at the incorporation of basic human rights, and in spite of the scrutiny of the UN Committee on the Rights of the Child. Committee Country Reports are referred to here, with a view to gauging whether the guidance contained in them might eventually persuade domestic jurists and policy-makers that origin deprivation ought to occur exceptionally rather than normatively as a consequence of social kinship. Essentially, if identity rights in international law engender a corresponding right to "repatriate" with one's genitors, this raises the possibility that a number of rights infringements might have arisen in respect of donor anonymity, closed birth records and non-disclosure vetoes. If one accepts that the "universal protection of relationships with significant others is in fact protection of the distinctness and the uniqueness in the individual" (Ronen 2004, 151) then it may be difficult to justify the complete loss of genetic connection as a means to enabling social kinship.

---

at http://www.un.org/en/documents/udhr/index.shtml accessed 10.07.09); See also Article 6 of The Universal Declaration on the Human Genome and Human Rights (1997) which states that '*No one shall be subjected to discrimination based on genetic characteristics that is intended to infringe or has the effect of infringing human rights, fundamental freedoms and human dignity*'. Available at http://portal.unesco.org/en/ev.php-URL_ID+13177 accessed 04.01.09.

[6] See for example The Children (Northern Ireland) Order 1995 Article 3 (1) where the need to promote child welfare paramountcy echoes the principle of child welfare primacy in Article 3 of the Convention on the Rights of the Child. Available at http://www.legislation.gov.uk/nisi/1995/755/article/3/made accessed 12.01.11.

[7] See for example Article 3 of The International Covenant on Economic, Social and Cultural Rights (1966) ('ICESCR') available at http://www2.ohchr.org/english/law/cescr.htm accessed 25.07.09; Article 5 of the UDHR; Article 7 of the International Covenant on Civil and Political Rights ('ICCPR') available http://www2.ohchr.org/english/law/ccpr.htm accessed 25.07.09.

4.2 The UN Convention on the Rights of the Child

Ratification of the Convention may nevertheless encourage domestic reform.—Stark (2005)

Ronen has described The Children's Convention as "the most authoritative legal text on children in international law", enshrining as it does the primacy of child welfare (148). Freeman has further argued that it represents a "celebration of children's rights (which) has reached a crescendo"(273). By emphasising the key concept of human dignity, and by highlighting the need to take into account the wishes and feelings of children, the Convention has to an extent focussed wider attention on important welfare notions such as "authenticity" and the "self-realization" of individual identity (Ronen, 149). As Stewart noted however "borderline or unusual conditions" were not taken into account by the drafters of the Children's Convention on the Rights of the Child: there was little or no discussion of the consequences of being rendered genetically kinless (223). During the drafting stage, several countries expressed concern that the inclusion of a provision referring to the "biological element" of identity might conflict with any domestic policies seeking to implement recent or future advances in reproductive technologies.[8]

As Detrick et al. (1992) noted at the time, several members of the Working Group involved in the initial drafting expressed considerable unease over the use of the term "family identity", preferring instead the phrase "family relations as recognised by law" (294). Such an amendment served to exclude any genetic affiliations not covered by the symbolic sanction of legal process. Persons conceived via anonymous gamete donation or relinquished adoptees whose records have been sealed will for example fall outside the scope of the Convention's protections. Arguably, the Convention epitomises the attitude of the international law community generally in seeking to define (if not respond) to the needs of young people trying to develop a "dynamic child-constructed identity" (Ronen, 148). Support for adolescents who wish to gain a sense of difference from other individuals has taken centre stage, rather than helping genetically kinless persons establish a sense of belonging (An-Na'im 2001, 2). Origin deprived persons may completely invert this developmental template, seeking to instead effect some measure of integration with an identifiable group, perhaps to replace lost onomastic identity.

It is perhaps unsurprising that, although the Convention is suggestive of a moral obligation to remove obstacles to "birth-kin repatriation" (Brower Blair 2000–2001, 642) no positive obligation to actively facilitate information release or contact with genetic relatives exists (Stewart, 224).[9] Although Article 21 deals

[8] During its drafting, only one signatory state suggested that a 'biological element' of identity ought to be included in the final draft of the Convention. See also Report of the Working Group on a Draft Convention on the Rights of the Child (1989) E/CN.4/1989/48.

[9] See also The Vienna Convention on the Law of Treaties U N Doc A/CONF. 39/27 Article 31 (1) which requires that Treaties ought to be interpreted in the light of their 'object and purpose'.

with adoption, it is aimed at the issue of inter-country adoption, framing the practice as on a similar footing to domestic adoptions and seeking to ensure that parental consents have been obtained where possible. The provision is effectively silent on the issue of whether biogenetic information ought to be made available at a later stage, largely leaving the matter to be decided by domestic legislators and jurists.[10] Arguably, this adds to the significant level of deference already existing around private family law matters, which is perhaps not dissimilar to the wide "margin of appreciation" that is often invoked in respect of interpreting and implementing other human rights provisions.[11]

Article 20 (1) does acknowledge that "special protection and assistance by the State" ought to be extended to any child "temporarily or permanently deprived of his or her family environment." The need to effect permanence of care-giving is also apparent in Article 20 (3)'s call for "desirability of continuity", although the provision is silent on more pragmatic issues such as kin reunion, familial contact or information provision. The concept of a right to preserve genetic heritage identity could fall within this provision's requirement on States to have "due regard" for the child's "ethnic, religious, cultural and linguistic background"[12] in respect of international or mixed race adoption. The concept of "due regard" possibly falls short of the more onerous "respect" standard which serves to remind States of their positive Convention obligations towards vulnerable groups however.[13] In relation to the need for genetic identity the Convention is much less straightforward. Its Preamble describes the family as being of "fundamental importance" and highlights the need to promote "full and harmonious development of ..personality"[14]; the Convention itself emphasises the protection of "authenticity" against a backdrop of "cultural

---

[10] See also The Hague Convention (1993) Art 31 which (together with Articles 15 and 16) states that any items of information '*shall be used only for the purposes for which they were gathered or transmitted*'. Available at http://hcch.e-vision.nl/index_en.php?act=convention.pdf accessed 01.08.09.

[11] On the 'margin of appreciation' see Handyside V. The United Kingdom—5493/72 [1976] ECtHR 5 (7 December 1976). The doctrine is referred to by enforcement bodies (such as the European Court of Human Rights) in determining whether signatory states have acted in breach of their Convention obligations. States are permitted varying degrees of discretion in carrying out such obligations at the level of domestic compliance, given that Convention rights will often be subject to a variety of differing interpretations by member states; factors such as religious or cultural differences will generally be taken into account. See further O'Donnell (1982).

[12] Article 21 (a) deals with the issues of parental consent, inter-country adoption and the prohibition of financial gain.

[13] See for example Article 8 of the European Convention on Human Rights which states that; '*1. Everyone has the right to respect for his private and family life, his home and his correspondence 2. There shall be no interference by a public authority with the exercise of this right except such as is in accordance with the law and is necessary in a democratic society in the interests of national security, public safety or the economic well-being of the country, for the prevention of disorder or crime, for the protection of health or morals, or for the protection of the rights and freedoms of others.*' On the concept of respect in family law see Eekelaar (2008).

[14] See The Preamble to the United Nations Convention on the Rights ('The Children's Convention') of the Child, Para 5.

context (and) personal meaning" (Ronen, 147). It does not create a clear duty to actively protect or enforce identity rights however but seems to at least express support for the psychological benefits that attach to having knowledge of one's familial origins. Articles 7 and 8 of the Children's Convention similarly emphasise the significance of "family relations" yet fail to provide a useful definition of what might actually constitute relatedness.[15] Article 9 (1) similarly permits kin separation where the best interests of the child so require, whilst Article 9 (3) provides for the maintenance of "personal relations and direct contact" unless the best interests principle demands otherwise.[16] McCarthy (2004) has further argued that the Children's Convention fails to define exactly which categories of relatedness might fall within its ostensibly wide remit of "family relationships" (12). Article 9 (3) for example refers only to the need for contact between child and parents, with no mention of siblings or other natal kinfolk.[17]

Its narrowly drafted scope is mirrored by much domestic legislation on this issue, where estranged genetic kin often also seem likely to fall outside the definitional remit of legal relatedness, if not that of "family life" itself. Article 9 (1) also permits the legal and physical separation of relatives to facilitate application of the "best interests" of the child principle, but limits this to circumstances where judicial review of any such decision is available as a remedy.[18] Where a child has been relinquished voluntarily pre-hearing, or via family case-conference, or where parents subsequently change their mind, it seems unlikely that the remedy of judicial remedy would be available however. Whether domestic decisions are likely to be altered by judicial review (for example to require social workers or adoptive parents to actively make provision for information release or sibling contact post-adoption or post gamete-donation) is difficult to say.[19] Equally, there are aspects of the relinquishment process that will often be beyond the scope of legal remedy, such as where domestic legislation puts no legal onus on social workers to notify birth relatives that a child has been relinquished or placed for adoption, or where the issue of contact has been left to the discretion of adopters.

[15] See Article 8 of The Children's Convention which states that (1) '*State Parties undertake to respect the right of the child to preserve his or her identity, including nationality, name and family relations as recognized by law without unlawful interference.*' (2) '*Where a child is illegally deprived of some or all of the elements of his identity, States Parties shall provide appropriate assistance and protection, with a view to re-establishing speedily his or her identity.*'

[16] Article 9 (4) covers the provision of information but is clearly aimed at situations involving civil or political rights violations such as '*detention, imprisonment, exile deportation or death*' rather than adoption or assisted conception. See also Articles 20 (1) and 22 (2).

[17] Article 9 (3) states that '*State parties shall respect the right of the child who is separated from one or both parents to maintain personal relations and direct contact with both parents on a regular basis, except if it is contrary to the child's best interests.*'

[18] See for example Re A [2001] NIFam 23 (24 October 2001); Re K [2002] NIFam 13 ; Re P, Freeing Without Consent [2004] NIFam 6.

[19] See for example Kearns v France ECHR 10 January 2008 (Application no. 35991/04).

It also seems unlikely that very young children at the centre of such processes would be either apt or able to bring proceedings given that issues such as contact and information release often tend to hinge upon the presence of "ties meaningful" (Ronen, 147). As a number of key cases suggest, the absence of "remembered relatedness" has often served to exclude natal ties e.g. between parent and relinquished newborn, siblings or other genetic relatives of the child.[20] The pre-requisite need to demonstrate meaningful ties to genetic kinfolk may present a considerable obstacle to accessing genetic identity. The subjective perceptions of the child (participatory rights of the child notwithstanding) might serve to exclude cultural heritage protections on the basis that such aspects of one's identity might not be "remembered." Arguably, to do so is to ignore the socio-cultural significance of biological kinship and sideline the psycho-social need to access accurate identity narrative. The Convention similarly places no legal obligation upon parents or social services agencies to actually inform children of their biological heritage or to provide them with any useful information on it. The apparent tensions between Articles 5, 12 and 13 (which together broadly outline the right of the child to "be heard" and to participate in all proceedings) seem reflective of the difficulties surrounding child protection aspects of identity formation generally (Ronen, 162).[21] Problems in balancing the competing principles of child protection and child participation for example have been repeatedly highlighted by The Committee on the Rights of the Child.[22]

That said, the Convention's provisions could yet perhaps be interpreted by domestic jurists and policy-makers in such a way as to frame genetic kinlessness as an exceptional rather than normative occurrence. The rights enshrined in Article 8 could at times be expanded upon in domestic hearings to look beyond the usual protected parameters of family, nationality and name. Courts might be open to the idea in certain cases of actually ordering some level of kin contact (if necessary, this could be in indirect form) or the release of information, so that these issues are not delegated completely to social workers or social parents. Article 10 might prove similarly useful, insofar as it appears to promote the reunification of estranged family members, by highlighting the need for regular contact between children and parents who "have been separated." Its wording suggests that it is aimed however at

[20] See for example Keegan v Ireland (1994) (App no. 16969/90) 18 EHRR 342; Kroon v The Netherlands (1995) 19 EHRR 263; X v Croatia (App no. 12233/04) (17.07.2008); Johansen v Norway [1996] EHRR 31 (App no. 17383/90) . See however Marckx v Belgium (1979) (application no. 6833/74) (1979).

[21] See further Freeman's (1996) comparative analysis of the lack of identity rights of donor gamete-conceived individuals as opposed to those of adult adoptees generally in open records jurisdictions such as the UK.

[22] See for example the Committee's Observations on Saint Lucia CRC/C/150 (2005) 10 para 66 which recommend that the views of the child be afforded greater consideration in decisions involving custody after parental separation; Islamic Republic of Iran CRC/C/146 (2005) 88 para 462 which notes that the 'best interests' principle of the child is often completely sidelined in favour of custody decisions based upon the age of the child, and how this often tends to discriminate against mothers.

reuniting relatives who have been separated by extreme events (territorial displacement, political conflict) or private issues (marital separation, familial breakdown) rather than by straightforward child relinquishment or child-protective state intervention aimed at securing emergency child care.

If the best interests principle were perhaps expanded upon to include the uniquely disadvantaged situations of many kinless individuals, the concept of child welfare paramountcy could possibly be taken to encompass not just childhood protections but longer-term psychological needs. In the context of the Children's Convention generally, child welfare is often subject to an over-riding "structured element of caution" (Ronen, 148). At the level of private family proceedings, such a policy may translate into complete loss of genetic ties, such as can occur with permanent vetoes on disclosure or contact.

4.3 UN Committee Guidance

> Despite the law's interpretations, the welfare principle remains potentially open to myriad meanings and is in that sense, indeterminate.—Sclater and Kaganas (2003)

Guidance provided by the various Human Rights Committees on issues such as the avoidance of genetic kinlessness is fairly limited. Although Committee Guidance is non-binding in nature, international scrutiny, if not some measure of censure, generally follows publication of Country Reports, critical Comments and Concluding Observations.[23] Several documents contain practical suggestions as to how signatory states might achieve greater adherence to the welfare-protective aims of the Children's Convention. As such, Committee Guidance does at least have the potential to persuasively embed international law principles within the domestic practices, policies and legislation of signatory states.[24] References to the child's need for identity are fairly commonplace; actual mention of genetic ties occurs much less frequently. A number of responses (to Country Reports) however do indicate the Committee's apparent disquiet over the inability of many children "born out of wedlock" to "know the identity of their father" within some jurisdictions.[25] The system of "anonymous birth" in Luxembourg for example is highlighted for criticism, as is the use of "baby nests" in Austria, with the Committee expressing the hope that such occurrences might eventually be addressed by signatory states introducing a range of "necessary measures" aimed at redressing

[23] See http://www2.ohchr.org/english/bodies/crc accessed 08.01.09.

[24] See also The Report of the Working Group on a Draft Convention on the Rights of the Child (1986) U.N. Doc E/CN.4/1986/39.

[25] St Vincent and The Grenadines CRC/C/118 (2002) 101 para 437. See also Haiti CRC/124 (2003) 95 para 426 which highlights Article 306 of Haiti's Civil Code, which makes it illegal for children born out of wedlock to learn the identity of their father.

the consequences of these forms of origin deprivation.[26] In respect of Italy, the Committee similarly expressed unease over the fact that adoptees cannot learn the identity of their biological parents even upon gaining the age of majority. Children born outside marriage may also be completely deprived of genetic parentage, unless they have been formally recognised as kin by their birth parents.[27] San Marino Registry Offices were noted for their failure to include any information whatsoever on the genetic parents of adopted children.[28] Slovenian authorities have also been reminded of the need to enact in domestic law "the right of the child to know his/her origin and access to information about his/her background."[29]

The notion of "best interest" primacy (under Article 3 of the Children's Convention) is itself not necessarily subject to straightforward, uniform application. In relation to Morocco, Syria and Pakistan for example, serious concern was expressed by the Committee that "custody" of a child might be determined by reference to the child's age rather than via the "best interests" principle.[30] Kyrgyzstan's new Family Code has also been highlighted by the Committee for failing to reform domestic laws on adoption secrecy and the lack of information provision to adoptees.[31] Antigua and Barbuda's lack of legal protection for the contact rights of separated parents and their children also warranted mention.[32] Likewise, the informal "god-parenting" (or "simple adoptions") that occur in some regions alongside legal adoption practice, were noted as being in urgent need of closer monitoring, as was some greater form of follow-up procedure to many inter-country adoption processes.[33] The Committee's "Days of General Discussion" have also attempted to examine a fairly wide range of issues that affect the rights of the child. The 40th session (2005) looked to the issue of "Children without Parental Care."[34] The Committee reiterated the message of the Preamble to the Children's Convention, that, "the child, for the full and harmonious development of his or her personality, should grow up in a family environment, an atmosphere of happiness, love and understanding."[35] Their Recommendations included a call for all relevant parties to

[26] Luxembourg CRC/C/146 (2005) 36 paras 184 and 185. See also Austria CRC/C/146 (2005) 47 paras 251, 252.

[27] Italy CRC/124 (2003) 36 at para 168.

[28] San Marino CRC/C/133 (2003) 9 para 41.

[29] Slovenia CRC/C/137 (2004) 104 para 566.

[30] Morocco CRC/C/132 (2003) 100 para 482; Syrian Arab Republic CRC/C/132 (2003) 116 para 554; Pakistan CRC/C/133 (2003) 37 para 210.

[31] Kyrgyzstan CRC/C/143 (2004) 50 para 282.

[32] Antigua and Barbuda CRC/C/143 (2004) 93 para 491.

[33] Madagascar CRC/C/133 (2003) 56 at paras 293. See also Papua New Guinea CRC/C/137 (2004) 94 para 504 on the use of informally 'adopted' girls as domestic servants.

[34] See CRC/C/146 Annex II (2005) available at http://bayefsky.com/getfile.php/id/48615273/misc/days accessed 30.12.08.

[35] Day of Discussion (2005) CRC Outline CRC/C/146 Annex II ibid para 4. Available at http://www.ohchr.org/english/bodies/crc accessed 21.12.08. See also CRC-'*Children Without Parental Care*' (2005) Recommendations [CRC/C/156, Chapter VI].

"show flexibility" towards the cultural aspects of substitutive care and highlighted the need to establish more effective monitoring mechanisms within signatory states. Some form of consultation with children and parents "throughout the process" was also called for. Concern was expressed over the lack of guidance available to signatory states in providing care for "parentless" children. It was argued that a "careful weighing of the child's best interests as the primary deciding factor" ought to be carried out before any action was taken to separate natal kin. Significantly, the wide variety of state policies surrounding social worker intervention ranged from "rarely, if ever taking action" to removing children permanently on the basis of "abuse and exploitation" with child poverty falling within this broad definition of harm. It is perhaps unfortunate that the 2005 Committee decided to avoid all discussion of "adoption-related issues" (which it also failed to define) in a bid to focus instead on the "main issues" of the day (i.e. child protection and substitute care). Some discussion of how states and judiciaries might seek to better protect the longer-term welfare of genetically kinless children might have been useful given the increasingly complex nature of social kinship processes. It is significant however that The Committee on the Rights of the Child noted that it "encourages the development of adoption, *Kafalah* and traditional foster care systems, such as family-based alternative care (e.g. extended family, grandparents."[36]

This statement seems to endorse the value of preserving biogenetic kinship links. As well as framing "children as partners" within the decision-making process, it suggests that the views of genetic parents ought to be actively sought, given that such "decisions have a far-reaching impact on the child's life and future." Calling for the establishment of a "special mechanism" which "values children as partners" (such as family group conferencing) the Committee seemed to indicate increased commitment to affording greater attention to the individual welfare of affected children. It remains difficult however to see exactly where the "identity rights" of newborn or infant adoptees might fit within this suggested framework. The need to demonstrate remembered relatedness (so as to trigger the protections of human rights law) might serve to exclude a significant number of genetically parentless individuals. Additionally, protection against the social stigma that can attach to "illegitimate" birth status seems unlikely to be achieved solely through the

[36] CRC-'*Children Without Parental Care*' (2005) Recommendations [CRC/C/156, Chapter VI] para 656. See also Article 20 of The Children's Convention which recognises the practice of *Kafalah* as an Islamic alternative to adoption; Sura No. 33 of the Koran prohibits (as '*haram*') adoption in Sharia law as a means of creating new filial bonds on the basis that it removes biological ties. *Kafalah* as a form of substitute parental care (maintenance, education, financial support) advocates that 'orphans' should be treated as genetic kinfolk; inheritance and onomastic identity rights are not created however. See also International Reference Centre for the Rights of Children Deprived of Their Family (ISS/IRC) 'Kafalah' (December, 2007) on how definitions and practices differ between Islamic states, with 'non-international Kafalah' (based upon a strict interpretation of Sharia law) predominating in countries such as Iran, Egypt and Mauritania, and a more adoption-like 'Conversion' form of Kafalah (in Tunisia and Indonesia) allowing 'full legal consequences of filiation', subject to religious requirements. Available www.crin.org/docs/Kafalah.BCN.doc accessed 01.01.09.

application of international law principles. As the International Reference Centre for the Rights of Children Deprived of Their Family ("IRC") noted recently in respect of incidences of child abandonment in Islamic states, "the reasons ...are very similar from one country to another; children born out of wedlock are the first victims of abandonment and they remain stigmatised throughout their life."[37] The transfer of such children to substitute family care under the doctrine of *Kafalah* seldom allows for international adoptive placement; there are further restrictions in respect of new parents' nationality and religious belief.[38] This emphasis on preserving the child's inherent biogenetic culture, arguably re-affirms the psycho-social importance of maintaining some element of blood-tied kinship; it also demonstrates that some ambivalence towards the genetic connection still cuts across key aspects of cultural relativism in this area. The Committee has also reiterated the need to preserve the links that exist between children and their "own distinctive communities" and stresses that child protection systems must take into consideration "indigenous culture, values and the child's right to indigenous identity." Although it is heartening to find mention of the importance of "identity rights" it is perhaps unfortunate that visible cultural or racial differences, or minority status seem pre-requisite in relation to enabling some level of international law protection. A basic need to gain authenticity of biogenetic background should perhaps warrant the same safeguards, irrespective of whether cultural heritage is an obvious issue. The reluctance of The Committee to engage with such matters possibly suggests that a considerable degree of social unease still attaches to questions of biogenetic heritage, biological "authenticity" and definitions of legal relatedness. Issues surrounding "illegitimacy" and infertility treatment fall within the realm of family privacy, into which jurists and policy-makers often seem reluctant to venture. The "best interests" principle however clearly demands that such issues must not be sidelined; definitions of child protection must attempt to include the lifelong psychological welfare of relinquished children. Decisions on how the harms of origin deprivation might be addressed or prevented ought to be subject to the same principles of international human rights law as those processes and policies which cause genetic kinlessness.

Surprisingly, jurisdictions that have actively incorporated the provisions of key human rights instruments may also come in for censure. Despite the enactment of the Human Rights Act 1998 (which incorporated much of the European Convention on Human Rights into its domestic law) the United Kingdom was reproached by The Committee in 2002 over the fact that in certain circumstances, "children born out of wedlock, adopted children or children born in the context of a medically assisted fertilization do not have the right to know the identity of their biological parents."[39] As such, it was recommended that "all necessary measures" should be

---

[37] IRC (2007) ibid at note 68, p. 2. See further Bargach (2002) and Ishaque (2008).

[38] Morocco is the exception to this; see further The Declaration on the Rights and Care of the Child in Islam. Available http://www.oic-oci.org/english/conf/fm/22/resolution22-c.htm#16 accessed 04.03.11.

[39] United Kingdom CRC/C/121 (2002) 23 para 123.

taken by the state to enable such children to gain access to information on their parentage "to the extent possible." Although donor anonymity has since been overturned by legislation in the United Kingdom, parents remain under no legal obligation to actually inform their children that they have been conceived through assisted reproductive techniques.[40] This seems at odds with the Guidance given in respect of separated or divorced parents, where the Committee stressed the right of affected children to "maintain contact" with both of their biological parents, in a bid to "maintain their identity."[41] Arguably, children in institutional care ought to have a juridical right to "maintain personal relationships" and have "direct contact with their parents and families" on a regular basis if they so desire.[42] As Oren (2001, 125) has argued however "the context of family law disputes shapes substance and procedure." Much of the jurisprudence involving genetic identity seems likely to continue to take place in the "private law" setting of parentage or parenting disputes. This does not necessarily remove the need for some form of genetic kinship "repatriation" to be effected in many cases, so as to bring about "just resolutions" in cases involving identity welfare. The psychological aspects of being deprived of genetic origin are likely to be similar, irrespective of how biological kinlessness might have arisen.

4.4 Preventing Non-ancestry?

> The potential for contact might be viewed as a resource that may or may not be appropriate to utilise in a particular family circumstance (Selina and Taylor 2009).

Although genetic connection is seldom referred to in explicit terms, a number of international law instruments do contain some degree of guidance on the issue of the need for accurate identity to be included within the definition of the longer-term best interests of the relinquished child. Article 16 of The Hague Convention on Protection of Children and Co-operation in Respect of Intercountry Adoption (1993) ("The Hague Convention")[43] for example requires signatory states to compile "identity" information on the "adoptable" child, including his or her "background, social environment, family history, medical history, including that of the child's family." Article 16 (1) (b) similarly requires that "due consideration" be given to the child's "ethnic, religious and cultural background" in the making of such reports. That said, under Article 16 (2) States are equally, if not more, obliged to take care "not to reveal the identity of the mother and father, if, in the State of

[40] See the Human Fertilisation and Embryology Act 1990 (as amended). Available at http://www.opsi.gov.uk/acts/acts1990/ukpagea_19900037_en_3#pb9-l1g34 accessed 01.08.10.

[41] Seychelles CRC/C/121 (2002) 41 paras 189–190.

[42] Estonia CRC/C/124 (2003) 9 at para 55 (f).

[43] The Hague Convention (1993). Available http://hcch.e-vision.nl/index_en.php?act=convention.pdf accessed 01.08.09.

origin, these identities may not be disclosed."[44] Article 31 further qualifies the use of any information collected under Articles 15 and 16 by adding that such data "shall be used only for the purposes for which they were gathered or transmitted." Arguably, there is scope within Article 31 for the transmission of information to affected parties, but only where the purpose of its collection complies with the terms of its release. Such a proviso gives much discretion however to signatory states who may decide that the information should be best treated as the property of the natal parents, or that triad members would be better off having it either destroyed or sealed permanently, in the interests of protecting parental anonymity, or ensuring child welfare. Again, adult-centric interests such as privacy (whether birth parent confidentiality or social parents' familial autonomy) appear capable of outweighing the identity rights of the relinquished child.

Stark has commented that this Convention also failed to set out a specific template that signatory states could either adhere to or modify to suit their own particular set of circumstances (57–68). Instead it provided a fairly general framework with "no substantive norm" attaching to either the notion of confidentiality or the practicalities of information release other than that signatory states should be aiming towards (rather than ensuring) the preservation of essential information. Equally, ratification of the Convention does not require states to change their domestic laws; this applies equally whether states have implemented a policy of open records or are adhering to policies that enshrine closed files or enable parental vetoes. As Horgan and Martin (2008) argue however, the concept of adoption is now generally "no longer seen as a one dimensional triangulation of interests, but as a constellation of interests which can often span different countries" with the child remaining as a fixed point, a "vulnerable party in a process conducted by adults" (157). As a result, the paramount aim of all concerned ought to be that of protecting the child's best interests.

The most recent European Convention on Adoption (2008)[45] appears to represent progress in this regard but is aimed primarily at regulating international rather than domestic adoption practices. Equally, whereas the 1967 Convention seemed

[44] Article 16 (2). See also The Convention on Jurisdiction, Applicable Law, Recognition Enforcement and Co-operation in Respect of Parental Responsibility and Measures for the Protection of Children (1996). Available at http://hcch.e-vision.nl/index_em.php?act=conventions.text&cid=70 accessed 31.07.09 (The United Kingdom is a signatory state, but at the time of writing has yet to ratify this Convention: a Declaration has also been entered in respect of domestic decision enforcement) which contains some guidance on promoting the best interests of the child at domestic level . See Article 4 (b) however which specifically excludes '*decisions on adoptions, measures preparatory to adoption, or the annulment or revocation of adoption.*' Article 6 (1) however does apply to 'refugee children', children who have been 'internationally displaced' and 'children whose habitual residence cannot be established.'

[45] Approved at the 118th Ministerial Session (Strasbourg) March 2008; adopted May 2008, opened for signatures November 2008. Available at http://conventions.coe.int/Treaty/en/Treaties/Html/202.htm accessed 08.05.09. This effectively replaced The 1967 European Convention on Adoption which appeared to support harmonization of domestic laws on international adoption; no means of enforcement was created by the 1967 nor was it included in the *Acquis Communautaire* of EU

aimed at achieving "absorption" of the adoptee into the new family "in every sense" (Horgan and Martin, 158) the 2008 Convention has instead been drafted to reflect the wide range of socio-cultural changes that have occurred within the fields of family law and family life, such as the increasing incidence of same-sex couples wishing to adopt children. It also, to an extent, seeks to embrace the spirit of the Children's Convention by highlighting the need to take into account the wishes of children in all proceedings affecting their future. The issue of making information available to adoptees seems to have been at least tentatively addressed, by way of the "participation rights" policy of "age-appropriateness."[46] Although children need not formally consent to their own adoption, it would seem that the views of older adoptees should at least be taken into account by domestic authorities. Arguably, a "nuancing of the severing of all links" (Horgan and Martin, 161) does now exist in certain circumstances, with the blood-tie retaining importance in respect of serving to maintain parental duties of fiscal support or prevent marriages between former genetic kin. Article 10 for also permits Contracting States to retain alternative models of adoption ("simple", customary or *de facto*) which will not result in a total severance of the blood-tie. Article 22 deals with disclosure of information and appears to address the issue of balancing the privacy (if not the "anonymity") rights of relinquishing genetic parents.[47] It seems slightly equivocal in tone in respect of the issue of gauging whether discretion (to disclose) might eventually create new policies of openness or engender a meaningful "right to know" for adoptees.

The wide degree of discretion afforded to domestic authorities in family life matters is particularly relevant here, with the task of deciding on information release being left to national bodies, who need only then "bear in mind" the provisions of Art 7 of the Children's Convention, rather than actually embed them in national law or policy. Unlike the 1993 Hague Convention, Article 14 of the 2008 Convention does acknowledge that adoptive placements can be subject to revocation.[48] Again, decisions on this issue are often left to domestic authorities but remain subject to the best interests of the child principle. As Horgan and Martin observe, the 2008 Convention complements the provisions of the older Hague Convention, at least in respect of inter-country adoption practice. Strasbourg

member states. Sweden formally denounced the 1967 Convention in January 2003, whilst the UK partially renounced its provisions in June 2005.

[46] Article 5(a) of the 1967 Convention set out the suggested minimum level of 'essential principles' that signatory states should aim for in respect of domestic adoptions. These permitted but did not require secrecy in relation to the provision of medical information.

[47] 47 see Article 22(3) which states that '*The adopted child shall have access to information held by the competent authorities concerning his or her origins. Where his or her parents of origin have a legal right not to disclose their identity, it shall remain open to the competent authority, to the extent permitted by law, to determine whether to override that right and disclose identifying information, having regard to the circumstances and to the respective rights of the child and his or her parents of origin. Appropriate guidance may be given to an adopted child not having reached the age of majority.*'

[48] See for example the Irish case A G v Dowse [2006] IEHC 65.

jurisprudence might provide further guidance for contracting states seeking to refine their policies on adoption or develop practicable legal norms that reflect the higher standards of the "requirements of modernity" (162). In sum, the 2008 Convention still ostensibly assumes that full integration into the substitute family will be the preferred outcome in many cases, allowing for the use of vetoes or wide-ranging parental discretion on issues such as disclosure or contact where necessary. At present, it does not seem likely that domestic hearings on issues concerning non-ancestry will be significantly affected by the revised provisions of the new Adoption Convention. Although it hints at the need for a juridical right to heritage identity (adoptees ought not to be deprived of their original nationality and may be subject to the terms of customary adoption, for example[49]) it says little on the issue of how knowable ancestry might actually be achieved, providing merely that, where parentage is in dispute, "adoption proceedings shall, where appropriate, be suspended to await the results of the parentage proceedings. The competent authorities shall act expeditiously in such parentage proceedings."[50] Similarly, signatory states are *encouraged* rather than bound to enact domestic laws which might be more favourable to the origin deprived child. Adoptees generally seem unlikely to be given an absolute, unqualified right to access their original, kin-identifying information. The new Convention thus seems to be aimed at preserving the status quo of normative secrecy, especially given that systems of anonymous birth registration have not yet been denounced as violating the identity rights of many adoptees.[51]

4.5 Non-ancestry as Loss of Cultural Heritage?

> The cultural integrity norm . . .requires diverse applications in diverse settings (Anaya 2000)

It seems unlikely that a single model of reparation or indeed repatriation could apply to the differing scenarios surrounding non-ancestry and child relinquishment. The loss of genetic identity may arise quite dramatically, as a result of civil or political strife, or via displacement from one's country of origin; it may also occur much more quietly, via domestic processes aimed at creating social kinship. Unsurprisingly, domestic interpretations of the "best interests" principle also vary widely, according to context. The families of the *Abuelas* argued for example (in respect of the illegally "adopted" children of the "Disappeared" in Argentina) that "knowing the realities of one's identity was in itself in the best interests of the child" (Oren, 137–150). Here, the quasi-adoptive parents were generally described by jurists as "appropriators", repressors or child kidnappers, rather than as

[49] See Articles 11 and 12 (2008 Convention).

[50] Article 16 (2008 Convention).

[51] See further Lowe (2006).

substitute care-givers. This was so even where they had, through lapse of time, effectively become the psychological parents of the children they were raising. The best interests principle was interpreted here as requiring that the children were removed from their de facto parents and returned to the "entire ecological nest or social network" of the genetic kin from whom they had been illegally taken. It was held that "living with secrets and lies has a terrible effect on a family, transforming it into a non-family" (Oren, 163). Given that the psychological aspects of origin deprivation must be taken into account in all such cases, the notion of "recapturing" the lost identity became a judicial imperative, aimed at "healing" the victims of the illegal adoptions.[52]

It is difficult however to see how the child's welfare needs were necessarily rendered paramount over those of their adult kin here however. Although the identity rights of the child were discussed in some detail, the main "balancing" exercises seemed to mainly involve the conflicting rights and interests of the two sets of parents as triad adults. As with a number of overturned adoptions in other jurisdictions, the socio-economic status of the birth kin (i.e. grandparents) was in many cases often much worse than that of the quasi-adopters.[53] In some cases, older children made it plain that they did not wish to be reunited with their birth relatives but preferred to remain kinned within their social families. As Oren argued, in such disputed parentage scenarios, the rights of the child, and the remit of the best interests principle is likely to be defined by determining which set of adults has a stronger "parental rights claim"(187). In the case of older children, such a formula may defeat their "participation rights" if the courts ignore their wishes.

Delays in reuniting genetic kin can also influence implementation of the best interests principle. In post-conflict Argentina this meant that calls to have children reintegrated with their estranged natal kinfolk were eventually replaced by more pragmatic requests for the restitution of their original "identity, knowledge of .... origins and name." Such an outcome did save the children from being removed from the only "psychological" parents they had ever known (Oren, 194). Categorising origin deprivation as a civil or political rights issue does serve to align the notion with powerful concepts in international human rights law. If, for example, the loss of genetic identity has arisen through political upheaval or enforced cultural assimilation it might, hypothetically at least, be regarded as tantamount to "inhuman or degrading treatment" as was argued in connection with the *Abuelas*. A number of civil or political provisions relevant to origin deprivation do seem to possess strong elements of justiciability. The duty upon states to actively enable identity protection under Article 24 of The International

[52] The Courts referred to attachment theory and to studies from the field of psycho-analysis ibid pp. 158–161.

[53] See also DeBoer v DeBoer 509 US 1301 [1995] (the 'Baby Jessica' case) and Baby Boy Richard v Kirchner 513 US 1138 [1995] which also involved overturned adoptive placement.

Covenant on Civil and Political Rights (1966) ("ICCPR")[54] mirrors for example, those found in Article 15 of The Universal Declaration on Human Rights (1948) (The "UDHR")[55] and Articles 9 and 4 (a) of The European Convention on Nationality (1997).[56] Such provisions emphasise the right to "have a name" from the time of birth and to either "acquire a nationality" or not be subsequently deprived of one. Article 9 of The Nationality Convention also refers to the recovery of lost nationality; "Each State Party shall facilitate, in the cases and under the conditions provided for by its internal law, the recovery of its nationality by former nationals who are lawfully and habitually resident on its territory." The wording of this provision seems to place a largely aspirational, non-interference obligation on signatory states to "facilitate" rather than actively enforce or monitor the right to recover lost nationality however. The right is further qualified here by provisos on territorial residence and prolonged habitation, not to mention the inclusion of a reference to national laws. Such limitations are usually more associated with resource-dependant, socio-economic human rights such as the state's duty to promote and protect respect for family life rights.[57] Cultural heritage identity rights do tend to be connected to, if not dependent upon, a wide range of "land-based" property aspects, such place of habitation or birth, or tribal status via length of residence on Native land for example.[58] Where identity is so aligned it perhaps weakens the argument that the right to identity can be easily conceptualised as a political entitlement. Where a child has been conceived or born in one jurisdiction, but adopted in another for example it will be difficult to establish a right to genetic identity information by simply arguing the "right" to name or nationality. Arguably, origin deprived persons in such circumstances might effectively be rendered almost "stateless", in addition to being deemed biologically "kinless." "Identity" rights are far from absolute, with key provisions tending to remain open to widely varying interpretations. In other words, possessing "a name" (Article 24 (2) ICCPR) is not necessarily the same as being afforded unfettered access to original onomastic information or genetic records. The right to "acquire nationality" (Article 24 (3) ICCPR) similarly does not equal those rights which might arises by virtue

[54] Articles 24 (2) and (3) of The ICCPR state that; '2. *Every child shall be registered immediately after birth and shall have a name. 3. Every child has the right to acquire a nationality.*'

[55] Article 15 of the UDHR states that '1. *Everyone has the right to a nationality 2. No one shall be arbitrarily deprived of his nationality nor denied the right to change his nationality*'.

[56] Article 4 (a) states that '*everyone has the right to a nationality*'.

[57] See Article 10 (1) of the International Covenant on Economic and Social Rights (The 'ICESCR') which states that '*The widest possible protection and assistance should be accorded to the family, which is the natural and fundamental group unit of society..*' and Article 16 (3) of the UDHR which describes the family as '*the natural and fundamental group of society and is entitled to protection by society and the State*'.

[58] See for example Re Bridget et al. (Minors) [1995] BO93520 (The 'Rost twins case') where the question of whether or not the children possessed Pomo ancestry appeared to turn on the issue of paternal residence on tribal lands.

of birth or ancestry. Additionally, a loss of nationality is often subject to domestically drafted sanctions (such as immigration laws).[59]

Stewart also highlighted the fact that Article II (e) of the Convention on the Prevention and Punishment of the Crime of Genocide (1948) ("The Genocide Convention"[60]) promotes the right to ascertain one's "national identity." This Article includes within its definition of genocide the act of "forcibly transferring children of the group to another group." The wording suggests some level of acknowledgement of the importance of ascertaining and preserving genetic identity; it also underscores the need for redress in many cases (223).[61] The harms of identity loss have also been acknowledged within a number of domestic jurisdictions. In respect of indigenous adoptees in the United States for example it was argued (during the drafting of domestic legislation aimed at preserving cultural heritage rights) that until quite recently "society was putting on them an identity which they didn't possess and taking from them an identity that they did possess" (Westermeyer, 1974).[62] Scenarios such as this perhaps suggest that the concept of identity must include both socio-cultural and civil or political aspects of birthright. The loss or absence of cultural heritage may affect title to land, language, religious belief or tribal status. Arguably, the hybrid nature of Article 27 of the International Covenant on Civil and Political Rights ("ICCPR") is relevant. The provision states that, "In those States in which ethnic, religious or linguistic minorities exist, persons belonging to such minorities shall not be denied the right, in community with the other members of their group, to enjoy their own culture, to profess and practise their own religion, or to use their own language."[63] The duty upon states to preserve "cultural integrity" especially in relation to the rights of the most vulnerable members of society such as minorities, "parentless" children or displaced peoples seems plain. As Anaya has observed in respect of indigenous cultural heritage rights, "While in principle the cultural integrity norm can be understood to apply to all segments of humanity, the norm has developed remedial aspects particular to indigenous peoples in light of their historic and continuing vulnerability (102)." If vulnerability perhaps were taken as the key

[59] Provided that this has occurred in a 'non-arbitrary' fashion; See Article 15 (2) UDHR.

[60] Available at http://www2.ohchr.org/english/law/genocide.htm accessed 12.07.09.

[61] This point, coupled with the argument that much harm can flow from loss of identity was central to the arguments of the *Abuelas'* families. Stewart appears to suggest that a right of repatriation should not however be 'absolute' given that it might defeat 'competing interests'. These are not defined here by Stewart but presumably might include conflicting parental rights such as privacy or familial autonomy or perhaps wider public interest, where the use of finite state resources would be involved e.g. in facilitating kin contact.

[62] Indian Child Welfare Programme '*Hearings before the Subcommittee on Indian Affairs of the Senate Committee on Interior and Insular Affairs*' (1974) 93d Cong 2d Session 46 per Dr Westermeyer J.

[63] Available at http://www2.ohchr.org/english/law/ccpr.htm#art27 accessed 07.07.09. See also Pritchard (1998) on The Human Rights Committee's General Comment No 23 (50) (Art 27) para 6.2 which suggests that Article 27 places a positive obligation on signatory states to actively protect 'the identity of a minority'.

element in gauging basic need, difficulty might still however surround the framing of identity rights as an element of cultural heritage, especially given the collective, "group rights" nature of many such entitlements (Hadden 2000). The wording of Article 27 is also such that individual identity rights issues could fall outside of its protective remit. It appears to assume that members of the group in question may possess essential knowledge as to the original identities of its "lost" members and an awareness of its own cultural background and nationality before being affected by non-ancestry. Many individuals are instead origin deprived through either via the legitimate exercise of parental discretion, or through private family law court decisions on information release or familial contact. As such, provisions such as Article 27 seem more aimed at "preserving" existing identities than interfering in private family law matters or overturning domestic court decisions.

That said, in some regions, kinship losses may be temporary, with identifying information later released, for example where an individual claimant can demonstrate a "good cause" exception to courts or agency workers.[64] Harms in this context still tend to quietly affect individuals as private law matters, rather than as dramatic, ethno-racial violations of collective, cultural heritage rights. They are also likely to be deemed "necessary and proportionate" given the benefits that social kinship tends to bring and the likelihood that parental interests such as privacy still often tend to outweigh relinquished children's identity rights in this context. Given that longer term child welfare requirements might include a need to restore narrative truth, there remains scope for overlap between the various categories of human rights. The psychological effects of genetic non-ancestry are arguably similar, whether individuals have been subject to the loss of genetic identity by gaps in domestic law or policy or through the less than robust workings of international legal systems.[65] Private law aspects of the rights most closely associated with family life tend however to rely significantly upon the effectiveness of national enforcement and monitoring systems.

Cerda has further argued, that rights deemed to be "new" are likely to lack the redressive urgency associated with older, more established rights (115–117).[66] By framing a right to genetic identity as a new right its promoters may also have left it vulnerable to being heavily qualified and resource dependent. Article 8 of The

[64] See for example Re Baby Boy K (1999) 701 NY S 2d 600 NY Fam Ct 1999 where medical grounds were deemed to satisfy the 'good cause' requirement and the use of an intermediary to pass essential information was sanctioned; See also however Re Adoption of Baby Boy S (1997) 705 A.2d.822 (NJ) Super.Ct.Ch.Div 1997.

[65] See for example the 'traumas and identity conflicts' arising in Serrano-Cruz Sisters v El Salvador Inter-American Court of Human Rights (March 1, 2005) Merits, Reparations and Costs. Judgment of March 1, 2005. Series C No. 120 para A (1) at p. 13. Available at: http://www.corteidh.or.cr/docs/casos/articulos/seriec_120_ing.pdf accessed 13.07.09. On the American Convention on Human Rights generally, see Robertson and Merrills (1997).

[66] Cerda (1990); see also Stewart (1992–1993) who suggests that the right to identity under the Children's Convention successfully conforms with Alston's (1984) template for 'new rights' and Alston (1984) on the 'creation' of new human rights.

Children's Convention for example seems aimed at redressing the "legal void" that surrounds displaced or abducted children to reconnect them with their original families (Stewart, 222). The Argentinean sponsor of Article 8 ultimately had to concede that meaningful implementation of such identity rights would depend largely upon domestic legal systems' willingness to interpret their Convention obligations pro-actively (Cerda 1990, 116). Article 8 (1) contains the proviso "without lawful interference" which would allow states to legitimately alter a child's "nationality, name and identity." Article 8 (2) is similarly worded, preventing only "illegal" removal of "elements of identity." Assistance provided by domestic governments to discover, retrieve or restore identity must similarly also be "appropriate" to the circumstances, although such support ought to be given "speedily" which does at least suggest awareness of how delays might attach to the domestic implementation.[67]

4.6 Enforcing Regionally Chartered Rights

> Psycho-social reparation...implies knowing the truth, official acknowledgement of the facts, justice being done, reunification and the reconstruction of relationships, experiences and affective ties with other people.[68]

The concept of a "right to know" is particularly relevant to the prevention of non-ancestry (Colliver 1995; Marks 2006). Many cases involving genetic identity losses appear to turn upon the issue of whether child welfare paramountcy should take precedence over conflicting parental interests such as privacy. Judicial discourses on the right to receive information tend to stress the importance of being able to obtain essential facts. They reinforce the argument that basic legal norms and human rights standards ought to be adhered to and actively implemented at the level of domestic decision-making, even if only as baseline minimal requirements.[69] As Colliver has further argued, "If national governments and institutions incorporate these standards into their laws or policies, they may be

[67] See also the Council of Europe Joint Declaration on Inter-Country Adoption February 2008. Available at http://www.coe.int/t/dc/files/themes/childrens_rights/declarationcommune_EN.asp accessed 11.08.09.

[68] See the testimony of expert witness Rosa América Laínez Villaherrera in I/A Court H.R., Case of Serrano-Cruz Sisters v. El Salvador at para 193. The Court considered, but did not find a violation of Articles 17, 18 and 19 of the American Convention on Human Rights (the rights of the family, right to a name, rights of the child respectively). It did however recommend the setting up of a body aimed at tracing lost natal kinfolk, and directed that '*the State must adopt all necessary measures to create a system of genetic information that allows genetic data that can contribute to determining and clarifying the relationships and identification of the disappeared children and their next of kin to be obtained and conserved.*'

[69] See also The Dissenting Opinion of Judge A A Cancado Trindade in Serrano-Cruz Sisters v. El Salvador. Merits, Reparations and Costs. Judgment of March 1, 2005. Series C No. 120 ibid para a (1) at p. 13. Available at: http://www.corteidh.or.cr/docs/casos/articulos/seriec_120_ing.pdf

enforced through the national courts and other institutional mechanisms" (43). In many "closed records" jurisdictions however, it is particularly difficult to overcome the obstacles created by domestic law or policy, even where signatory states have subsequently ratified international Conventions or introduced their own national system of rights protection. Despite the enactment of The Canadian Charter of Rights and Freedoms (1982) ("The Canadian Charter")[70] for example, the province of Quebec's "clean-break" model of adoption survives due to its closed record, parental non-disclosure veto system. Quebec adoptees are often unable to access an original birth certificate, irrespective of whether a parental veto has lodged or not. They are entitled to receive basic non-identifying information as to their "background" (birth weight, birth date, hospital of birth) but have no right to access any information that might be deemed "identifying." This includes such basic information as original birth surname, and whether or not they may have natal siblings or other genetic kinfolk. Cultural integrity may be similarly compromised as mixed race adoptees are often unable to receive information on which tribal band they might be descended from.[71]

Reunions do occur perhaps "illegally" via private investigators in privately funded searches, or through strictly handled intermediary contact by social welfare agencies, where social workers are given access to confidential files and carry out discreet searches. Where a birth mother refuses to engage in contact with or accept any correspondence from an adoptee, social workers are not allowed to make a second attempt at contact. The effect of the non-disclosure veto extends here to other kinfolk: birth fathers are unlikely to be contacted and informing birth siblings of the existence of the adoptee is similarly not permitted.[72] Denial of access to an original, unaltered birth certificate constitutes a form of origin deprivation that might well infringe a number of international law principles, such as the right to identity, best interests of the child, cultural heritage integrity or the right to be afforded respect for family life. As a number of Canadian NGO's have observed, Quebec's system is at odds with the various provisions of The Canadian Charter. Section 15 (1) for example, guarantees that, "Every individual is equal before and under the law and has the right to the equal protection and equal benefit of the law without discrimination and, in particular, without discrimination based on race, national or ethnic origin, colour, religion, sex, age or mental or physical

accessed 13.07.09 wherein he criticizes the Court for not taking the opportunity to create a useful precedent in respect of the child's right to identity (para 13).

[70] The Charter represents a domestic Bill of Rights, now enshrined in the Canadian Constitution via the Constitution Act 1982. See also The Canadian Human Rights Act 1977.

[71] See Quebec's Civil Code Sections 35–39. Available at http://www2.publicationsduquebec.gouv.qc.ca/dynamicSearch/telecharge.php?type=2&file=/CCQ/CCQ_A.html accessed 11.08.90. See further Quebec's Code of Ethics on International Adoption http://www.quebecadoption.net/adoptiona/organizations/codea.html accessed 10.07.09. This is so even where correspondence from birth kin has revealed that indigenous ancestry is present. See further http://www.bastards.org/bb/3.SealedHistCanada.html accessed 01.02.09. See also Canada's Child Welfare Act RSA C-81 (1984) as amended.

[72] See http://www.pfmtl.org accessed 02.06.09.

disability."[73] This section provides that equality of treatment and freedom from discrimination should apply to all Canadian citizens irrespective of birth status. Section 3 of The Canadian Human Rights Act (1985) includes "family status" as one of the prohibited grounds of discrimination.[74] Adult adoptees in a number of other Provinces (British Columbia[75] or Ontario[76] for example) have a much greater degree of access to their original birth certificate; arguably Quebec law and policy prejudices its adoptees on the basis of birth status. This also seems at odds with the spirit of The Quebec Charter of Human Rights and Freedoms (1976) which makes particular reference to the protection of socio-economic rights (e.g. employment, education, religion, voting, access to goods and services).[77] Section 10 for example provides that "Every person has a right to full and equal recognition and exercise of his human rights and freedoms without distinction, exclusion or preference."[78] Although the Quebec Charter carries "quasi-constitutional" status (The Canadian Charter will prevail in matters of federal jurisdiction) this provision largely mirrors the Canadian Charter's list of prohibited grounds for discrimination. Although "family status" does not appear in Section 10 of the Quebec Charter, "civil status" has been included, with an additional ground of "social condition." Section 10 further stresses that "discrimination exists where such a distinction, exclusion or preference has the effect of nullifying or impairing such a right." Arguably, adoptees could challenge non-disclosure vetoes, on the basis that their effects might be deemed to constitute a fundamental interference with Charter rights. The right to identity could perhaps find further protection within the remit of Section 4, which safeguards "dignity, honour and reputation." Section 49 similarly permits the victims of "unlawful" rights violations to "obtain the cessation of such interference and compensation for the moral or material prejudice resulting therefrom." Section 44 also provides that "every person has a right to information." Unfortunately it immediately limits this right by including the phrase "to the extent provided by law." The Charter's drafting also enables a balancing exercise to be effected in favour of preserving parental privacy. Although Section 5 guarantees the

---

[73] The Charter has been incorporated into Part I of The Constitution Act 1982. Available at http://law.justice.gc.ca/en/charter/ accessed 03.08.09.

[74] Section 3 of the Canadian Human Rights Act (1977) lists the prohibited grounds as '*race, national or ethnic origin, colour, religion, age, sex, sexual orientation, marital status, family status, disability and conviction for which a pardon has been granted.*' Available at http://laws.justic.gc.ca/PDF/Statute/H/H-6.pdf accessed 09.08.09.

[75] See The Adoption Act SBC 1995 c.48 which permits adult adoptees (at age 19) and their birth parents to gain access to their birth certificate and Adoption Order, subject to there being no 'No-Contact' Declaration (veto) registered on the file.

[76] See Ontario's Access to Adoption Information Act 2008 (Ontario). Available at http://www.ontla.on.ca/web/bills/bills_detail.do?locale=en&BillID=1899 accessed 11.08.09.

[77] Current version (29.10.2008). Available at http://www.canlii.org/en/qc/laws/stat/rsq-c-c-12/latest/rsq-c-c-c12.html accessed 09.08.09.

[78] The Charter's Preamble also declares that '*all human beings are equal in worth and dignity and are entitled to equal protection of the law.*' See further Gosselin v Quebec (Att Gen) [2002] 4 SCR 429, 2002 SCC 84 on violations of the Charter's provisions.

"right to respect for ...private life", Section 9 copper-fastens parental welfare paramountcy by proclaiming that "every person has a right to non-disclosure of confidential information."[79]

In contrast to Article 8 of The European Convention (which cites the notions of "home, family and private life" together as almost overlapping rights and interests) the concepts of home and privacy rights have been teased apart here and are addressed discretely via separate provisions. Article 7 states that "a person's home life is inviolable" while Article 5 provides "a right to respect for private life." The right to "family life" is not afforded its own article, but appears as a plainly juridical concept later on in the contexts of property i.e. leased family dwellings (Article 14), social welfare assistance (Article 45), marriage obligations (Article 47) and children's disabilities (Article 48). It is difficult to see how an adoptee might successfully plead that unjustified interference has occurred in respect of either their identity or family life rights under the Quebec Charter.[80] Canadian NGO's have noted that the use of the non-disclosure veto does not appear to comply with the requirements of Section 15(1) of the Canadian Charter, Section 3 of The Canadian Human Rights Act or indeed Article 3 of The Children's Convention, in preventing discrimination towards adoptees.[81] This is perhaps ironic given The Supreme Court of Canada's definition of discrimination as, "a distinction, whether intentional or not but based upon grounds relating to personal characteristics of the individual or group, which has the effect of imposing burdens, obligations or disadvantages on such individuals or group not imposed on others, or which withholds or limits access to opportunities, benefits and advantages available to other members of society."[82] NGO's in other provinces have also highlighted the fact that closed records systems perpetuate discrimination, and have recently urged adopted adults to file formal complaints with The UN Committee on the Rights of the Child and with Canada's Ministry of Heritage.[83] Whether a successful case of discrimination will ever be brought against the state under either of these two regional Charters, remains to be seen. In theory, sealed records vetoes contravene

[79] S 52 also provides for general derogation from The Canadian Charter.

[80] Quebec's opposition has been well-documented with the Province's federal government being the only one to avail of the Charter's controversial '*notwithstanding*' clause to effectively achieve a degree of non-compliance in respect of other matters.

[81] See http://www.parentfinders.org accessed 12.08.09. The pressure group lobbies for legislative reforms on closed records and anonymity in reproductive and genetic technologies and drafts Recommendations on how compliance with the various provisions of human rights law might be better achieved. See for example http://www.gov.mb.ca/legislature/hansard/3rd-36th/la_005/la-005.html accessed 12.08.09.

[82] Andrews v The Law Society of British Columbia [1989] 1 SCR 143 pp 174–5. See also the Canadian Human Rights Tribunal case of Shirley (Starrs) McKenna v Dept of the Sec of State (October 8 1993) T D where it was found that the provisions of the Citizenship Acts 1974–76 had resulted in discrimination against the adopted children of a Canadian citizen (in respect of its bar on *jus sanguinis* transmission of citizenship and 'Canadian heritage') p. 22. Available at http://www.chrt-tcdp.gc.ca/search/view_html.asp?doid=205&lg=_e&isruling=0 accessed 09.08.09.

[83] See htpp://www.parentfinders.org/News_Newfoundland.htm accessed 12.08.09.

many aspects of human rights law, at the level of both international and regional instruments. In practice, government agencies are likely to continue to rely upon the careful wording of provisions such as Section 9 of the Quebec Charter or Quebec's Civil Code. Together these form a domestic framework that trenchantly protects the privacy rights of triad adults over and above the welfare of adoptees. It is difficult to see how origin deprivation could be prevented here without the enactment of veto-free domestic laws or policies aimed specifically at inverting the current situation, as has occurred in other jurisdictions. Given the high regard that attaches to concepts such as family sanctity and adult privacy here, it also seems unlikely that such reforms will happen without some form of direct legal challenge from relinquishing parents argued on the basis of a rights violation.

4.7 Conclusion

> If knowledge of one's origins destabilizes the child's family life which of the child's interests takes precedence and must be protected? (Ronen, 2004).

A "rights-based approach to children's issues" has at least enabled a "decisive shift" in domestic judicial attitudes towards the notion of children's rights in general (McCarthy, 1) . Although the need to effect child protection predominates in most proceedings involving children, there is a gradual move towards allowing children some degree of participatory input, especially where court decisions are likely to impact upon their long-term psychological welfare (Sinclair and Franklin 2000).[84] The balancing exercise is perhaps slowly starting to afford greater weight to the identity needs and rights of relinquished children, although the jurisprudence on this area is far from consistent. Cases involving cultural heritage or national identity are likely to involve calls for "displaced" children to be repatriated; private law cases involving adoption or donor anonymity remain subject to tough veneers of judicial deference. Margins of appreciation for domestic policies, high regard for family sanctity and adult privacy often combine to render the genetic connection legally meaningless.

Arguably, the loss of genetic kinship through non-disclosure of essential information may amount to a significant interference with a number of fundamental rights, permanently orphanising a significant section of the population. Domestic legislation in many jurisdictions has created a distinct group of "veto-victims" who may be prevented from ever accessing accurate genetic identity or achieving genetic kinship with estranged relatives. The Committee on the Rights of the Child has however in some cases avoided entirely the issue of the disclosure veto, opting to not raise their use "as a matter of concern" in their Concluding

[84] See further Sinclair (2004); Kirby and Bryson (2002).

Observations on Country Reports.[85] As Brower Blair notes (643), "current practices of the nations of the world regarding release of identifying information are not sufficiently uniform to support an absolute right to information under customary international law or general principles of law." She cites two key factors: a lack of international focus on the issue of disclosure and the use of balancing exercises to determine when parental interests ought to outweigh the rights of the child. Whether the persuasive influence of international law principles might eventually bring about "the required level of recognition" remains to be seen (660–661).[86] The current human rights framework does offer at least some scope for reform however. The Council of Europe's recently drafted Joint Declaration on Intercountry Adoption (2008) appears to acknowledge the difficulties faced by "parentless" children. In its Preamble it stresses the paramountcy of child welfare by declaring that the "higher interests of the child must take precedence over any other consideration." It stresses also the failings in the system. By amalgamating, for example, international law policies on intercountry adoption with those on preventing child trafficking, a large number of "abandoned and orphaned children" fell outside the remit of the law and were denied any "right to a family."[87] To some extent the Joint Declaration also recognises the need to preserve genetic kinship, by noting that The Hague Convention was aimed at keeping children in the care of their original families where possible, with Intercountry adoption described as an option of last resort.[88] It is also significant insofar as it places a positive obligation upon signatory states "to respond to the psychological distress of all their abandoned and orphaned children" (para 4). It calls for example for consideration to be given "at European level" to the setting up of an "adoption procedure based on the exchange of good practice between States."

Practical suggestions also include registration from birth for all children; "encouraging adoption in another European country where there is no national solution", reduction of delays and assistance in the "transmission of information." This last suggestion is particularly important given that it could actively help promote "the right of the child to know its origins" provided that meaningful, identifying information is gathered properly and then made generally available at a later stage. Similarly, if the "best interests" principle were to be interpreted with more consistency by domestic judiciaries it might yet come to include a juridical right to avoid non-ancestry, by enabling a more repatriative norm of access to genetic kinship via information or contact. Such a development could serve to invert the current model of passive preservation to one of active achievement, with national authorities placed under a positive obligation to qualify parental anonymity rights, modify the absolute nature of

[85] See Concluding Observations: Canada UNCRC/UN/Doc.CRC/C/15/Add 37 (1995). See also UN Doc CRc/C/11/Add.3 (1994). Available at http://www.unhchr.ch/tbs/doc.nsf/(Symbol)/CRC.C.9.WP.1 accessed 11.08.09.

[86] As Brower Blair notes however (660) Irish adoption disclosure law was subject to some measure of reform following representations to the Committee on the Rights of the Child by the Constitutional Review Group and the Ministry of Health and Children.

[87] They often remained in institutional care facilities or in long term foster care.

[88] See further Smolin (2004).

non-disclosure vetoes or actively facilitate information release or kin contact, unless the best interests of the child clearly demands otherwise. Where child protection issues are a factor, the principle of child welfare paramountcy could still be invoked (e.g. to place a veto on contact or information release) but this ought to occur only as an exception event and preferably as a short term measure rather than as a normative feature of social kinship creation.

In sum, if the right to identity in international law were to create a corollary right to actually *identify* one's genitors in domestic law, then the use of such potentially adult-centric devices as donor anonymity, closed records and non-disclosure vetoes could at least be called into question. The concept of genetic identity should not simply be dismissed as an unimportant or irrelevant issue in court proceedings involving the cauterisation of genetic ancestry. This is especially so given that the notion of usefully "creating and re-creating" one's own identity during adolescence does not necessarily apply automatically to genetically kinless individuals. As Oren has argued (187) "competing versions of "best interest of the child" are also competing versions of children's rights." Framing access to genetic identity as a means to achieving "psychological integrity" (Oren, 175) may yet come to be regarded as a particularly important aspect of child welfare paramountcy discourse, in both international and domestic law.

References

Alston, P. (1984). Conjuring up new human rights: A proposal for quality control? *American Journal of International Law, 78*, 607–615.

Anaya, S. J. (2000). *Indigenous peoples in international law*. New York: Oxford University Press.

An-Na'im, A. A. (Ed.). (2001). *Cultural transformation and human rights in Africa*. London: Zed.

Bargach, J. (2002). *Orphans of Islam: Family, abandonment and secret adoption in Morocco*. Oxford: Rowman and Littlefield.

Brower Blair, M. (2000–2001). The impact of family paradigms, domestic constitutions and international conventions on disclosure of an adopted person's identities and heritage: A comparative perspective. *Michigan Journal of International Law, 22*, 587–672.

Cerda, J. S. (1990). The draft convention on the rights of the child: New rights. *Human Rights Quarterly, 12*, 115–117.

Convention on the Rights of the Child. (1989, November 20/1990, September 20). Office of the United Nations High Commissioner for Human Rights [Web. 21 July 2009].

Detrick, S., Doek, J., & Cantwell, N. (Eds.). (1992). *The United Nations convention on the rights of the child: A guide to the "Travaux Preparatoires*. Dordrecht: Martinus Nijhoff.

Donnelly, J. (2003). *Universal human rights in theory and practice*. Ithaca: Cornell University Press.

Eekelaar, J. (2008) Respect. In L. D. Wardle, & C. S. Williams (Eds.), *Family law: Balancing interests and pursuing priorities* (pp. 1–12). Buffalo, NY: William S. Hein.

Freeman, M. (1996). The new birth right? Identity and the child of the reproductive revolution. *The International Journal of Children's Rights, 4*, 273–297.

Hadden, T. (2000) The pendulum theory of individual, communal and minority rights. *Critical Review of International Social and Political Philosophy, 3*, 77–90.

Horgan, R., & Martin, F. F. (2008, September). The European Convention on adoption 2008: Progressing the children's rights polemic. *International Family Law*, 155–162.

Ishaque, S. (2008). Islamic principles on adoption: Examining the impact of illegitimacy and inheritance related concerns in context of a child's right to an identity. *International Journal of Law, Policy and the Family, 22*, 393–420.

Kirby, P., & Bryson, S. (2002). *Measuring the magic? Evaluating and researching young people's participation in public decision-making*. London: Carnegie Young People's Initiative.

Marks, S. P. (Ed.). (2006). *Health and human rights: Basic international documents* (2nd ed.). Cambridge, MA: Harvard University Press [Harvard Ser. on Health and Human Rights].

McCarthy, P. (2004). Making the most of international law on the right to identity: An analysis of article 8 of the United Nations Convention on the rights of the child. *Cork Online Law Review, 3*, 1–32.

O'Donnell, T. A. (1982). Margin of appreciation doctrine: Standards in the jurisprudence of the European Court of Human Rights. *Human Rights Quarterly, 4*, 474–507.

Oren, L. E. (2001). Righting child custody wrongs: The children of the 'disappeared' in Argentina. *Harvard Human Rights Journal, 14*, 123–195.

Pritchard, S. (1998). *Indigenous peoples, the United Nations and Human Rights*. Annandale: Federation.

Robertson, A. H., & Merrills, J. G. (1997). *Human rights in the world: An introduction to the study of the international protection of human rights* (4th ed.). Manchester: Manchester University Press.

Ronen, Y.'i. (2004). Redefining the child's right to identity. *International Journal of Law, Policy and the Family, 18*, 147–177.

Sandra, C., et al. (Eds.). (1995). *The right to know: Human rights and access to reproductive health information*. Pittsburgh: University of Pennsylvania Press.

Sclater, S. D., & Kaganas, F. (2003). Contact: Mothers, welfare and rights. In A. Bainham, et al. (Eds.), *Children and their families: Contact, rights and welfare*. London: Hart.

Selina, T., & Taylor, B. J. (2009). Post-adoption face to face contact with birth parents: prospective adopters' views. *Child Care in Practice, 15*(1), 21–38.

Shue, H. (1980). *Basic rights: Subsistence, affluence and U. S. Foreign Policy*. Princeton: Princeton University Press.

Sinclair, R. (2004). Participation in practice: Making it meaningful, effective and sustainable. *Children and Society, 18*, 106–118.

Sinclair, R., & Franklin, A. (2000). *Young people's participation: Quality protects research briefing no 3*. London: Dept of Health (Govt Report).

Smolin, D. M. (2004) Intercountry adoption as child trafficking. *Valparaiso University Law Review*, 39(2), 281–325.

Stark, B. (2005). *International family law: An introduction*. London: Ashgate.

Stewart, G. A. (1992–1993). Interpreting the child's right to identity in the U.N. Convention on the rights of the child. *Family Law Quarterly*, 26, 221–233.

The Children (Northern Ireland) Order 1995 Article 3 (1). Available at http://www.legislation.gov.uk/nisi/1995/755/article/3/made. Accessed 12.01.11.

The European Convention on Nationality, European Treaty Series No. 166. (1997, November 6). Council of Europe [Web. 20 July 2009].

The Hague Convention of 29 May 1993 on Protection of Children and Co-Operation in Respect of Inter-Country Adoption. (2008, September). Hague Conference on Private International Law [Web. 13 July 2009].

The International Covenant on Civil and Political Rights. (1966, December 16/1976, March 23). Office of the United Nations High Commissioner on Human Rights [Web. 30 July 2009].

The International Covenant on Economic, Social and Cultural Rights ('ICESCR'). (1966). Available at http://www2.ohchr.org/english/law/cescr.htm. Accessed 25.07.09.

The Universal Declaration on the Human Genome and Human Rights. (1997). Available at http://portal.unesco.org/en/ev.php-URL_ID+13177. Accessed 04.01.09.

Universal Declaration on Human Rights. (1948). Available at http://www.un.org/en/documents/udhr/index.shtml. Accessed 10.07.09.

Chapter 5
Strasbourg Jurisprudence: 'Remembered Relatedness'

5.1 Introduction

> even if it has been adopted, a child who is unable to gain access to any type of information about its family origins is made to endure a form of suffering and that suffering may leave scars.[1]

Legal ties do not necessarily reflect genetic realities. Whereas courts may show reluctance to sever the ties of legal relatedness, biological bonds are often simply deemed to be non-existent or irrelevant and therefore unlikely to warrant meaningful protection even where they may have been unjustifiably interfered with. In a number of cases involving genetic ties, there appears to be some degree of reluctance on the part of jurists to interfere with the sanctity of private family life.[2] Parents may for example be afforded a fairly broad amount of discretion over such matters as the release or withholding of information or the enablement or vetoing of kin contact. In respect of the European Convention on Human Rights ('ECHR'), signatory States are often also afforded significant leeway in respect of how they interpret and implement its provisions. Arguably, at the level of domestic proceedings, difficult decisions on private issues of family life are perhaps subject to a sort of 'double margin of appreciation' due to this added layer of judicial deference.

This chapter will examine the nature and significance of European Convention-defined 'family life rights' as they relate to the issue of the blood-tie. Several precedents seem to come close to almost endorsing a juridical right to genetic identity via natal kinship while others perhaps at least provide some measure of useful guidance as to how the harms of origin deprivation might best be prevented or addressed. In some of the cases referred to here, the obligation to preserve natal

[1] *Odièvre V France* [2003] (App No 42326/98) Joint Dissenting Opinion per Wildhaber J, Sir Nicolas Bratza, Bonello J, Loucaides J, Cabral Barreto J, Tulkens J And Pellonpää J, Para 8. See Callus (2004), pp. 658–669; Scheiwe (2003), p. 144.

[2] See further Welstead and Edwards (2008), pp. 339–341.

A. Diver, *A Law of Blood-ties - The 'Right' to Access Genetic Ancestry*,
DOI 10.1007/978-3-319-01071-7_5, © Springer International Publishing Switzerland 2014

heritage seems to be a key element while in others it merits very little or no discussion. In relation to preventing origin deprivation, Article 8 of the ECHR holds much potential, given its relevance to the concepts of privacy and family life rights.[3] As the case law makes plain however, the right to be afforded respect for family life seldom amounts to an 'absolute' right to enjoy all aspects of it.

Similarly, if it is accepted that Article 8 family life essentially falls within the remit of socio-economic human rights (alongside privacy, home life and correspondence) then the concept has been grouped with a number of other entitlements that seem unlikely to achieve meaningful justiciability at the level of domestic implementation (e.g. the right to adequate housing, social care or health). As such, several aspects of family life might also be subject to the wider-reaching limitations that flow from such rights of 'negative obligation' on the part of signatory States (i.e. which are likely to be subject to the more easily-achieved 'non-interference' rather than active implementation).[4] Arguments on the finite nature of resources tend perhaps to also frame family life rights as aspirational rather than as urgently binding state obligations; it costs money to set up, maintain and monitor systems that will facilitate contact, searches and reunion.

Maintaining invisible, non-legal bonds of relatedness will necessarily take second place to ensuring that vulnerable children and infants are afforded placement permanency, especially as a means of protecting them from harmful or disinterested birth kin. That said, the issue of longer-term psychological harm through a loss of genetic ancestry has been discussed in a number of recent cases, and tied to the issue of the child's best interests. Arguably the requirement that relatedness must be 'remembered' to merit legal protection, has to some extent been challenged by the notion that genetic heritage should not simply be overlooked or permanently discarded.

5.2 Genetic Realities, Legal Fictions

The European Convention on Human Rights and Fundamental Freedoms (1951) ('The European Convention') has long been regarded as one of the key instruments of international human rights law, especially given its capacity to act as a 'living

---

[3] Article 8 states that '1.*Everyone has the right to respect for his private and family life, his home and his correspondence 2. There shall be no interference by a public authority with the exercise of this right except such as in accordance with the law and is necessary in a democratic society, in the interests of national security, public safety or the economic well-being of the country, for the prevention of disorder or crime, for the protection of health or morals, or for the protection of the rights and freedoms of others*'..

[4] Arguably, positive domestic statute-based obligations to actively effect child protection perhaps more accurately fall within the remit of civil or political rights, given that there is often an 'absolute' duty upon governments in respect of preventing the 'degrading treatment' of child abuse.

instrument' that continually evolves to reflect societal and cultural changes.[5] In the wake of its incorporation into the domestic law of the United Kingdom (via The Human Rights Act 1998) Choudry suggested that the European Convention also effectively entrenched a number of key elements from the United Nations Convention on the Rights of the Child ('The Children's Convention') such as the principles of child welfare primacy and 'best interests.'[6] Article 8 (2) qualifies the remit of the right to 'family life' however, given that the notion of 'respect' for one's family life permits a number of justifications to be argued in connection with any potential infringements. 'Proportionality' of outcome for example is likely to allow for the 'heavier' privacy rights of parents to 'trump' those of their children, as can occur in domestic proceedings involving family and private life and in respect of triad adult vetoes on information disclosure or contact. Weightier interests include for example anonymity in relinquishment or gamete donation, or confidentiality over the details of conception.

That said, much of the Strasbourg jurisprudence does examine the remit and notion of 'family life' rights. Although case law on the need for blood-tied relatedness is relatively uncommon, a number seem to offer some guidance as to how domestic courts might be under an obligation to protect genetic ancestry. In *Keegan v Ireland* (1994)[7] for example, the nature of the link between the unmarried father and genetic child was referred to, albeit briefly. The applicant father argued that his Article 8 right to be afforded respect for his family life had been breached by the State's failure to grant him automatic guardianship of his daughter, who was subsequently placed for adoption. The respondent State's arguments seemed to focus to an extent on the behaviours of the triad adults, highlighting how a 'sporadic and unstable' relationship had existed between the unmarried parents, and declaring that it lacked the

> minimal levels of seriousness, depth and commitment to cross the threshold into family life within the meaning of Article 8. . .neither a mere blood link nor a sincere and heartfelt desire for family life were enough to create it.[8]

Significantly, the Irish Supreme Court did seem to take into account the issue of genetic relatedness as 'one of the factors to be weighed in the balance in assessing the child's welfare'.[9] It was eventually found that a violation of the father's Article 8 rights had occurred, given that he'd had no means of challenging the adoption as an unmarried father. Given however that 'the scales concerning the child's welfare had tilted inevitably in favour of the prospective adopters' through the passage of time, the adoptive placement was not overturned, but it was held that the father's

[5] *Tryer v United Kingdom* (25 April 1978) Series A no 26 para 31.

[6] This is so despite the fact that judicial balancing exercises may serve to afford priority to parental interests such as privacy. See Choudry (2003), p. 119.

[7] *Keegan v Ireland* (1994) (no 16969/90) 18 EHRR 342; See further *Kroon v The Netherlands* (1995) 19 EHRR 263, 283.

[8] *Ibid* para 42.

[9] *Ibid* para 47.

'right to a hearing' under Article 6(1) of the European Convention had also been violated. The state's actions therefore amounted to an 'interference' which was 'not necessary in a democratic society;'

> The essential problem was that Irish law permitted the child to be placed for adoption shortly after her birth without the applicant's knowledge or consent. The placement not only jeopardised the applicant's ties with the child, but also set in motion a process which was likely to prove irreversible, thereby putting the applicant at a significant disadvantage in his contest with the prospective adopters for the custody of the child. The Government had advanced no reasons relevant to the welfare of the child to justify this.[10]

If, as perhaps occurred in *Keegan*, a child's 'right' to genetic kinship is dependent upon the quality of the relationship between her biological parents, then to an extent her identity rights are subject to adult-centric issues and factors completely beyond her control. Some examination of the significance of the relationship between father and daughter, even in respect of whether he might have had the requisite *locus standi* to complain under the European Convention on her behalf might have perhaps proven useful here. A little discussion on the nature of the child's need or right to original identity might also have been welcome in terms of providing guidance on how domestic systems might address the child's need for genetic kinship in such cases. In terms of 'fair hearing' and the processes connected with the severance of the blood-tie, the decision to an extent presages some of the recent case law on birth parent involvement.

As Lowe and Douglas have recently noted, although unmarried fathers now seem to have some degree of 'protection in the adoption process', it remains to be seen whether others affected by the system

> can claim a breach of their Article 8 rights by the severance of the legal ties with the whole family resulting from the adoption. It is certainly not beyond argument that the complete severance of legal ties with the whole family is a disproportionate effect of adoption.[11]

A not dissimilar situation had previously arisen in the case of *K v The United Kingdom* (1986).[12] Here the applicant father had cared for his infant son for the first eleven months of his life before the local authority sought to have the child placed for adoption. In this case, unlike *Keegan*, the instigation of Guardianship (and then High Court Wardship) proceedings by the father was permitted by domestic legislation. Infringement of the father's Article 8 rights was found to have occurred however due to the lack of consultation with him before the child was placed with foster carers. In relation to the definition of 'family life' it was held that this depended upon 'the real existence in practice of close personal ties'. Under this interpretation, such a relationship could constitute family life where, for example, caring duties had occurred or where perhaps a child was of sufficient age and understanding to actually remember the absent parent, or other genetic relatives.

---

[10] *Ibid* para 55.

[11] Lowe and Douglas (2007), p. 821.

[12] *K v The United Kingdom* (1986) (European Commission) Application no 11468/85.

This test of 'remembered relatedness' would clearly serve to exclude newborn infants and perhaps many young children from the protective remit of family life. The European Commission did however observe that 'the primacy of a natural's parent's relationship with his or her child as family life is clearly implicit in Articles 8 and 12.'[13] It is also perhaps significant that the interests of the child were found capable of being sufficiently important to perhaps 'override those of their natural parents.' Although this stops far short of suggesting that a right to genetic kinship might exist, it does at least focus greater attention on the needs of the origin deprived child in such circumstances.[14]

Similarly, in *Kroon v The Netherlands* (1994), the key issues surrounded the 'biological realities' (rather than the legal fictions of domestic law) of the paternity of the applicant's son.[15] The Dutch government contended that an extra-marital relationship would not give rise to the concept of family life relatedness and that any child born as a result could not plead any *ipso jure* family tie to his genetic father.[16] The Court noted however that the Article 8 concept of family life need not be 'confined solely to marriage-based relationships' and found that a bond 'amounting to family life' did in fact exist between the child and his biological father in this instance.

The 'exceptional' nature of such a situation was stressed however, as was the fact that four other children had since been born to the unmarried, non-cohabiting parents. The fact that the child's 'legal' father (the since-divorced husband) would not have objected to any proposed adoption of the child by his genetic father also appears to have been a key factor here.[17] In finding that Article 8 rights had been infringed the Court seems to have made it clear that in this instance the 'biological and social reality' of the case ought to take precedence over 'legal presumption.'[18] As Cretney et al. have also argued, child welfare might be 'better served by knowing the truth' especially given the 'current emphasis on rights and genetic heritage.'[19] The unique rights and vulnerabilities of the child seem however to an extent to have been subsumed here into the interests of the adults, especially in respect of the question of what might amount to 'unjustified interference' with family life.

More recently the Court seems, in certain contexts, to have begun to afford greater significance to the blood-tie and to perhaps find value in its protection. In

---

[13] *Ibid.*

[14] *Ibid* para 1. The example of an over-riding interest given here is the basic health of the child. In terms of process, and in determining whether the applicant had received a 'fair hearing,' the European Commission declined to make a decision on whether the issue of contact had involved a "determination of civil rights" within the meaning of Article 6 (1), explaining that a full examination of the law and facts of the case would be necessary.

[15] *Kroon v The Netherlands* (1994) 19 EHRR 263 para 11.

[16] *Ibid* para 29.

[17] *Ibid* para 30.

[18] *Ibid* para 40.

[19] Cretney et al. (2003), p. 528.

Clemeno v Italy (*2008*)[20] for example, it was held that the severance of all ties with the biological family of a child amounted to an unjustified interference with the family's Article 8 rights. The case is particularly difficult however because of the circumstances surrounding the child's adoption: she had been removed from her original family in the wake of suspected rape and sexual abuse by some members of her original family. Her birth father was eventually acquitted of the offences; because the domestic court had not sufficiently considered the child's interests in maintaining her genetic links it was held that the State had failed to take appropriate steps to enable 'reconstitution' of the original family.

Much significance was also placed upon the fact that contact had not been permitted between the child and her mother and brother, neither of whom had been charged with any offence. This effectively meant that the 'breaking of every link with them had been total and final'.[21] The case is important not just in relation to the question of whether there is a positive obligation upon signatory states to enable the maintenance of some form of contact between estranged family members, but in respect of the importance that the Court seemed to place upon preservation of genetic kinship, and their apparent obligation to 'repair' severed family ties.

Similarly, in *Jucius and Juciuvienė v Lithuania* (2008)[22] the Court looked at the case of two orphaned sisters. The applicants (the girls' aunt and uncle) had been caring for them for three years since their parents' deaths, but the domestic authorities ordered that residence be transferred to their paternal grandparents. Through the lapse of time, the girls came to live apart under the separate care of their respective guardians. The views of the older girl (then aged six) and of the Child Rights Protection Institution (who favoured the girls staying with their current carers) were taken note of by the domestic family courts, but not actually acted upon. The issue of whether the girls' tender ages perhaps precluded them from demonstrating the requisite degree of 'family life' for Article 8 protections was discussed here as was the applicants' argument that the children's wishes should have warranted 'particular attention' under domestic legislation, alongside the principle of 'non-separation of siblings'.[23]

The Court ultimately found that 'close personal ties'[24] had existed between the girls and their aunt's family, as demonstrated by the fact that one of the girls

---

[20] *Clemeno v Italy* (2008) Judgment 21.10.2008 [Section II] No 19537/03. See also Chamber Judgments Press Release of 21.10.08 available at http://cmiskp.echr.coe.int/tkp197/view.asp?action=html&documentId=842420&portal=hbkm&source=externalbydocnumber&table=F69A27FD8FB86142BF01C1166DEA398649 accessed 04.01.11.

[21] Para 61; See Judgment (in French) available at http://cmiskp.echr.coe.int/tkp197/view.asp?item=1&portal=hbkm&action=prof&highlight=clemeno%20%70%2019537/03&sessionid=17673504&skin=hudoc-en accessed 04.04.11.

[22] *Jucius and Juciuvìene v Lithuania*—application no 14414/03 (25-11-2008) available at http://cmiskp.echr.coe.int/tkp197 accessed 03.01.11.

[23] *Ibid* para 24.

[24] As previously defined in *K and T v Finland* (2001)25702/94 (150) ECHR (2001).

vehemently refused to leave her aunt's home to take up court-ordered residence with her grandparents.[25] The exclusion of the applicants from all matters concerning the younger girl's upbringing (which occurred in the wake of her removal to the grandparents' home) was also deemed to be an unjustified interference.[26] Perhaps most significantly, it was also noted that 'future relations between parent and child should not be determined by the mere effluxion of time'.[27] Although the need for some form of future kin contact between the sisters was not actually considered, the Court did imply that a duty to create 'proper conditions' (which would enable them to effectively 'communicate with each other') arose, presumably to restore and strengthen the bonds of family life that had existed between them as siblings.[28]

Such reasoning could suggest that a positive obligation does exist in respect of domestic authorities' duties to actively preserve the ties of genetic relatedness, at least in situations involving siblings separated by child-protective adoption practice. Unreasonable delays might give rise to litigation in similar cases; failure to facilitate some form of contact between genetic kin might also potentially involve Article 8 rights, especially where such delays have led to a weakening of the ties of family life. The children in this case were however not the applicants, nor were they unable to demonstrate 'remembered relatedness'; they were also legally related rather than simply attempting to establish that their genetic connection merited legal protection in the absence of statutory or policy frameworks. Had they for example been separated from their original relatives or from each other at birth or in early infancy on the basis of child protective adoption practice or to effect permanence, the Court might not have arrived at such a conclusion.

That said the case does at least serve to highlight how unjustified interferences with Article 8 rights may arise when genetic ties are severed or weakened in a bid to enable the creation of replacement ties of social kinship. It also illustrates the difficulties associated with achieving workable, best interests-led solutions to apparently intractable problems of family life, especially where matters such as contact, psychological welfare and exclusively defined legal parenthood arise. As in other similar cases involving genetic kinship or identity, the courts are perhaps being tasked with an impossible undertaking: to achieve a solution that might adequately balance the rights of all concerned while protecting vulnerable children from the harms of bad parenting or origin deprivation. Arguably, the sharing of parental responsibility (if not residence) in such cases might to some extent enable a wider variety of outcomes, but this option is unlikely to be regarded as a workable solution where adults are in direct, acute conflict with each other. Nor does it seem likely that domestic authorities would have the public support or the resources to monitor the details of family life so closely. In such cases it might be useful

---

[25] *Jucius and Juciuvière v Lithuania* at para 27.

[26] *Ibid* para 30.

[27] *Ibid* para 32.

[28] *Ibid* para 32.

however if the child's need for information and contact were to be framed more explicitly as a priority concern, in respect of best interests rights and needs when faced with the risk of losing genetic kinship.

Not entirely dissimilar issues arose in *X v Croatia* (2008)[29] where the applicant mother was completely divested of her power to object to the adoption of her daughter. Although she suffered from schizophrenia, she had cared for the child for the first two years of her life and had engaged in regular contact visits with her afterwards. The bond between them was held to amount to 'family life' within the remit of Article 8. Although adoption of the child was justified in domestic law as a legitimate means of child protection under the 'best interests' principle, the European Court did find that the relationship between mother and daughter ought to have been assessed at an earlier stage of the domestic proceedings. Given the profound consequences, failure to inform the mother of the adoption proceedings constituted a violation of her Article 8 rights.[30] The Court stressed that Article 8 does not contain any 'explicit procedural requirements'[31] but noted that 'this [was] not conclusive of the matter.'[32] Local authorities ought to include the views of and interests of parents in such decisions, so that appropriate remedies might be sought.

Failure to involve parents in the decision-making process, especially where the consequences are so serious, may well amount to failure to afford due respect to the interests protected by Article 8. Where this has occurred, 'interference resulting from the decision will not be capable of being regarded as necessary.'[33] Had the child been a younger infant however, or had her mother not engaged in contact to the extent that she did, it is possible that the Court would have found no significant degree of family life existing between mother and daughter. The decision seems to acknowledge that significant harm can attach to the loss of genetic relatedness. Equally, it is the mothers' rights to genetic kinship which were deemed worthy of protection here rather than those of the child.

If a connection exists however between parent and child that can give rise to some level of rights protection, then the issue of whether it is legal, genetic or emotional ought to be secondary to that of whether harm will flow from the severance of it. By the same token, it might yet be argued that such connections tend to be bilateral in nature. Although children may be too young to verbalise or otherwise demonstrate 'remembrance', they may still be most at risk of suffering

[29] *X v Croatia* [2008] (app no 12233/04) (17.07.2008).

[30] See also The Motion of The Council of Europe Parliamentary Assembly Doc 11742 (9th October 2008) para 4 which noted that the use of mental incapacity to exclude parental responsibility in X v Croatia also potentially violated Articles 6 ('fair trial') and 13 ('remedy'), that judgments in family proceedings in England are often not made publicly, in conflict with Article 6 (para 9) and requesting an investigation into 'forcible' adoptions (in respect of Portugal, Croatia and the United Kingdom) at para 12. Available http://assembly.coe.int/Main.asp?link=/Documents/WorkingDocs/Doc08/EDOC11742.htm accessed 10.01.11.

[31] *X v Croatia op cit* n 29 para 48.

[32] *Ibid*.

[33] *Ibid* para 48.

the harms of origin deprivation. It is unlikely that family law courts will be quick to find that a 'two-way right' exists in respect of genetic relatedness, especially where such a finding could result in abusive parents or formerly anonymous gamete donors claiming an absolute right to retain or gain parental responsibility. The main significance of cases such as these is perhaps that they serve as useful reminders of how biological realities may be totally obscured by legal fictions. They perhaps also indicate a degree of reluctance on the part of courts to address the issue of how such decisions might impact upon the children concerned. Although there is no onus upon judges to discuss rights issues from the adopted child's point of view or indeed to refer to the potential risks that attach to rendering children genetically kinless, some reference to the best interests principle might be useful, in respect of providing guidance for domestic jurists facing similarly difficult issues.

5.3 Family Life: Privacy as 'Weightier' Right?

The concept of the right to private life features frequently in judicial discourses that concern genetic kinship, especially in connection with the rights of parents. In *Marckx v Belgium* (1979)[34] for example, the Court had to determine whether the 'natural tie' between a mother and her 'out of wedlock' child could give rise to 'family life' within the meaning of Article 8. Under Belgian law, no legal bond existed between them based upon 'the mere fact of birth'.[35] Maternal affiliation could only be achieved if the child's mother chose to 'recognise' the child as being related to her or legally adopt her.[36] In any event, the child would not have had the same legal status as her other 'legitimate' siblings, in respect of inheritance succession or filial maintenance.[37] Under Belgian legislation, family life rights seemed to depend upon the presence of 'a family unit;' as was noted;

> The 'tranquility' of 'legitimate' families may sometimes be disturbed if an 'illegitimate' child is included, in the eyes of the law, in his mother's family on the same footing as a child born in wedlock, but this is not a motive that justifies depriving the former child of fundamental rights.[38]

---

[34] *Ibid.*

[35] *Ibid* para 14.

[36] Under Article 334 of the Belgian Civil Code available http://eur-lex.europa.eu/n-lex/info/info_be/index_en.htm (accessed 23.03.13).

[37] *Ibid* para 16. Limits were also placed upon the amount of property transferable on death or via *inter vivos* dispositions from mother to child. This provision left mothers with the choice of either recognising affiliation to the child (and thus limiting the inheritance) or renouncing the child completely, so that testamentary freedoms associated with 'stranger dispositions' might be enabled instead. The distinction between the rights of 'heir presumptive' and 'heir exceptional' was similarly found to be in breach of Articles 8 and 14, given that family life may include 'interests of a material kind' (para 52) Oddly, maternal grandparents could however, if they so chose, exercise their 'right of access' to the child.

[38] *Marckx v Belgium* (1979) (application no 6833/74) (1979) para 48.

A breach of Article 14 of The European Convention (which prohibits discrimination, in conjunction with the exercise of another Convention-protected right) was argued here on the basis that no 'objective or reasonable justification' could attach to the exclusion of single-parent, or unwed families from the protective remit of Article 8.[39] Domestic law had served to legally separate mother and child, to categorise them as 'non-relatives', and to render the child legally 'motherless'.[40] The Court also found that, in respect of the child's ties to her maternal kinfolk, that 'such relatives may play a considerable part in family life'.[41] As such, the State was under a positive obligation to act in a way that might permit 'these ties' to 'develop normally.'[42] The Court did not define the extent of such 'ties', which perhaps renders uncertain exactly what degree of relatedness must be present before familial connections are judicially recognized or protected.

The concept of 'normal development' of such ties is similarly ambiguous, but this phrase is perhaps telling: if 'development' of such ties is needed, then they probably will not arise spontaneously, simply by virtue of the natal connection. Arguably, this case does not provide an overly useful precedent for the protection of such bonds. It does however highlight how socio-cultural attitudes towards genetic non-ancestry might give rise to discriminatory treatment of such individuals, through the denial or loss of property rights or social status. The Court did not consider the argument that this case may also have involved Article 3 rights on the grounds that aspects of the parties' treatment, although perceived by them as humiliating, still fell short of the human rights definition of 'degrading treatment'.[43]

The Court similarly did not discuss the child's need for or right to identity, nor was the significance of the loss of paternity discussed. In his dissenting judgment however, Sir Gerald Fitzmaurice J did make a number of observations on family life, stating that it

> can exist whatever the civil status of those resident in the common home, provided there is an inter-relationship between them by blood, adoption, or even, conceivably of amity, convenience or long-continued habit.[44]

He goes on to suggest that the interpretation of Article 8 family life favored by the Court in this instance ought (due to its listing alongside the concepts of privacy, home life and correspondence) not necessarily to be used to determine more difficult issues of legal relatedness in other contexts. Questions involving civil status, affiliation and patrimonial or property rights for example may belong 'to a different order, class or category' of rights. Given also that the European Convention contains an entire article devoted to the protection of the right to marry and

[39] *Ibid* para 36; Breach of Article 14 ECHR ('non-discrimination') may only be pleaded in connection with another Convention Article.

[40] *Ibid* para 37.

[41] *Ibid* para 45.

[42] *Ibid.*

[43] *Ibid* para 66.

[44] Dissenting Opinion of Judge Sir Gerald Fitzmaurice para 13.

found a family (Article 12), the Convention draughtsmen may well have intended that Article 8 was simply not meant to 'introduce a whole code of family law'.[45] If so, its interpretation of the concept of family life may not be entirely incompatible with the Belgian legislation on illegitimacy and property inheritance; equally, the mother-child relationship might still arise per se 'as a result of birth.'[46]

That said, the Dissenting Opinion does argue that unmarried mothers ought to be afforded a right to choose whether or not they should be 'saddled' with the 'consequences of the birth.'[47] This case is still regarded as good authority on the definition of family life and the scope of Article 8 rights.[48] As Cretney et al. have noted, it also seems to confirm that states should not simply be seen to have avoided interference with such rights but must engage in some form of 'positive action to promote family relationships.'[49] That said, as a rights enabler, Article 8 itself perhaps 'presupposes the existence of a family.'[50] Whether a relinquished newborn would be able to automatically avail of its protective remit to plead a right to genetic ancestry and identity remains to be seen.

In *Mikulić v Croatia* (2002)[51] the issues of the need for identity and respect for private life were discussed in some detail. Here, a putative father's non-compliance with a court order for DNA testing (and the State's lack of enforcement procedures) gave rise to unease over the child's need to receive information 'necessary to uncover the truth about an important aspect of their personal identity.'[52] The Court found that the requirements of Articles 6 (1) and 8 of the Convention had not been complied with and stressed the positive obligations that States are under in respect of actively protecting private and family life rights.[53] Delays during domestic court proceedings had also 'left the applicant in a state of prolonged uncertainty as to her personal identity.'[54] In respect of the rights of third parties, it was found that a fair balance had to be struck between such competing rights, which should still have enabled the applicant to 'have her uncertainty as to her personal identity eliminated.'[55] The decision appears to at least acknowledge that the need for genetic identity is often acutely felt by origin-deprived individuals. Arguably, it

---

[45] *Ibid* para 15.

[46] *Ibid*. See also para 29; '*of course it is there by blood*'.

[47] *Ibid* para 30.

[48] See Hale et al. (2009), p. 31.

[49] Cretney et al. (2003), n 19, p. 507.

[50] Lowe and Douglas (2007), n 11, p. 820.

[51] *Mikulić v Croatia* (2002) (no 53176/99) 54 ECHR; see also *Human Rights Case Digest* (2002) (13) 1–2, pp. 55–95.

[52] *Ibid* para 64.

[53] *Ibid* para 57. Lack of domestic remedy (in respect of delays) also amounted to a violation of Article 13.

[54] *Ibid* para 66.

[55] *Ibid* para 56.

may provide useful guidance for anyone seeking to protect, preserve or, most significantly, access identifying information about their history.

In relation to the concept of kinship identity rights, however, a key case is that of *Odièvre v France* (2003) which concerned the applicant's inability to 'find out her origins'. It was argued that the French anonymous birth registration system breached the provisions of Article 8 especially in respect of her private life rights.[56] As in the *Marckx v Belgium* case, the principle of *mater semper certa est*[57] had not been enshrined in the French Civil Code, allowing unmarried mothers the option of complete relinquishment of all parental responsibility and permanent anonymity.[58] This served to stop the applicant from accessing any personal information ('biological truth') or establishing any 'emotional ties' with her genetic relatives, including any siblings and half-siblings.[59] The applicant contended that French domestic law therefore created an unfair 'legal fiction' which negated the very facts of her birth and existence. The process of anonymous birth, it was argued, amounted to 'an act of violence' against her.[60] She further argued that 'the irrational fear of certain adoptive parents and adoption agencies . . .had encouraged the belief that children without a past were easier to adopt'.[61]

The respondent Government's counter argument focused on the need to protect 'distressed mothers' and prevent potential harm to abandoned infants, who under this system, were immediately made available for clean-break adoption.[62] Any interference with the child's Article 8 right to private life was thus 'proportionate' and therefore justified. Citing *Gaskin v United Kingdom* (1989)[63] it was argued that

---

[56] *Odièvre v France* op cit n 1 para 3.

[57] The legal presumption that '*The mother is always certain*', presuming there is no counter-evidence to the contrary; since the development of assisted conception techniques , the presumption has however lost its status as *praesumptio iuris et de iure* and is now instead rebuttable.

[58] Via Law no 93-22 (1993) and Articles 341 and 341-1 of the Civil Code which together created an 'estoppel defence' to proceedings which had been brought to establish maternity; this has since been replaced by a 'veto' system of maternally consented-to, mediated disclosure; *ibid* para 15. See also The Report of the Conseil d'Etat '*Status and Protection of the Child*' (1990) on the need for consensual release of original information and the circumstances in which adoption professionals might be released from their duties of confidentiality; Mattei's Report '*Children From Here, Children From Elsewhere-Adoption Without Borders*' (1995) which advocated the preservation of anonymity but the gathering of non-identifying information; Fabius (1998) which proposed reforming the law on anonymous birth to enable adoptees' access to their information on reaching the age of 18, given that this would not affect 'the parental ties they already enjoyed'. It was also suggested that the reforms might also be extended eventually to persons conceived via 'medically assisted procreation'. *Ibid* para 16.

[59] *Odièvre v France ibid* para 28.

[60] *Ibid* para 31.

[61] *Ibid* para 32.

[62] See however Thery (1998) which suggested repeal of Article 341-1 based on the 'serious consequences of anonymous births' given the deprivation of maternal and paternal filiation which results; Dekeuwer-Defossez (1999) which supports anonymous birth and the right of mothers to give birth 'discreetly' (via permanent veto).

[63] *Gaskin v UK* (1989) 12 EHRR 36.

a clear margin of appreciation existed where there was conflict between two private interests i.e. the respective rights of the child and mother. The margin was perhaps further widened here by the fact that 'no European consensus on the issue of a child's access to information about its origins existed'.[64] The Court also distinguished the applicant's situation from those of the claimants in *Gaskin* and *Mikulić*, stating that,

> access to information about one's origins and the identity of one's natural parents is not of the same nature as that of access to a case record concerning a child in care or to evidence of alleged paternity.[65]

As such, the abandoned child could not overturn the express wishes of the relinquishing mother; the competing interests of both parties fall within the remit of Article 8 privacy but could legitimately be weighted in favour of protecting the health and anonymity rights of the mother. The Court also highlighted the need to protect the private and family life interests of other third parties such as the birth father, siblings and other biological relatives.[66] The exact dangers posed to them by the searching adoptee were not elaborated upon however; the fact that the applicant was eventually able to trace some of her natal relatives (from the non-identifying information provided[67]) perhaps aided the Court in finding that her Article 8 rights had not been breached by the French legislation.[68]

In his Concurring Opinion Rozakis J observed that, because she had gained the age of majority at the time of her search, the applicant had lost the legal protections that might have accompanied the 'privileged position' of the child under international law, such as the need to achieve personal development. Her Article 8 interests must then be regarded on an equal basis to those of the potentially affected third parties.[69] Such reasoning perhaps suggests that, as a child, the applicant might have been afforded some right to access confidential information on the basis of child welfare paramountcy principles. Whether these rights might arise under the directly applicable European Convention, or via the less persuasive best interests principle of The Children's Convention is not made clear however. Nor was discussion forthcoming in respect of how such an apparently 'privileged' rights position might actually be used to realize a meaningful, positive right to access genetic identity, during or after one's childhood.

This point highlights one of the key difficulties for many origin deprived persons. As children they are subject to the weighing out and balancing of their rights and interests against the conflicting (and generally heavier) rights of triad adults. In many cases, the State's obligation to achieve child protection will also

---

[64] *Ibid* para 37.

[65] *ibid* para 43.

[66] *Ibid* para 44.

[67] The non-identifying information provided by domestic authorities is perhaps interesting for its level of detail, noting as it did the relinquishing mother's 'heavily made up brown eyes'. *Ibid* para 12.

[68] *Ibid* para 52.

[69] Opinion of Rozakis J at para 3.

necessarily place a bar on their accessing genetic kinship via information or contact. On gaining adulthood, adoptees lose the international law protections afforded to them as vulnerable children seeking to engage in the process of identity formation. It is difficult therefore to ascertain an exact point at which the interests of the origin-deprived adoptee might be found to take precedence over those of a relinquishing parent, if ever. Given also the wider social interests involved in preserving anonymity post-relinquishment[70] it seems unlikely that domestic interference with the right to genetic identity in such circumstances will ever constitute a violation of Article 8.

The Joint Dissenting Opinions in *Odièvre* however do argue fairly clearly that a violation of Article 8 has occurred. They describe the majority's definition of family life ('filiation') as 'too narrow' but agree that the Court's focus was correctly kept on private rather than family life, especially given that the concept of dignity via identity development falls more within the remit of private life than within family life. They stress that French legislation impeded the balancing of the conflicting rights and that the maternal veto created an absolute bar to any requests for information from the adoptee;

> The mother thus has a discretionary right to bring a suffering child into the world and to condemn it to lifelong ignorance....the rights of the child...are entirely neglected and forgotten.[71]

As such, the absolute nature of the veto might also 'paralyze' the Article 8 rights of third parties such as those of the birth father and siblings. The Dissenting Opinion also criticizes the majority for focusing on how the applicant had been successfully adopted, and for suggesting that her search for her natal relatives was 'superfluous.' The need to protect the adoptive parents was also described as a matter of conjecture rather than established fact.[72] In sum, the right to an identity was described within this Dissenting Opinion as 'an essential condition of the right to autonomy.' The fairest scrutiny ought to be used when weighing up such competing interests.[73] Psychological welfare must not be sidelined;

> even if it has been adopted, a child who is unable to gain access to any type of information about its family origins is made to endure a form of suffering and that suffering may leave scars.[74]

---

[70] Such as the judicial focus on abortion rates *ibid* para 3. See however the Concurring Opinion of Greve J (opening remarks) wherein he disagrees with the view that abortion figures can automatically be related to a lack of confidential child abandonment facilities.

[71] Dissenting Opinion para 7, per Judges Wildhaber, Sir Nicolas Bratza, Bonello, Loucaides, Cabral Bonetto, Tulknes and Pellonpää.

[72] *Ibid* para 8.

[73] *Ibid* para 11. See also *Pretty v the United Kingdom* (2002) (no 2346/02) 61 ECHR 2002 and *Bensaid v United Kingdom* (2001) no 44599/98 47 ECHR which were cited in support of this view. They further argued that the absence of a common denominator in relation to domestic law (i.e. in respect of maternal anonymity) did not justify the automatic granting of a wide margin of appreciation.

[74] *Odièvre op cit* at n 1; Dissenting Opinion, para 8.

Such logic accords with the Recommendations of the Parliamentary Assembly of The Council of Europe (2000) on international adoptions,[75] that the right to learn the truth about one's origins ought to be made available to adoptees, at least upon attaining the age of majority. National legislation should therefore contain no clauses that might serve to prevent this aim, given that a narrow 'rights-based approach' might be driven more by the desires of adults than by the needs of vulnerable children.[76]

5.4 Lifelong Psychological Welfare

In *Johansen v Norway (1997)* it was found that Article 8 rights may be infringed where an irreversible decision on state intervention is made in respect of an infant.[77] Here, the mother, (who had previously been addicted to drugs and alcohol and was a victim of ongoing domestic violence) was denied contact after the child was placed in long term foster care. The decision was made on the grounds that the she might have disrupted the foster-care placement, or would perhaps attempt to abduct the child. Genetic relatedness counted for little, it was argued, given that a lack of contact was capable of rendering the child 'a stranger to her' mother.[78] The applicant's contention that 'contact could have contributed to a stable development of the child's identity had it been allowed to continue'[79] was deemed by the European Court to be one of several relevant considerations in respect of determining the issue of 'necessity' under Article 8(2).[80]

It was held that no Article 6 violation had occurred in respect of taking the child into care, but that the lack of contact did deprive the applicant of her parental rights in contravention of Article 8.[81] The margin of appreciation had been 'overstepped' by the national authorities[82] in respect of the mother's parental rights, especially given that lack of contact (a decision over which the mother had no control) had been cited as a key factor in the decision to continue the substitute care placement

[75] Parliamentary Assembly of the Council of Europe '*International Adoption: Respecting Children's Rights*' Doc 9377 Recommendation 1443 (26 January 2000, 5th Sitting) para vii—available at http://assembly.coe.int/Main.asp?link=/Documents/AdoptedText/ta00/EREC1443.htm accessed 02.01.09. It was further proposed that all member states should move to ratify and observe the principles of The Hague Convention on Adoption (1993) as a matter of urgency, to monitor the 'psychological well-being' of internationally adopted children.

[76] See further Herring (2009) available http://www.newlawjournal.co.uk/nlj/content/family-revoking-adoptions accessed 10.08.11.

[77] *Johansen v Norway* [1997] 23 EHRR 33.

[78] *Ibid* at para 27.

[79] *Ibid* at para 75.

[80] *Ibid* para 81.

[81] A violation of Article 6 (1) was not found to have occurred.

[82] *Ibid* para 84.

and proceed with adoption.[83] The violation of the mother's Article 8 rights served therefore to deprive her absolutely 'of her family life with the child.'[84] The lack of contact was also

> inconsistent with the aim of reuniting them. Such measures should only be applied in exceptional circumstances and could only be justified if they were motivated by an overriding requirement pertaining to the child's best interests.[85]

Despite noting the 'particular importance' of the best interests of the child principle, this decision is grounded upon adult-centric, parental rights. Arguably, it sets out a template for cases involving pre or post-adoption kin contact, rather than providing guidance for decision-makers faced with the possibility of having to sever or weaken genetic kinship ties. The Court seemed keen to provide some measure of reassurance for substitute parents; as the cited psychologist's report noted, adopters 'should be confident that it is they who take decisions about important events in the child's life. This must be taken into account if a right of access or visit is established.'[86] The Dissenting Opinion is note-worthy, given its argument that the European Court should avoid 'playing the role of a court of appeal and limit itself to reviewing whether the applicant's interests were duly protected in the decision-making process.'[87] Morenilla J went on to argue that the difficulties associated with the applicant's history meant that the child welfare authorities ought to have been exempted from the normal requirements of Article 8, which would otherwise perhaps give rise to a 'repatriative' duty of mother-child reunion.[88] Equally, where 'life, health and development' of the child are at risk, society must be free to intervene to protect the best interests of the child even where this disrupts 'in an irreversible manner the natural bonds between' mother and child.[89]

That said, the placing of a child in substitute care is a difficult decision and not one which the European Court often seems eager to either overturn or interfere with. Scrutiny of domestic decisions to sever the links between mother and child seems to be key, especially if one accepts that kin reunification ought to remain an eventual possibility, unless the best interests of the child principle clearly demands otherwise. In *P C & S v United Kingdom* (2002)[90] the Strabourg Court did discuss to some extent the issue of the newborn child's interests. The applicant's arguments sought to focus judicial attention on the harm suffered by the child, for example by not having been breastfed.[91] Had this separation not occurred,

---

[83] *Ibid.*

[84] *Ibid* para 78.

[85] *Ibid.*

[86] *Ibid* at para 29.

[87] Dissenting Opinion Morenilla J at para 2.

[88] *Ibid* para 3.

[89] *Ibid.*

[90] *P C and S v United Kingdom* (2002) 35 EHRR 1075.

[91] *Ibid* para 102.

> it would have allowed S. to retain the comfort and security of knowing her parents loved her, avoided any damaging sense of abandonment and reinforced S's sense of family and personal identity.[92]

The fact that 'there had already been a degree of natural bonding through contact visits'[93] was also raised, but this point was outweighed by the wishes of the adopters who 'were opposed to direct contact'.[94] The State contended that the child's need to achieve the 'establishment and reinforcement of her new home'[95] should take precedence, thereby effectively delegating decisions on post-adoption contact to the social parents.[96]

It was found that the removal into care of a newborn baby a few hours after her birth was a non-substantive violation of both the child's and parents' family life rights under Article 8. (This was so despite the mother's history of child abuse and mental illness.) Significantly, the severance of these original links was described as being akin to cutting a child off 'from its roots' and would only be justifiable in exceptional circumstances or where the best interests principle obviously required it.[97] Similarly, although Article 8 contained no explicit procedural requirements, where interference with a fundamental right might arise, decision-making processes must remain fair and should afford, to the parents at least some degree of 'due respect to the interests safeguarded by Article 8.'[98]

Because the child's mother had had no legal representation during the Care or Freeing Order hearings, a breach of Article 6(1) had also occurred. In sum, the removal of a child at birth now seems to require some form of 'especial justification'.[99] As Cretney et al. have noted however, the adopted child did not receive compensation here, by virtue of her being unaware of the proceedings and having 'no legal ties' to the natal parents post-adoption.[100] Arguably, had she been older, or had perhaps previously enjoyed some manner of direct contact with her original parents, this aspect of the decision may have been differed.

Similarly, in *Gorgülu v Germany* (2004) the Court looked at the issue of whether the 'long term effects' of separation from one's genetic parent should have at least been considered by the domestic authorities.[101] It was held that Article 8 had been violated in respect of the genetic father's 'access rights' but not in relation to the decision-making processes. The German Court of Appeal had through its decision 'rendered any form of family reunion and the establishment of any kind of further

---

[92] *Ibid* at para 101.

[93] *Ibid* at para 103.

[94] *Ibid* at para 112.

[95] *Ibid.*

[96] Indirect ('letterbox') contact was allowed once a year. *Ibid* para 134.

[97] *Ibid* at para 118. See also *Gnahoré v. France,* (2000) no. 40031/98, § 59, ECHR 2000-IX.

[98] *Ibid* para 119. The Court cited *W v United Kingdom* [1987] ECHR 9749/82 paras 62 and 64.

[99] Lowe and Douglas (2007) at n 11, p. 764.

[100] Cretney et al. (2003), n 19, p. 823.

[101] *Gorgülu v Germany* (Application No 74969/01) [2004] 1 FLR 894 para 46.

family life impossible.'[102] It did so by unjustifiably refusing to permit contact between the child and his putative father, on the grounds that,

> any contact with his natural father would be a physical and psychological strain for the child. Suspending access for a certain time would allow ..(for).. inner repose and emotional balance.[103]

It was noted also that a parent cannot be entitled under Article 8 to 'have such measures taken as would harm the child's health and development.'[104] Arguably, a court-sanctioned refusal by adoptive parents to permit contact between genetic relatives could fall within this remit, especially perhaps where sibling contact might be at issue. In addition to implying that children need to not be denied access to their genetic 'roots', the Court referred both to family reunification and to 'further' family life. Again, the fact that the father had engaged in some level of contact with his child no doubt assisted the Court in declaring that some degree of familial connection still existed between them and was worthy of legal protection. Although no overt reference appears to have been made in the case to the welfare primacy principles of the Children's Convention, the Court did cite the need to effect equitable outcomes based upon achieving some measure of 'fair balance between the interests of the child and those of the parents.'[105] Adherence to equitable principles such as fairness is perhaps reminiscent of other jurisdictions where the practice of making reference to quasi-contractual notions of 'flawed consent' may be used to gauge whether or not disputed relinquishments were legally valid.[106]

In *Gnahoré v. France* (2000) the Court emphasized that in child protection cases the child's interests must 'come before all other considerations.'[107] In such circumstances these interests may however be comprised of 'two limbs,' the first of which consists of the need to ensure that the child's health and development are not harmed by the actions of his parents. The second is of no less importance and concerns the need for the child's

> ties with its family to be maintained, except in cases where the family has proved particularly unfit, since severing those ties means cutting a child off from its roots. It follows that the interest of the child dictates that family ties may only be severed in very exceptional circumstances and that everything must be done to preserve personal relations and, if and when appropriate, to "rebuild" the family.[108]

---

[102] *Ibid* para 48.

[103] *Ibid* at para 48.

[104] *Ibid* para 43. See also *Johansen v Norway*; *Scozzari and Giunta v. Italy* (2000) [GC] nos. 39221/98 and 41963/98, § 169, ECHR 2000-VIII.

[105] *Ibid* para 43.

[106] See for example Clayton (2007–2008), pp. 249–266.

[107] *Gnahoré v France* [2000] ECHR 40031/98 para 59.

[108] *Ibid* para 59. See also *Kutzner v Germany* 46544/99 66 ECHR [2002] where a ban on contact was found to have infringed Article 8, insofar as it would act 'to alienate' children from their parents and from each other.

Although this statement refers primarily to the placing of children into substitute care (and, by extension, to the right of natal parents to have contact with their children whilst they are in care) it may also be significant for those seeking to argue a right to genetic connection. It seems to potentially imply that biological relatedness *per se* might be capable of attracting legal protections, on the basis that 'personal relations'[109] should where possible be defended against the effects of the passage of time. Where familial bonds are capable of being 'rebuilt', this suggests that they must at some stage have pre-existed, on some legally *or* socially significant level. Genetic ties could perhaps be found to fall within this principle, with the exception being 'where the family has proved particularly unfit'.[110]

By highlighting the child's need to know, and more crucially, to have access to his 'roots,' the Court seems to accept that psychological development and longer-term welfare needs often require some manner of active kinship enablement, such as contact facilitation, accurate record-keeping or the unsealing of confidential files. This possibly also represents a move beyond the norms of 'non-interference' with family life rights by member states. Thus, where concerns over child protection issues have rendered kin contact impossible, this does not necessarily completely negate the child's need for meaningful ancestry information. Despite finding that no violation of Articles 8 or 6(1) had occurred in this case, the judgment still perhaps contains some useful guidance on the importance of preserving or accessing biological identity to achieve long-term psychological welfare.[111]

In respect of the issue of legal representation, the partly Dissenting Opinions make a number of interesting points.[112] The means-tested system of access to legal aid meant some applicants had to present a prima facie case on appeal. Applicants able to pursue cases on points of law post-refusal might also be disadvantaged given that a negative inference was likely to arise. Given that in such cases points of law may involve Convention rights, the placing of applicants in such a position gave rise to concern. Arguably, such a 'pre-hearing' of the case could perhaps also occur in other contexts where, for example, birth parents are not legally represented; case conferences pre-adoption, where issues of post-adoption contact are perhaps being discussed or pre-agreed with a view to either 'freeing' the child or arranging a placement with prospective adopters, might also potentially call into question the concept of having a 'fair hearing' of the issues. In other words

> the least-well off members of society should not therefore by denied access to justice and a category of litigants forfeit the substance of their right to access to a court, as guaranteed by Article 6 of the Convention.[113]

---

[109] *Ibid.*

[110] *Ibid.*

[111] The applicant father had argued that a lack of legal aid had led to his case being pre-judged; that his son ought not to have remained in care once criminal proceedings (ill-treatment) had been dropped; that his right to contact had been illegally restricted.

[112] Joint Partly Dissenting Opinion, Tulkens J and Loucaides J para 1.

[113] *Ibid.*

In relation to the issue of contact (three visits in 7 years) it was noted that care orders have a 'dual aspect', namely the interests of the child and the 'ultimate aim' of 'reunion of parent and child'. Because of this, Article 8 sets out two requirements: firstly that 'only very exceptional circumstances' may lead to the 'break-up of a family' and secondly that 'everything must be done to maintain' contact and ensure that 'the rights of both parties' (i.e. the child as well as the birth parent) are protected, especially when reunion is no longer the key aim. The difficult role of foster parents is acknowledged, described here as having to cope with the 'triangular relationship between the child and both his families, with equal rights for all.'[114]

Significantly, the severance of links with the original family is framed as 'a form of 'social ill-treatment' for the child.[115] The applicant's reluctance to co-operate with the authorities (in respect of undergoing therapy for example) is also mentioned, insofar as a decision to do so ought not to be a 'wholly decisive' factor during initial proceedings on the question of barring or permitting contact.[116] The role of the European Court (and the use of the doctrine of margin of appreciation) in respect of scrutinizing such questions is also noted;

> while the Court is obviously precluded from substituting its own assessment for that of the relevant national authorities-which are best placed to assess the situation-it is nonetheless under an obligation under para 2 of Art 8 to assess the relevance and adequacy of the reasons asserted for the interference and whether due weight has been given to the respective interests. It cannot avoid that assessment merely by invoking the margin of appreciation, just as reference to the interest of the child can be no substitute for proper argument.[117]

Although non-binding in nature, these persuasive arguments contain guidance on important issues such as the child's need for some measure of natal kin contact and on the remit of family law courts where Article 8 rights arise. Where domestic authorities decide for example to sever all ties between the child and his original family (given that the interests of siblings could be affected as well) this outcome might need to be viewed as an exceptional event, or clearly justified by more detailed arguments involving explicit references to the best interests of the child.

In *Lebbink v The Netherlands* (2004) the issue of genetic ties was raised by the applicant father, who argued that his Article 8 rights had been violated by the state's bar on contact with his daughter.[118] The Court was unwilling to accept that biological connection could create *ipso jure* ties worthy of Article 8 protections. They stressed that the blood-tie amounted merely to a 'potential relationship which could develop between a child born out of wedlock and its natural father.'[119]

---

[114] *Ibid* para 2.

[115] *Ibid.*

[116] *Ibid.* They cite *Olsson v Sweden (no 1)* [1988] ECHR 10465/83 paras 81, 83.

[117] *Ibid.*

[118] *Lebbink v The Netherlands* (2004) (app no 45582/99).

[119] *Ibid* para 36. See also *Nylund v. Finland (1999)* no. 27110/95, ECHR 1999-VI).

The Court did however show a degree of willingness to take into account such factors as 'the nature of the relationship between the natural parents and the demonstrable interest in and commitment by the father to the child both before and after its birth.'[120]

In this case, the fact that he had been present at her birth, had taken care of her on a number of occasions and had at least some measure of involvement in her upbringing meant that 'certain ties' were found to exist between them in addition to their biological link. These bonds gave their relationship 'sufficient constancy and substance'[121] to merit Convention protections. This was so in spite of the fact that her father had at first declined to 'recognize' her as his daughter, as domestic law had required (albeit out of respect for her mother's wishes).[122]

In *Sahin v. Germany* (2003)[123] not dissimilar issues arose. Here the interests of the child were deemed to be best protected by the barring of all direct contact with her father, due mainly to her mother's strong degree of antipathy towards him, and the child's opposition to contact. Again, their genetic connection alone was insufficient to give rise to Article 8 rights: the fact that the applicant had lived with the child for the first two years of her life was cited in support of his argument on discriminatory treatment however. Issues such as the lack of a psychologist's report and the distinction that domestic law drew between legitimate and illegitimate births were held to be relevant however. The partly Dissenting Opinion suggested also that an Article 8 violation could have arisen. The domestic court had failed to hear the child's views and 'refused to allow the father any further involvement in proceedings', basing their decision on 'mere hypothesis.' It had made no effort to balance the 'various interests at stake' or address the 'difficulties of the triangular relationship by finding and imposing a compromise solution.'[124]

The harshness of the eventual outcome was such that it had led to a 'permanent alienation' between father and daughter.[125] Again, the focus here was mainly on the rights of the father, with the best interests of the child being largely defined by reference to the interests of the mother. Equally, the ties deemed worthy of protection were described as 'emotional' rather than simply genetic. The second Dissenting Opinion stressed however the procedural aspects of Article 8, declaring that domestic courts should 'assess the difficult question of the child's best interests on the basis of a reasoned and up-to-date psychological report.' If possible, the child should also be heard by the court and by a psychologist.[126] The discriminatory aspects of the Civil Code had 'influenced the whole court proceedings' and contributed to the violation of Article 8, with the unmarried father having the

---

[120] *Ibid.*

[121] *Ibid* para 37.

[122] *Ibid* para 40.

[123] *Sahin v Germany* (2003) no. 30943/96, § 12, ECHR 340.

[124] *Ibid* para 108.

[125] *Ibid* para 109 per Rozakis J and Tulkens J.

[126] *Ibid* para 113. (Citing *Elsholz v Germany* (2000) no 25735/94 ECHR 2000-VIII.)

burden of convincing the court that contact would be in the child's best interests (such a presumption only applied to cases where the child was legitimate). Taken together, these Dissenting Opinions at least highlight the potential for discrimination that can attach to birth status. They also serve as a reminder of how the psychological aspects of family life ought not to be overlooked, in determining the best interests of children and deciding whether family life rights have been affected by decisions based mainly on adult interests.

In *Sommerfeld v. Germany* (2003)[127] the best interests principle was also discussed. Here, the child had made clear her aversion to contact with her birth father, despite having lived with him for five years during her early childhood. Considerable emphasis seemed to be placed by the Court upon her mental well-being, so that the lack of a new psychological report (on the possibilities of establishing contact) potentially amounted to insufficient protection of the applicant father's interests. Although no violation of the father's Article 8 rights had occurred, he had still been discriminated against, for example, in respect of being able to bring an appeal.[128] The partly Dissenting Opinion noted the duty of domestic courts to uphold the best interests of the child principle and argued that many decisions on disputed contact seem to turn upon the express wishes and welfare of the child. Given the lack of fresh evidence, the outcome here may have resulted from the general 'disadvantages of the legal situation of children born out of wedlock at the material time'.[129]

It was argued also however that Article 8's implicit procedural requirements in relation to parental rights of access demanded that a detailed, up to date psychological report should have been carried out to more accurately ascertain the child's wishes and enable a meaningful degree of participation in the decision-making process. A 'more thorough approach' may have allowed for the 'development of relations' between the child and her genetic father.[130] Suggesting that a violation of Article 8 had in fact occurred, it was proposed that the domestic legislation directly denied the applicant father his inherent procedural rights. Such guidance does not suggest that genetic ties will automatically amount to family life under Article 8; it does however seem to point to a need for domestic authorities to ensure that such ties are not simply severed (or let wither) without decision-makers having regard for issues such as the psychological aspects of such relationships. Although concerned here mainly with the rights of the applicant father, the child's right to participation was highlighted. Arguably, psychological evidence ought not to be easily discounted in cases where origin deprivation might occur.

[127] *Sommerfeld v Germany* (2003) no. 31871/96, §§ 11-12, ECHR 2003-VIII.

[128] *Ibid* para 111 see also s.63a of the Act on Non-Contentious Proceedings (2003). See also *Elsholz v Germany* (2000) no 25735/94 ECHR 2000-VIII.

[129] Dissenting Opinion para 4 per Ress J, Pastor Ridruejo J and Türmen J.

[130] *Ibid* paras 3–4. The Dissenting Opinion referred also to Gardner's empirical work on 'Parental Alienation Syndrome'—see further Gardner (2001), pp. 61–106.

5.5 A Right to Relinquish?

Where a genetic parent has decided to revoke their relinquishment, the issue of the child's best interests may be particularly relevant to the interpretation of Article 8. In *Kearns v France* (2008)[131] for example an Irish mother travelled to France to avail of the anonymous birth registration system.[132] One day after the birth, she signed placement consent forms, following a half-day interview by social services.[133] Under domestic legislation, a two month 'cooling off' period existed to allow mothers the chance of a change of heart; 'absent' birth parents thus had six months to reclaim a child.[134] Five months after the birth, the mother sought to revoke her consent,[135] arguing that it had not been freely given and that the French law contravened Articles 13 and 14 of the European Convention. In the meantime, the child's genetic father, with whom she'd had an extra-marital affair, brought domestic proceedings in Ireland to have his parental rights recognized. The Lille *Tribunal de Grande Instance* found that the infant had 'no legally established parentage' given the lapsed time limits and the mother's original, explicit wish that she 'should never be able to have legal ties to her.'[136]

There was no provision in domestic law for third parties to attempt a revocation or appeal against maternal consent; once a child was placed with foster carers with a view to adoption, a legal bar arose in respect of overturning relinquishment and in relation to 'any declaration of filiation or recognition.'[137] The court stressed that domestic legislation therefore struck a 'delicate balance' between the competing rights of mothers (to give birth anonymously) and the interests of affected children ('stability and certainty, both psychological and legal'[138]). Although time limits for revocation were short, they did exist; a domestic remedy was thus available. The Douia Court of Appeal found however that the secondary, six-month time limit (which applied to 'second parents' i.e. genetic fathers) did not apply here, due to the anonymous registration system itself. This legislative provision may have misled the birthmother, given that she did not speak French, and the fact that the sole interpreter was a post-natal ward nurse. The validity of consent was thus negated,

[131] *Kearns v France* (App no. 35991/04) [2008] 1 FLR 888. The birth mother was married and had conceived as a result of an extra-marital affair.

[132] '*Accouchement sous X*'. See Article 348-3 of France's Civil Code available http://lexinter.net/ENGLISH/civil_code.htm (accessed 10.10.12).

[133] Under Article L. 224-5 of France's Social Action and Families Code; a folder containing a letter and photographs was provided for the child. Available http://www.proz.com/kudoz/french_to_english/law_general/3844397-code_de_l%E2%80%99action_sociale_et_des_familles.html (accessed 11.12.12).

[134] See Article L 224-6 (2) *ibid*.

[135] *Kearns v France op cit* n131 para 18.

[136] *Ibid* para 21.

[137] *Ibid*.

[138] *Ibid*.

the placement ordered to be annulled, and the 'establishment of parental ties' ordered to be restored between applicant and child.[139]

A subsequent appeal to the Court of Cassation found that parental consent to relinquishment was not required, given that legal parentage of the child had never been established. By opting for an anonymous birth, the applicant mother had self-severed all genetic links with the child, enabling the adoption to proceed.[140] The European Court considered Article 21 of the Children's Convention and Article 5 of the European Convention on the Adoption of Children (1968) including the (then draft) revised version of Article 5.[141] The timing of consent became a key factor, given that it came only one day after the birth, with six weeks being suggested as the minimum time period for reflection upon the consequences. The question of whether there was an 'absence of pre-existing family life' between child and mother was relevant, with the Court finding that their relationship *did* fall within the legal sphere of family life to qualify for the protections of Article 8.[142] The Court stressed however that the principle of *mater semper certa est*[143] did not apply in France, as it would have had the baby been born in the mother's native Ireland. This fact was significant given the applicant's decision to travel to France for the purpose of anonymous relinquishment, the implication being that she clearly intended to legally exclude both herself and the genetic father from the child's records.

The European Court found ultimately that no Convention violations had occurred. Given the lack of consensus amongst member States in such matters (in respect of the need for a post-relinquishment period of reflection for example) the Court was justified in affording a wide margin of appreciation to such issues, especially where achieving balance between 'competing private and public interests or Convention rights' was at stake.[144] It was noted however that 'in striking a balance between these different interests the child's best interests should be paramount.'[145] The child's welfare needs in this situation were, it was found, best addressed via speedy adoption, despite the applicant mother's contention that,

> with regard to the child, the painful, psychological effects of adoption should not be overlooked, seeing that before reaching adulthood, many children or adolescents experienced the suffering of being abandoned and pursued the aim of restoring contact with their

---

[139] Under Article 334-8 of the Civil Code op cit n 133 at para 22.

[140] Under Article 627 (2) of the New Code of Civil Procedure; this was held to have been the appropriate rule all along. *Ibid* para 25.

[141] France has signed, but not ratified the 1968 Convention. Although not directly applicable, Article 4 of The Hague Convention on the Protection of Children and Co-operation in Respect of Inter-Country Adoption (1993) was also referred to (as persuasive) insofar as it requires that persons whose consent is needed for adoption must have received such counselling as might be necessary and be informed of the effects of their consent; birth mother consent may only be given post-partum. *Ibid* para 35.

[142] *Ibid* para 72 (Citing *V S s Germany* (App no 4261/02) (Admissibility Decision, 22 May 2007).

[143] '*Birth creates legal affiliation between mother and child*'.

[144] Para 74 (Citing *Evans v UK* (2007) 22 BHRC 190).

[145] Para 79 '*biological mother, the child and the adoptive family*'.

> biological parents. Such suffering could only be exacerbated if the child discovered that, a few months after the birth, the natural mother had unsuccessfully sought to take him or her back.[146]

The case is significant for several reasons. The concept of identity is briefly discussed but only insofar as it is something possessed by birth mothers and which must be kept secret if they have opted for anonymity.[147] The issue of the child's loss of nationality and cultural heritage was not discussed here nor, to any real extent, were the respective claims of either her genetic or legal fathers in Ireland, both of whom had separately sought to establish some legal claim to her parentage. It would have perhaps been useful if the Court had briefly examined the issue of which hypothetical claim might have been regarded as the stronger, that of legal father (by virtue of his marriage to the birth mother) or that of her unmarried, genetic father, especially given his application to the Irish Supreme Court. Arguably, an opportunity to perhaps clarify or expand upon definitions of such concepts as the child's best interests, the scope of family life under Article 8, or the positive obligations of domestic authorities in such cases, was missed.

The wide margin of appreciation therefore allowed for absolute maternal relinquishment and anonymity, despite the applicant mother's subsequent change of mind. Although the remit of the European Court is not that of an appellate court some guidance on the more difficult procedural aspects of such cases (extent of relinquishment, time limits, consent frameworks) or some acknowledgement of the importance of preserving accurate genetic identity information again might have been useful here. The case does not seem to sit well with some of the earlier jurisprudence on the need for genetic connection, especially given the growing focus on involving parents more fully in decision-making processes (via information, or the availability of legal representation) and the emphasis in some cases upon the possibility of reunion.

That said, had the Court perhaps suggested that the mother's Article 8 rights had been infringed, for example through the domestic time limits on revocation or via the process of obtaining her consent without some form of legal representative being present, it is perhaps still likely that the best interests of the child principle would have required the child to remain with her adoptive parents given her psychological attachment to them. Jeopardizing the child's welfare by exposing her to a convoluted parentage dispute involving two putative fathers (genetic and legal) and a mother who had taken extreme steps to preserve her own anonymity might have led to an unfortunate outcome.[148] Had the best interests of the relinquished child been more fully discussed here however, some judicial acknowledgement of the psychological need for accurate identity narratives might have occurred. The issue of loss of original nationality might have also merited some

---

[146] Para 49.

[147] Para 21.

[148] Such decisions are not unheard of in other jurisdictions however. See for example the Irish case *N & Anor v Health Service Executive & Ors* [*2006*] IESC 60 (the 'Baby Anne' case).

discussion, if only to allow for issues of cultural heritage rights to enter the debate.[149] As a newborn infant, the child would also have lacked the necessary degree of 'remembered relatedness' to give rise to Convention-protected family life rights in such cases. Equally, where States have long customs of legally enshrined parental anonymity or relinquishment, it seems unlikely that an argument based only upon the best interests of the child principle would be able to challenge or overturn such deeply-imbedded socio-cultural veto traditions.

5.6 Matters of Procedure

A number of Strasbourg cases have highlighted the importance of the procedural aspects of such cases, focusing particularly on the involvement of the original parents in the decision-making process. In *W v United Kingdom* (1987)[150] for example it was noted that 'the natural family relationship is not terminated by reason of the fact that the child is taken into public care.'[151] Failure to include parents in the decision-making process 'to a degree sufficient to provide them with the requisite protection of their interests,' may amount to 'a failure to respect their family life.'[152] Equally, procedural delays pre-hearing might bring about a 'de facto determination of the issue.' This should not occur given that

> an effective respect for family life requires that future relations between parent and child be determined solely in the light of all relevant considerations and not by the mere effluxion of time.[153]

Active consideration of the best interests of the child principle, especially perhaps where rights-balancing exercises are involved, may also be necessary to satisfy the requirements of Article 8. In *Anayo v Germany* (2011) for example an absent genetic father was refused contact with his children by the German Court of Appeal.[154] This decision was deemed to have breached the provisions of Article 8 (2) because 'sufficient' reasons for the refusal had not been provided. Significantly, adequate consideration had not been given to the fundamental issue of whether contact with their genetic father might have been in the children's best interest. German legislation seemed to assume that a 'social and family relationship' would have arisen had the father 'lived with the children in domestic community for a

[149] See further CRC/C/FRA/Q/4 (24 February 2009) List of Issues from The Committee on the Rights of the Child which includes queries on domestic child protection, international adoption and the monitoring of adoption agencies (see Part IV para 5). Available http://daccess-dds-ny.un.org/doc/UNDOC/GEN/G09/408/26/PDF/G0940826.pdf (accessed 01.03.11).

[150] 10 EHRR 29.

[151] *Ibid* para 59.

[152] *Ibid* para 64.

[153] *Ibid* para 65.

[154] *Anayo v Germany* [2010] ECHR 2083 (21 December 2010).

lengthy period of time.'[155] The fact that his particularly difficult circumstances (i.e. the 'legal and practical reasons' of complying with immigration laws[156]) had completely precluded him from establishing such a relationship with the children had not been taken into account by the national authorities however.[157]

The European Court stressed that even though there was 'no uniform approach' evident amongst Member States to the issue of contact disputes in such circumstances (i.e. where a genetic father was in dispute with the legal father) this did not necessarily mean that domestic courts were prevented from examining the issue of the best interests of the child.[158] The national court's decision was geared towards effecting compliance with domestic legislation, which was itself aimed at protecting 'existing family relationship between a legal father and a child, who are actually living together with their wife and mother respectively'.[159] Such a stance afforded 'existing family' ties of kinship priority over those that might arise only by virtue of genetic connection. The fact that the biological father had no way of ever applying to become the legal father of the children (or of challenging the legal paternity of the mother's husband) under domestic law was a key factor given the potential for 'intended' family life rights to arise in such cases.

As Sloan has further observed, a close relationship falling somewhere short of familial life might well still fit within the remit of Article 8 private life and thus merit protection against unjustified State interference.[160] The decision reached in *Anayo,* though somewhat equivocal, at least indicates 'an admirably flexible approach'[161] on the part of the judiciary in respect of the question of whether potentiality of relatedness ought to simply be rendered void *ab initio* by domestic legislative or policy frameworks. Given the wide-ranging impacts of such a template for gauging, recognising or vetoing blood-ties (insofar as the affected parties here could include mother, legal father, genetic father, siblings as well as the twins themselves) an appropriate degree of 'strict scrutiny' oversight by the Strasbourg judiciary seems entirely appropriate. That said, the Court did stress that, in terms of jurisdiction and procedural decision-making,

> it is for the domestic courts, who have the benefit of direct contact with all the persons concerned, to exercise their power of appreciation in determining whether or not contacts between a biological father and his children are in the latter's best interest.[162]

Under the German civil code, in such circumstances 'the legal parents' motives for refusing contact did not necessarily have to be based on considerations relating

---

[155] *Ibid* at para 67. See however *Zaunegger v. Germany* ECHR 22028/04 (December 2009).

[156] *Anayo ibid* at para 69.

[157] *Ibid* para 71.

[158] *Ibid* para 68.

[159] *Ibid* para 70.

[160] Sloan (2011), p. 316.

[161] *Ibid.*

[162] *Ibid* para 66.

to the children's best interest.'[163] Perhaps the most significant aspect of the *Anayo* decision was that the Court both highlighted and afforded recognition to the 'potentiality' of relatedness that existed between the genetic father and his children, albeit on the basis of his clearly expressed intention to commit to family life. To an extent, this represents a departure from earlier jurisprudence which often required the presence of a more established social relationship before Article 8 rights might be considered to have arisen. Whether this decision marks a move towards a more child-centric reasoning in such rights-balancing cases remains to be seen, but it at least serves as a useful reminder that children (and their welfare needs) are at the very centre of many such cases; they may be vulnerable to having their best interests almost completely overlooked at the level of domestic proceedings. It seems unlikely that a new rights-bearing category of 'potential family life' has been declared here; in certain contexts however, absolute bars on genetic kinship (via loss of contact or vetoes on identifying information) might perhaps be capable of amounting to unjustified, perhaps unjustifiable, rights violations under Article 8, as an aspect of family or private life.[164]

Equally, domestic decision-making processes in such scenarios ought now to include, if not actively address, those issues that might touch and concern the best interests of the children affected. Whether 'potentiality' of relatedness and 'intention' to forge kinship bonds can be pleaded by other members of the triad (such as separated siblings or disappointed adoptive parents) or by veto-affected origin-deprived individuals who are no longer in childhood, is more difficult to predict. In terms of being adequately involved in decision-making processes, it was held in *R and H v United Kingdom* (2011)[165] that the applicant parents' complaints had both a 'substantive and a procedural dimension.'[166] The Court found that the domestic proceedings amounted to an interference with the applicants' family life within the meaning of Article 8, but that this was fully in accordance with the national law and in the pursuit of a legitimate aim (i.e. to protect the child's rights and well-being). Whether the proceedings were necessary in a democratic society and 'struck a fair balance between the competing interests,' was also discussed.[167] Although Article 8 contains no explicit procedural requirements, 'decision-making processes involved in measures of interference must be fair and such as to afford due respect to the interests safeguarded by Article 8.' Parental involvement, viewed holistically, must be such that parents were afforded 'requisite protection.'[168]

---

[163] *Ibid* para 69.

[164] See for example *Marckx v Belgium* (1979) (application no 6833/74) (1979); *Keegan v Ireland* (1994) (App no 16969/90) 18 EHRR 342; *Kroon v The Netherlands* (1995) 19 EHRR 263; *X v Croatia* (app no 12233/04) (17.07.2008); *Johansen v Norway* [1996] EHRR 31 (app no 17383/90).

[165] *R and H v United Kingdom* (2011) 35348/06 [2011] ECHR 844.

[166] *Ibid* para 57.

[167] *Ibid* para 72.

[168] *Ibid* para 75.

This is especially so in cases involving adoption given that it is 'an interference of a very serious order to split up a family.'[169]

If however it was clearly in interests of the child to be adopted (and the child's chances of being successfully placed would, as was argued here, be maximized by the presence of a court order freeing them for adoption) then 'the interests of the biological parents must inevitably give way' to those of the child.[170] Even if prospective adopters agreeable to post-adoption contact could not be found, 'the benefits of adoption outweighed those of post-adoption contact.'[171] Although a wide margin of appreciation attached to domestic authorities' decision-making powers (i.e. on taking children into care) 'stricter scrutiny' by the European Court would be required where limitations were to be placed upon such key family life matters as contact[172];

> For these reasons, measures which deprive biological parents of the parental responsibilities and authorise adoption should only be applied in exceptional circumstances and can only be justified if they are motivated by an overriding requirement pertaining to the child's best interests.[173]

That said, Article 8 does not oblige domestic authorities to 'make endless attempts at family reunification.' It requires them instead to 'take all necessary steps that can reasonably be demanded to facilitate the reunion of the child and his or her parents.'[174] The Dissenting Opinion of Kalaydjieva J argues however that 'a reasonable determination of a child's best interests' could only be made after considering the child's present situation, 'the extent to which it calls for separation from the biological parents' and 'the extent to which these interests would be better met with the candidate-adoptive family (including the future contacts with the biological parents).'[175] Given how long the freeing order proceedings had taken here, and taking into account the fact that an adoption order hearing would only have been possible after their completion, it was difficult to see how a

---

[169] *Ibid* para 76; citing *X v Croatia* [2008] (app no 12233/04) (17.07.2008) para 47, it was noted that local authorities must be afforded 'a measure of discretion in this respect.'

[170] *Ibid* para 77.

[171] *Ibid* para 79; It was also noted that the Trust's efforts to accommodate the parents' worries went some way towards mitigating the effects of the Northern Ireland High Court's decision to grant the freeing order without consulting with the original parents.

[172] 'Such further limitations entail the danger that the family relations between a young child and one or both parents would be effectively curtailed.' Citing *Elsholz v. Germany* [GC], no. 25735/94: [2000] ECHR 371, 49, ECHR 2000-VIII; *Kutzner v. Germany*, no. 46544/99 [2002] ECHR 160, § 67, ECHR 2002 I) para 81.

[173] *Ibid*, citing *Aune v. Norway*, no. 52502/07, [2010] ECHR 1994 66, 28 October 2010; *Johansen v Norway* [1997] 23 EHRR 78; *P., C. and S. v. the United Kingdom*, no. 56547/00, [2002] ECHR 604, 118, ECHR 2002 VI).

[174] *Ibid* para 88; citing *Pini and Others v. Romania*, nos. 78028/01 and 78030/01, [2004] ECHR 275, § 155, ECHR 2004 V

[175] Available http://www.bailii.org/eu/cases/ECHR/2011/844.html accessed 28.02.12.

'consequential examination of the current and the future situation of the child better met her urgent needs.'[176]

The child's interests were comprised of 'inseparable aspects'; domestic authorities should have assessed these as they stood at different points in time. The applicants were later deemed eligible (by a different Social Services Trust) to act as foster carers for other children in their family, and found capable of parenting their remaining child. These facts were not relevant however to the issue of the adoption itself, only to the question of post-adoption contact frequency. As such the nature of the freeing order was to aid in the search for prospective adopters rather than necessarily supporting a 'proper assessment of the child's best interests or the applicants' rights under Article 8 as to the moment of the decision for adoption.'[177]

In *Y C v The United Kingdom* (2012)[178] the refusal of the domestic courts to order an assessment of the child's mother (as sole carer) and their failure to take into account all relevant considerations when making a placement order did amount to a serious interference with her Article 8 rights.[179] Determining whether or not this interference was necessary in a democratic society, required consideration of two aspects of the proceedings, namely whether (in the light of the case as a whole) the court's justifying reasons were 'relevant and sufficient' and whether the decision-making process was 'fair and afforded due respect' to her Article 8 rights.[180] Where a 'permanent severance of family ties' via adoption was at issue, the child's best interests would be the paramount consideration.[181] In gauging these matters, two points were key. Firstly, the child's original family ties should be maintained unless the family had 'proven to be 'particularly unfit'; secondly, the child's development required a 'safe and secure environment.'[182] Only in 'very exceptional circumstances' might original family ties be severed, whilst everything had to be done 'to preserve personal relations and, where appropriate, to "rebuild" the family.'[183] If however the maintenance of these ties would be harmful to the child's health and development, then parents would not be able to rely upon Article 8 rights.

Again the need for parents to be involved fully in the decision-making process was stressed, given that adoption was a possibility.[184] The lower court's failure to order the assessment was mitigated by several factors however: an assessment

---

[176] *Ibid.*

[177] *Ibid.*

[178] *Y C v United Kingdom* (App 4547/10) ECHRR.

[179] The court had not made explicit reference to the various provisions of the 'welfare checklist' contained in s 1 (2) of the Adoption and Children Act 2002. See para 103.

[180] *Ibid* para 133.

[181] *Ibid* para 134.

[182] *Ibid*, citing *Neulinger and Shuruk v Switzerland* (2010) [GC] app no 41615/07 available http://hudoc.echr.coe.int/sites/eng-press/pages/search.aspx?i=003-3194121-3556272 (accessed 03.01.13).

[183] *Ibid.*

[184] *Ibid* para 139.

would most likely have led to placement delays, the applicant's case had subsequently been reviewed at the Court of Appeal stage, and the County Court judge had made reference to the reports and verbal evidence of the guardian ad litem and social worker involved in the case. As such, no violation of Article 8 was found to have occurred.[185] The Dissenting Opinion of De Gaetano J stressed however that the child in question was eight years old, and had lived with his mother until he was seven; the placement order had been made in an 'unorthodox' fashion and the applicant had been refused leave to appeal to the Supreme Court in 'a cavalier way.' Taken together these last two factors could be deemed to amount to a violation of Article 8, whilst the margin of appreciation that had been afforded here raised a 'serious issue,' given that the various dicta did not deal in any meaningful way with Convention issues.[186]

Previous Strasbourg authorities had called for 'careful assessment' of the impact of the proposed measures,[187] while the English Court of Appeal had itself set down 'stringent requirements' for the contents of such decisions.[188] In making the placement order which dispensed with the mother's consent to the adoption, no reference had been made to the 2002 Act's 'welfare checklist' nor did the County Court 'expressly spell out relevant and sufficient reasons' for its decision, as domestic courts should be required to do in such cases.[189] Perhaps most significantly, the doctrine of margin of appreciation should not be allowed to 'absolve' the European Court from its supervisory duty towards domestic courts, in making sure that their decisions meet Convention criteria 'in form or in substance.'[190]

That said, the margin of appreciation appears to widen where the issue at hand is one that has attracted a lack of consensus. In *Ahrens v Germany* (2012)[191] for example, a child's genetic father was prevented by the domestic legislation from challenging the child's legal paternity, on the basis of the child's best interests, and the need to protect the stability of the family unit. The Court considered whether the private life rights of the genetic father had been compromised, noting that he had 'a protected interest in establishing the truth about an important aspect of his private life' as the child's father.[192] Had the issue been 'a particularly important facet' of his existence, the margin of appreciation afforded to the member state would

---

[185] *Ibid* para 150.

[186] *Ibid* para 1.

[187] See for example *Saviny v Ukraine* (2008) no. 39948/06 EHRR; *Kutzner v Germany* (2002) no 46544/99; *Kurochin v Ukraine* (2010) no 42276/08.

[188] *E H v London Borough of Greenwich & Ors* [2010] 2 FLR 661, [2010] 2 FCR 106, [2010] Fam Law 577, [2010] PTSR CS23, [2010] EWCA Civ 344.

[189] *Y C v United Kingdom* (App 4547/10) ECHRR para 6.

[190] *Ibid* para 8. De Gaetano J further notes that the doctrines of margin of appreciation and *quatrième instance* 'should be applied evenly across the board and irrespective of whether the legal or judicial system concerned can trace its lineage back to the post-Soviet era or to the meadow at Runnymede.'

[191] *Ahrens v Germany* (App 45071/09) ECHRR (22 March 2012).

[192] Para 71.

normally have been narrow. The Court distinguished this case from the strict-scrutiny requirements of *Anayo* on the basis that the genetic father there had sought establishment of paternity only for the purposes of gaining contact rights; here, the proceedings would have been 'fundamentally different' with a more far-reaching objective'[193] of altering the child's birth status as an acknowledged descendant of her legal father. The genetic father's interests were not greater than those of the stable, 'existing family relationship' that had survived the revelation of the mother's infidelity.[194]

Equally, a wide margin of appreciation attached to issues involving legal status, so that the domestic court's decision on refusing to allow the paternity challenge fell within this. In terms of being discriminated against under Article 14 of the European Convention (on the basis that the mother, and legal father could bring a challenge and that any child bringing a challenge would not be hindered by the presence of their social relationships) the Court found that the applicant's rights were justifiably curtailed with a view to 'protecting the child and her social family from external disturbances.'[195] As such, no rights violation had occurred.

The case raises an interesting question in terms of the margin of appreciation afforded to domestic jurists: analysis of the young child's best interests may require that issues such as contact with an absent genetic parent are factored into the equation. The rights of older adoptees to ascertain genetic realities, even where health and welfare issues might be highly relevant, do not however necessarily fall within this margin, given the lack of consensus amongst member states.

The case of *S H and Others v Austria* (2010)[196] was perhaps quite telling in respect of attitudes to genetic identity and the need for blood-ties generally. The Court noted that respect for private life may encompass the right to 'establish and develop relationships with other human beings'[197] and found that a domestic legislative ban on certain types of IVF techniques amounted to unjustifiable discriminatory treatment. A fairly substantial discussion on the potential risks of assisted reproduction techniques ('selection' of children, exploitation or humiliation of female ova donors) was provided in the judgment, although there was only very limited reference to the donor-conceived child's need for information 'about their actual descent.'[198] The Court did accept however that respect for private life generally requires the ability to 'establish details of . . .identity. . .such information [being] of importance because of its formative implications for. . .personality.'[199] Such a right was essentially only a highly qualified one, given that member states' domestic laws (such as those underpinning the French anonymous birth system)

193 Para 67.

194 Para 72.

195 Para 89.

196 *S H and Others v Austria* (2010) App no 57813/00.

197 *Ibid* para 58.

198 *Ibid* para 82.

199 *Ibid* para 83.

were capable of achieving a 'proper balance' between competing public and private interests in such cases.[200]

Significantly, the disappointed parents in this case argued successfully that having to 'resort to' the 'difficult and painful decision' of adopting a child instead (after failed IVF attempts abroad) rather than being permitted to attempt the conception of 'children of their own' had caused them 'great emotional distress.'[201] The Court declined to comment on the issue of their 'suffering' but did make an award in respect of non-pecuniary damage. Bainham's observation is entirely apposite here: 'social parenting, comes about not as the option of first choice but because….the natural family has failed.'[202]

At the Grand Chamber stage, it was further noted that,

> the splitting of motherhood between a genetic mother and the one carrying the child differs significantly from adoptive parent-child relations and has added a new aspect to this issue.[203]

Issues such as 'human dignity [and] the well-being of children thus conceived'[204] should be taken into account, not least because of the lack of consensus in Austrian society to such issues. Perhaps most significantly, the Grand Chamber stressed that even though there was no breach of Article 8 in this case, the law needed to be kept under review by signatory states, given the 'particularly dynamic development[s] in science and law'[205] that were occurring. In a separate Opinion, De Gaetano J called for 'proper acknowledgement of the parameters of Article 8' and stressed that the concept of human dignity underpinned the European Convention. As such, neither Article 8 nor Article 12 should be interpreted as creating a right to conceive 'at any cost.'[206] The Joint Dissenting Opinion further suggested that not only had a violation of Article 8 occurred but that its role in relation to questions of procreation now appears to be an 'enhanced' one. Articles 12 (1) and 15(1) (b) of the International Covenant on Economic, Social and Cultural Rights (1966) ('ICCPR') were also cited to stress 'the right of everyone to enjoy the benefits of scientific progress….[and] the highest standard of physical and mental health.'[207]

In affording a very wide margin of appreciation here, the Grand Chamber perhaps departed 'dangerously' from the Court's earlier case law by failing to

---

[200] *Ibid* para 84.

[201] *Ibid* para 98.

[202] Bainham (2008), p. 338.

[203] *S.H. And Others v. Austria*—57813/00 [2011] ECHR 1878 (3 November 2011) para 105 available at *http://www.bailii.org/eu/cases/ECHR/2011/1878.html* (accessed 31.01.13).

[204] *Ibid* para 113.

[205] *Ibid* para 118.

[206] Separate Opinion para 2. In relation to the notion of European consensus, he added that 'history teaches us that [it] has in the past led to acts of gross injustice both in Europe and beyond.' *Ibid* para 4.

[207] *Ibid* para 9 per Tulkens J, Hirvelä J, Lazarova Trajkovska J and Tsotsoria J.

'contribute to harmonizing across Europe the rights guaranteed by the Convention.'[208] By focusing on the question of margin of appreciation rather than on the central issues, a 'minimalist' approach had been taken which provided little in the way of guidance on crafting laws for 'atypical family relations' in domestic contexts. It was further noted that the applicants' freedom to engage in gamete donation outside of their own jurisdiction meant that the Government's arguments on the need to protect the best interests of the child (and their concerns for vulnerable egg donors) were rather undermined.[209]

From the perspective of the child so conceived, the rights-issues of human dignity, identity, health and well-being clearly could have merited fuller discussion. It is perhaps at least a useful development that such matters have been acknowledged as potentially giving rise to discrimination, controversy and 'danger'.

5.7 Conclusion

Taken together, the 'blood-tie' cases referred to in this chapter may provide some level of useful, if limited, guidance on a number of issues, such as a gradual, tentative recognition of the importance of genetic identity. Cases such as *Sommerfeld* and *Sahin* for example, discuss the issue of psychological well-being, including the concept to some extent in their definitional analyses of the child's 'best interests'. The discussion in *Sahin* of the potential for discriminatory treatment, and the finding of an Article 14 violation, seems to support this argument. *Gnahoré* also seems to offer a useful observation on the child welfare paramountcy principle, separating the child's interest into two 'limbs' and possibly suggesting that severance from one's genetic 'roots' should occur only in exceptional circumstances, rather than normatively. The case perhaps serves also as a reminder of how parents, social or genetic, ought to refrain from acting in any way that might harm the health and development of their children. The child's need for genetic identity was similarly touched upon in *Clemeno* and *Jucius,* with the interests of other relatives being at least discussed. Such discourses may support the minority reasoning in *Odièvre* on the importance of preserving 'origin' and finding 'biological truth'.

That said, parental interest still seems to be the paramount factor in most cases. The limited discourse on maternal rights in *X v Croatia* and the stance taken in *Lebbink* on the 'potentiality' for legal relatedness seem to indicate a continuing judicial reluctance to identify which rights, if any, might attach to the need for knowable genetic connections. Arguably, creating an 'absolute' right to genetic kinship (i.e. to receive accurate information or engage in kin contact) could potentially reconfigure the child-protective role of domestic authorities and

[208] *Ibid* para 10.

[209] *Ibid* para 13.

provoke public outcry. Changing the current legal definitions of parenthood or parentage might prove particularly contentious, especially in regions where international adoptions and assisted reproduction have long been regarded as esteemed or accepted means of achieving parenthood or overcoming infertility.

The principle of child welfare primacy has itself served to link the child's right to identity, to the right to exist within a family.[210] In many cases however, the family no longer simply comprises of 'a dyad of parent and child being assessed, but a triad of child, parent and extended family.'[211] Although interpretations of the best interests principle must remain sensitive to adult concerns, harsh limitations upon kin contact or information provision should not be allowed to adversely impact upon the child's long term psychological welfare. International law principles of equality and non-discrimination may yet become particularly relevant, perhaps by encouraging Convention-signatory states to implement domestic laws and policies that are aimed more specifically at preventing origin deprivation, and, where possible, actively enabling eventual reunion.

Given the ever-changing 'requirements of modernity'[212] and the increasingly complex domestic definitions of legal relatedness, decision-making is perhaps best not left almost entirely to parents, social workers or the passage of time. In some cases domestic courts are perhaps failing to honour their *parens patriae* obligations towards vulnerable children, by too readily delegating or postponing the more difficult family life decisions rather than providing some form of guidance as to how the welfare paramountcy rights of vulnerable children might be better protected. Judicial regard for family sanctity and parental privacy should not get in the way of a positive obligation to promote and actively protect children's rights throughout their lifetime. Arguably, traditional 'reluctance to see our children as having rights to anything, let alone a right to protection'[213] may be in decline, as a result of international scrutiny of domestic practices. There remains much scope however for international law definitions of family life to perhaps be further expanded upon and to perhaps take more account of the potential for unlawful discrimination that can exist in these areas of law. The identity 'rights' of the newborn infant are often fairly fictive, or perhaps prone to being overlooked in much the same way as their long-term psychological welfare needs may be sidelined or left to chance. As a result, many origin deprived persons wishing to realize the 'right' to genetic identity by accessing sealed records or engaging in kin contact must seek out black letter domestic remedies on an individual basis.

Some jurisdictions have introduced legislation aimed specifically at preserving genetic ties, irrespective of whether family life exists or not, on the basis of

---

[210] See Fortin (2003), p. 48.

[211] Hegar and Scannapieco (1999), p. 81.

[212] See Horgan and Martin (2008), pp. 155–162, p, 162 available at http://www.coe.int/t/e/legal_affairs/legal_co-operation/family_law_and_children%27s_rights/Documents/Rosemary%20Horgan%20Article.pdf accessed 01.05.11.

[213] Lyon et al. (2003), p. 639.

protecting parental or tribal rights or promoting cultural heritage.[214] Such provisions can however leave children exposed to inadequate or abusive parenting. Neither of these two templates for dealing with genetic connection ('at all-costs' preservation or complete severance) seems to elevate child welfare to a meaningful position of paramountcy over that of adult-centric interests. The need to involve affected parties more fully in decision-making processes (to ensure Article 6 ECHR compliance) perhaps places a useful burden upon domestic jurists and practitioners to ensure that all stages of proceedings involving children provide opportunities for 'fair hearing'; pre-hearing, informal agreements and discussions aimed for example at resolving contact issues may have to highlight more clearly how the best interests of the child will be protected against the losses and potential harms of origin deprivation.

The chapter which follows will examine domestic jurisprudence on disputes over legal parentage and parenthood, and the child's 'right to know' in a bid to gauge whether a juridical right to genetic connection might yet be found to be exist. It will also discuss the extent to which domestic judiciaries might be obliged to include more fully the psychological aspects of origin deprivation in their best-interests-led, decision-making processes.

References

Bainham, A. (2008, July). Arguments about parentage. *Cambridge Law Journal, 67*(2), 322–351.

Callus, T. (2004). Tempered hope? A qualified right to know one's genetic origin: Odièvre V France. *Modern Law Review, 67*(4), 658–669.

Choudry, S. (2003, June) The Adoption and Children Act 2002: The Welfare Principle and The Human Rights Act 1998—a missed opportunity? *C Fam, 15*(2), 119.

Clayton, B. J. (2007–2008). A day late and a dollar short: Should Utah's unmarried dads get one more chance to claim their newborns? *Journal of Law and Family Studies, 10*, 249–266.

CRC/C/FRA/Q/4 (2009, February 24) List of issues from The Committee on the Rights of the Child available http://daccess-dds-ny.un.org/doc/UNDOC/GEN/G09/408/26/PDF/G0940826.pdf. Accessed 01.03.11.

Cretney, S. M., Masson, J. M., & Bailey-Harris, R. (2003). *Principles of family law.* London: Sweet & Maxwell.

Dekeuwer-Defossez, F. (1999, September 14). *Modernising family law: proposals for a law adapted to the realities and aspirations of our times.*

Eekelaar, J. (1994) The interests of the child and the child's wishes: The role of dynamic self determination. In P. Alston (Ed.), *The best interests of the child: Reconciling culture and human rights* (pp. 42–61). Oxford: Clarendon.

Fabius, L. (1998). *Rights of the child, uncharted territory.* Parliamentary Inquiry Committee.

Fortin, J. (2003). *Children's rights and the developing law.* Cambridge: Cambridge University Press.

---

[214] See for example The Indian Child Welfare Act 1978 A 25 U.S.C.A. 1901 (United States) See further McGillivray (1983–1985), p. 449; Hollinger (1988–1989), p. 453. See also Eekelaar (1994), pp. 42–61.

Gardner, R. (2001) Should courts order PAS children to visit/reside with the alienated parent? A follow-up study. *American Journal of Forensic Psychology*, 61–106.
Hale, B., Pearl, D., Cooke, E., & Monk, D. (2009). *The family, law and society*. Oxford: Oxford University Press.
Hegar, R. L., & Scannapieco, M. (1999). *Kinship foster care: Policy, practice and research*. New York: OUP.
Herring, J. (2009, April 6). Revoking adoptions. *New Law Journal*.
Hollinger, J. H. (1988–1989). Beyond the best interests of the tribe: The Indian Child Welfare Act and the adoption of Indian children. *University of Detroit Law Review, 66*, 451–501.
Horgan, R., & Martin, F. (2008, September). The European Convention on adoption 2008: Progressing the children's rights polemic. *International Family Law, 3*, 155–162.
Lowe, N., & Douglas, G. (2007). *Bromley's family law*. Oxford: Oxford University Press.
Lyon, C. M., Cobley, C., Petrie, S., & Reid, C. (2003). *Child abuse*. London: Jordans.
McGillivray, A. (1983–1985). Transracial adoption and the status Indian child. *Canadian Journal of Family Law, 4*, 437–467.
Parliamentary Assembly of the Council of Europe. (2000). *International adoption: respecting children's rights*. Doc 9377 Recommendation 1443 (26 January 2000, 5th Sitting) para vii. Available at http://assembly.coe.int/Main.asp?link=/Documents/AdoptedText/ta00/EREC1443.htm
Scheiwe, K. (2003). Giving birth anonymously-a suitable export from France to Germany? *International Family Law,* 144.
Sloan, B. (2011). Unmarried fathers and the frustration of family life. *Cambridge Law Journal,* 314–317.
Thery, I. (1998, May 14). *Couples, filiation and parenthood today-the challenges posed to the law by changes in family and private life*. Report to the Minister of Justice and the Minister for Employment and Solidarity.
Welstead, M., & Edwards, S. (2008). *Family law*. London: Oxford.

Cases

Strasbourg

Ahrens v Germany (App 45071/09) ECHRR (22 March 2012)
Anayo v Germany [2010] ECHR 2083 (21 December 2010)
Aune v. Norway, no. 52502/07, [2010] ECHR 1994 66 (28 October 2010)
Bensaid v United Kingdom (2001) ECHR no 44599/98 47
Clemeno v Italy (2008) ECHR Judgment 21.10.2008 [Section II] No 19537/03.
E H v London Borough of Greenwich & Ors [2010] 2 FLR 661, [2010] 2 FCR 106, [2010] Fam Law 577, [2010] PTSR CS23, [2010] EWCA Civ 344
Elsholz v Germany (2000) no 25735/94 ECHR (2000-VIII)
Evans v UK (2007) 22 BHRC 190)
Gaskin v UK (1989) 12 EHRR 36
Gnahoré v. France, (2000) no. 40031/98, § 59, ECHR 2000-IX
Gorgülu v Germany (Application No 74969/01) [2004] 1 FLR 894
Johansen v Norway [1997] 23 EHRR 33
Jucius and Juciuviène v Lithuania - application no 14414/03 (25-11-2008)
K and T v Finland (2001)25702/94 (150) ECHR (2001)
K v The United Kingdom (1986) (European Commission) Application no 11468/85
Kearns v France (App no. 35991/04) [2008] 1 FLR 888.
Keegan v Ireland (1994) (no 16969/90) 18 EHRR 342
Kroon v The Netherlands (1995) 19 EHRR 263, 283

Kutzner v. Germany, no. 46544/99 [2002] ECHR 160, § 67, ECHR 2002 I) para 81
Lebbink v The Netherlands (2004) (app no 45582/99)
Marckx v Belgium (1979) (application no 6833/74) (1979)
Mikulić v Croatia (2002) (no 53176/99) 54 ECHR;
Neulinger and Shuruk v Switzerland (2010) [GC] app no 41615/07
Nylund v. Finland (1999) no. 27110/95, ECHR 1999-VI)
Odièvre V France [2003] (App No 42326/98)
Olsson v Sweden (no 1) [1988] ECHR 10465/83 paras 81, 83
P C and S v United Kingdom (2002) 35 EHRR 1075
Pini and Others v. Romania, nos. 78028/01 and 78030/01, [2004] ECHR 275, § 155, ECHR 2004 V
Pretty v the United Kingdom (2002) (no 2346/02) 61 ECHR
R and H v United Kingdom (2011) 35348/06 [2011] ECHR 844
Sahin v Germany (2003) no. 30943/96, § 12, ECHR 340
Saviny v Ukraine (2008) no. 39948/06 EHRR
Scozzari and Giunta v. Italy (2000) [GC] nos. 39221/98 and 41963/98, § 169, ECHR 2000-VIII
S.H. And Others v. Austria - 57813/00 [2011] ECHR 1878 (3 November 2011)
Sommerfeld v Germany (2003) no. 31871/96, §§ 11–12, ECHR 2003-VIII
Tryer v United Kingdom (25 April 1978) Series A no 26
VS s Germany (App no 4261/02) (Admissibility Decision, 22 May 2007)
W v United Kingdom [1987] ECHR 9749/82
X v Croatia [2008] (app no 12233/04) (17.07.2008)
Y C v United Kingdom [2010] (App 4547/10) ECHRR
Zaunegger v. Germany [2009] ECHR 22028/04 (December 2009)

England/Wales

E H v London Burough of Greenwich & Ors [2010] 2 FLR 661, [2010] 2 FCR 106, [2010] Fam Law 577, [2010] PTSR CS23, [2010] EWCA Civ 344

Ireland

N & Anor v Health Service Executive & Ors [2006] IESC 60

Chapter 6
Never Knowing 'One's Past': Genetic Ancestry Vetoes as Discrimination?

6.1 Introduction

> Genetics alone does not create a bond; in fact, it is played down in order to create a biological bond based on gestation and care. Blood, food and the experience of carrying the foetus are the biological substance shared with the child.[1]

Although the 'biogenetic substance shared in reproduction has the constitutive power to create kinship,' it does not necessarily always serve to 'dictate social relations.'[2] Where reproductive technologies have enabled conception, the ties of relatedness might not be 'modelled on nature' but upon parents' 'relational actions' of 'desire, gestation and care.'[3] Such a 'revisioning' of the law on family life has seen domestic law and policy makers in a number of jurisdictions struggle to define or perhaps re-define the right to identity, as relevant to the concepts of best interests, non-discrimination and equality. In respect of the child's 'right to know', it has been argued that the act of truth-telling may not necessarily always be in the best interests of the child, due to the psychological harms that could result.[4] Others have suggested that parental rights to privacy (especially where they might for example wish to avoid the stigma of infertility) must negate the child's 'right' to access genetic identity.[5] As a number of regions begin to unseal previously closed birth records and place legal bars upon donor anonymity, it may be argued that the 'putative authority of biology'[6] can no longer simply be dismissed in favour of protecting parental rights to privacy. As Bainham has argued,

---

[1] Bestard (2004), p. 255.

[2] Ibid p. 253.

[3] *Ibid* p. 255.

[4] See for example Braude et al. (1990), pp. 1410–1412; Snowden and Mitchell (1981); Rowland (1985), pp. 391–396; Nachtigall (1993), pp. 1846–1851.

[5] See for example Walker and Broderick (1999), pp. 38–44.

[6] Mykitiuk (2002), pp. 771–781 at p. 772.

A. Diver, *A Law of Blood-ties - The 'Right' to Access Genetic Ancestry*,

DOI 10.1007/978-3-319-01071-7_6,

> advances in technology have made it increasingly difficult for the law to justify an approach to establishing paternity which is characterised by fiction and presumption.[7]

Where however a right to identify genetic access ancestry is contingent upon birth status, birthplace, date of birth, parental consent (or the absence of a parental veto on disclosure) or perhaps subject to differing regional interpretations of the child welfare paramountcy principle, then it may be the case that institutionalised frameworks of discrimination still exist in respect of protecting adult-centric interests. The recent decision of the Court of Appeal for British Columbia in *Pratten* highlights the on-going difficulties associated with implementing a meaningful 'right' to genetic identity via knowable ancestry on the basis of arguing inequality of treatment.[8] In its discussion of whether donor anonymity might serve to perpetuate 'unlawful' discrimination against one section of the population (i.e. donor children) the Court provides a telling insight into the wider, underpinning aspects of such cases, namely whether there is any juridical, positive obligation upon states to let origin deprived persons know the details of their 'past.'[9]

Other jurisdictions offer a different approach to the issue. Following the abolition of donor anonymity in the United Kingdom, for example, a number of English decisions indicate an increasing acknowledgement of the importance of preserving an authentic genetic identity from the perspective of the child, arguably as a free-standing right that both arises and exists free from the expressed wishes of a parent and in the absence of any legal connection between the estranged donor and child.[10] The Irish High Court has also recently declared that blood-ties tend to create a significant connection that merits legal recognition, framing this as a Constitutional entitlement and effectively discarding the long-established doctrine of *mater cemper certa est* in the process.[11]

This chapter will argue that there is at least some degree of 'societal consensus'[12] in respect of the genetically relinquished, origin deprived child's need for identifiable ancestry, despite the fact that 'two distinctive orders of the social and

[7] Bainham (2008), pp. 322–351 at p. 324.

[8] *Pratten v British Columbia (Attorney General)* [2012] BCCA 480, per Frankel HJ, at para 46 (citing *R.v. Malmo-Levine*; *R. v. Caine*, SCC 74, (2003) 3 S.C.R. 571 at para. 113) available at http://www.canlii.org/en/bc/bcca/doc/2012/2012bcca480/2012bcca480.html (accessed 21.02.13) on the duty to remedy disadvantages created by state-sanctioned dissociation of adoptees from their original biological parents (at para 37). The judge found that adoptees were a group which had historically, if not perhaps currently, been subject to a 'negative social characterization' (para 38).

[9] *Ibid* para 7.

[10] See for example *Re G* [2013] EWHC 134 (Fam) on whether sperm donor fathers may apply for court-ordered contact with their children.

[11] *M.R & Anor-v-An tArd Chlaraitheoir & Ors* [2013] IEHC 91('*the mother is always certain*').

[12] See *Pratten op cit* n 8, para 46, on the concept of fundamental justice being underpinned by 'significant societal consensus that is fundamental to the way in which the legal system ought to fairly operate, and it must be identified with sufficient precision to yield a manageable standard against which to measure deprivations of life, liberty or security of the person' (citing *R. v. Malmo-Levine*; *R. v. Caine*, 2003 SCC 74 (CanLII), 2003 SCC 74, [2003] 3 S.C.R. 571 para. 113).

the biological' still seem to exist in relation to such concepts as family, kinship and legal relatedness.[13] In terms of avoiding discriminatory outcomes and embedding the principles of equality of treatment and human dignity, domestic law-makers may yet find that adult-centric vetoes on the release of identifying information tend to fall short of international law standards, not least in terms of promoting and protecting the best interests of the affected child and also potentially in terms of protecting the rights of their estranged relatives. As the European Court of Human Rights noted recently in respect of domestic decision-making processes, there will be situations where 'the interests of the biological parents must inevitably give way' to those of the child.[14] It was similarly declared that only in 'very exceptional circumstances' should original ties be severed, given that there was often a positive obligation on the part of domestic authorities to actively work towards the maintenance or 'rebuilding' of lost or broken family connections.[15]

Although the genetic tie between gamete donor and child tends generally not to be underpinned by legal relatedness (or a social acknowledgement of familial connections[16]) from the child's point of view, there may still be some degree of 'potentiality' for relatedness. A sense of origin deprivation can occur, especially where, as in *Pratten*, an absolute veto on paternal parentage has been enshrined in law and then supported by a policy of non-storage of information. Given the wide variety of 'fragmented ...processes' for reproduction and the diverse range of 'alternative family forms'[17] that have sought judicial oversight of domestic legislation, it may be argued that the 'multiple influences of assistive reproductive technology' have impacted significantly upon a number of 'iconic biological truths.'[18] Blood-ties continue to matter; the practice for example of donor 'matching' the characteristics of lesbian co-parents perhaps indicates a desire to create the impression of some degree of biogenetic linkage, as does the use of the same sperm donor in later insemination sessions to strengthen blood-ties between siblings.[19] As Freeman has observed however, 'arguments for secrecy rarely invoke the interests of the child.'[20]

---

[13] Franklin (2003), pp. 65–85 at p. 67.

[14] *R and H v United Kingdom* (2011) 35348/06 [2011] ECHR 844 para 77.

[15] *Y C v United Kingdom* (App 4547/10) ECHRR para 134.

[16] *Lebbink v The Netherlands* (2004) (app no 45582/99); on DNA testing see also *Mikulić v Croatia* (2002) (no 53176/99) 54 ECHR and *Human Rights Case Digest* (2002) (13) 1–2 pp. 55–95; *Gnahoré v France* [2000] ECHR 40031/98.

[17] Mykitiuk (2002) n 6 p. 772.

[18] Franklin (2006), pp. 547–555 at. p. 548.

[19] Jones (2007), p. 140.

[20] Freeman (1996), pp. 273–297 at p. 286.

6.2 Vetoed Ancestry

The analogous position of adoptees and donor gamete children has been much discussed,[21] especially in respect of why vetoes on information disclosure (whether they are enabled by donor anonymity or sealed birth records) are still deemed to be a legal necessity in some jurisdictions.[22] The 'complex emotional responses'[23] of parents have been examined and cited as a possible reason for maintaining bars on information disclosure. The argument that secrecy serves to protect triad members from social stigma (e.g. promiscuity, infertility, illegitimacy, 'adultery'[24]) has however to an extent been replaced by an acceptance that a considerable degree of 'dishonesty' tends to attach to the active concealment or falsifying of essential information on genetic or onomastic identity.[25]

Parental privacy rights often remain the most significant barrier to accessing genetic ancestry however, not least in terms of hiding parental infertility or relinquishing all ties of legal relatedness and obligations of parenthood.[26] Arguably, these are supported by the various norms of secrecy that have traditionally served to promote the 'social and psychological'[27] construct of socially-created families, especially in relation to gamete donation.[28] As Frith has noted, the 'desire to mimic the natural family' may also have featured as an important aspect of the parental decision to simply not tell children of their genetic background.[29] Also, because gamete donated children tend to 'blur into the general population in a way that adopted children do not'[30] it has been suggested that truth-telling might not always be in their best interests.[31] Bonney's work for example (on parental attitudes

---

[21] See for example Brandon and Warner (1977), pp. 335–341; Brewaeys (2001), pp. 38–46; Blyth (2002), p. 324; see also the lobbying activities of Canadian NGO *Parentfinders* which included the issue of donor anonymity; available at http://www.parl.gc.ca/35/Archives/committees352/srta/english_committee.html accessed 09.08.11.

[22] See further Jones (2007); Grace et al. (2008), pp. 301–314.

[23] Brandon and Warner (1977), n 21, p. 338.

[24] See Brandon and Warner (1977), on the 'social reservations' associated with the decision to not tell: 'in principle the child's unusual biological origin should be forgotten'; distaste over revealing intimate conception details; concern over promises of donor anonymity; mixture or pooling of donor sperm.

[25] Swanson (1993–1994), p. 179. See also Nachtigall et al. (1997), p. 83; Leiblum and Hamkin (1992), p. 267.

[26] See for example Walker and Broderick (1999), pp. 38–44.

[27] Shenfield (1997), pp. 371–372.

[28] Cahn (2000–2001), pp. 1–27 at p. 15.

[29] Frith (2001), pp. 818–824; see also Bestard (2004) n 1 on the philanthropic altruism of egg donors in Spain, where the right to complete maternal anonymity has been enshrined in domestic law.

[30] Brandon and Warner (1977) n 21, p. 337 on how 'apparent' legitimacy, 'technical adultery' and 'psychological infidelity' may be associated with gamete donation.

[31] See Snowden and Mitchell (1981); Rowland (1985), pp. 391–396; Nachtigall (1993), pp. 1846–1851; Braude et al. (1990), pp. 1410–1412.

to donor insemination in Germany) found that some gamete recipient parents would simply opt to never inform their children 'about the biological roots of their existence'[32] on the basis that 'the child might attach more importance to biological than social parenthood.'[33] As Cahn has argued however, the term 'genetic essentialism . . .(biogenetic fundamentalism) is not an accurate label'[34] for the psychological need to find or re-establish one's original biological connections. Equally, in respect of the 'genetic gap' that exists between social and biological relatedness, there is an 'assumption that genes make us who we are. . .[which] is both too true to ignore and too partial to be enough truth by itself.'[35] Parental honesty over the issue of genetic ancestry could help avoid the longer term, 'psycho-pathogenic' consequences of having the 'donor's ghost' remain as a 'mental-emotional presence,'[36] perhaps damaging familial bonds, or playing an unwelcome, 'unconscious and phantasmatic role.'[37]

Haimes' summary is as valid now as it was several decades ago: 'one of the paradoxes of the prescription to be open is that it comes from sources which promote institutional secrecy.'[38] Legal vetoes on accessing ancestry may be grounded in fairly blunt terms (the need or obligation to protect parental privacy, honour promises of confidentiality, prevent unwanted or unseemly inheritance claims, shielding children from the truth of their origins[39]) but it can still be argued that children so affected are being treated 'differently from children in traditional nuclear families'[40] and from those who had the good fortune to be born in jurisdictions that do not seal birth records or permit donor anonymity. Arguably, the worst aspects of 'closed' adoption practice are mirrored by donor anonymity. Adoption and 'artificial reproduction' both serve to enable the formation of family groupings for individuals who would not otherwise be so 'kinned with.'[41] Parents

[32] Bonney (2002), pp. 201–208 at p. 206; the situation was compounded by policies on destroying medical records.

[33] *Ibid*; see also Haimes (1988), pp. 46–61 on how psychological 'imbalances' can occur where only one parent is related to the child biologically.

[34] Cahn (2000–2001) n 28, p. 7.

[35] Franklin (2003) n 13, p. 83 noting that moral difficulties often tend to attach to scientific advances especially where modern reproductive technologies are concerned.

[36] Bonney (2002) n 32, p. 207.

[37] Grace (2008), pp. 257–262 at p. 261. Additionally, a social father could perhaps subconsciously perceive himself as having been 'cuckolded' by the process, with the norms of 'paternal function' becoming divided between himself and the invisible male genitor p. 258.

[38] Haimes (1988) n 33, p. 55.

[39] See further Turner and Coyle (2000), p. 2041.

[40] Cahn (2000–2001) n 28, p. 5.

[41] Haimes (1988) at n 33. The term Artificial Insemination by Donor ('AID') fell into disuse during the 1980's given its similarity to the acronym for Auto-Immunity Deficiency Syndrome ('AIDS'). The term Donor Insemination ('DI') fails to encompass the wide range of modern infertility treatment techniques (e.g. egg donation and surrogacy). The gentler term 'assisted birth' focuses less on conception and more on gestation and parturition; arguably this also hides or removes

gain children and children are permitted entry to a new family circle. This is a slightly misleading generalisation in relation to the child so conceived. Adopted children do have genetic relatives, as do donor-gamete children, in spite of laws and customs which serve to render such ties void.

This element of 'biological truth' may be completely overlooked by domestic judiciaries, or perhaps deemed to merit active maintenance or rehabilitation. As Haimes further noted, the wide range of problems that surrounds a lack of knowledge over genetic parentage will not necessarily find a solution within the changing templates for adoption practice that arose in the latter half of the twentieth century. Despite the new atmosphere of 'openness'

> 'uneasiness. . .still persists about adoption as an institution. Moves to make the practice of adoption more open have, in effect, disguised but not eliminated this uneasiness.'[42] Wider societal developments, such as the fast-changing concepts of family life, best interests and privacy rights further complicate the issue.

Judicial challenges to donor anonymity laws and policies based solely or mainly on the need to prevent psychological harm seem unlikely to succeed as long as there is a prevailing emphasis upon adult-centric interests such as privacy and the 'right to reproduce.'[43] The conflicting 'hierarchy of needs' that continues to attach to bans and vetoes on biological heritage has its roots in a long, adult-centric tradition of parental rights-paramountcy.[44] Thus if parents have opted to never tell their child about the gamete donation, there is little that can be done to compel disclosure. As long as this aspect of assisted reproduction remains a grey area, policy-sanctioned silences seem likely to continue this enablement of the removal of genetic identity.[45] That said, domestic and international courts do now generally seem to at least accept that delaying the discovery of truth in respect of adoption 'may not promote the child's welfare.'[46] This is not necessarily so in respect of genetic secrecy arising from gamete donation however, where the basic legal presumption seems to be that both of the social parents are 'related' to their child, despite the presence of donated genetic ancestry. Often no legal obligation attaches to the act of parental 'telling', with only the principles of 'good practice' gently stressing that there is a need for biological truth to eventually be forthcoming.[47] The 'transplant model'[48] (wholesale rights severance) of closed adoption practice arguably seems to have

biological genitor(s) from the reproductive process. In a bid to focus on the persons involved rather than the various processes, the phrase 'gamete donor children' will generally be used here.

[42] Haimes (1988) n 33, p. 46.

[43] Swanson (1993–1994) n 25, p. 181.

[44] Dennison (2007–2008), p. 25. Dennison further argues that forms of 'fertility tourism' might arise if donor anonymity were to be banned in the United States.

[45] See further Herring (2009) available http://www.newlawjournal.co.uk/nlj/content/family-revoking-adoptions accessed 10.08.12.

[46] Bainham (2008) n 7, p. 346.

[47] Ibid p. 347.

[48] Ibid p. 349.

been imported into many policies aimed at the regulation of assisted reproduction.[49] As such, greater recognition should be afforded to the unique vulnerabilities associated with non-ancestry. As Freeman has argued, 'identity as what we know and what we feel is an organising framework for holding together our past and our present and it provides some anticipated shape to future life.'[50]

6.3 Legal Loopholes

A gap may be said to exist between the recognition of children's rights in international law and the best interests-led duties of parents, wider society 'and science in particular.'[51] Blyth and Farrand have observed that the provisions of The UN Convention on the Rights of the Child ('the UNCRC') provide a potentially useful but ultimately ineffective framework for implementing meaningful change at national level. Implementational delays, together with high levels of impunity mean that there is often little pressure to endorse or embrace reform.[52] That said, some examples of useful guidance are emerging from case law involving issues of disclosure across various jurisdictions.

English jurisprudence is perhaps particularly relevant given the abolition of donor anonymity in that region.[53] That the biological reality of the blood-tie clearly matters is borne out by the reasoning underpinning cases such as *Re R* (2003).[54] Here it was decided that a key purpose of the Human Fertilisation and Embryology Act 1990 ('HFEA 1990') was 'to regulate treatment services so as to control the acquisition of paternity by those who were not genetically related to children'.[55] The common law presumption of legitimacy (*'pater est quem nuptiae demonstrant'*[56]) was significantly amended by the provisions of the HFEA 1990 (itself amended by the 2008 Act and the 2010 Regulations), linking legal paternity to the issue of parental consent to infertility treatment.

In *Re H & A (Children)* (2002)[57] the putative father of 5 year-old twin girls sought parental responsibility and contact, asking the Court to order paternity

[49] See further McWhinnie (1992), p. 37; see further Baran and Pannor (1989).

[50] Freeman (1996) n 20, p. 273.

[51] *Ibid* p. 290. See also O'Donovan (1988), p. 27.

[52] Blyth and Farrand (2004), pp. 89–104 at p. 101.

[53] See The FHEA (Disclosure of Donor Information) Regulations 2004 available at http://www.legislation.gov.uk/uksi/2004/1511/contents/made (accessed 21.02.13).

[54] *Re R [2003] EWCA Civ 182* Per VC Hale LJ.

[55] Lind (2003), pp. 327–340 at p. 333. In this case, the question of the child's long-term welfare did appear to influence the Court's thinking on whether to 'afford' legal paternity to the estranged (at the time of implantation) partner of the donor-inseminated mother.

[56] '*a child born to a married woman has been legally fathered by her husband and is therefore 'legitimate'*.

[57] *Re H & A (Children)* (2002) EWCA Civ 383.

testing. The mother was married to another man but had concealed her affair from her husband. Genetic connection was a central issue, with the mother's husband (under the legislation, the legal father) declaring his 'inability to continue as primary carer' and threatening to 'forsake' the family if it transpired that he was not the genetic father.[58] Previously, the lower court had referred to the importance of abiding by the 'best interests' principle and being mindful of Article 7 of the Children's Convention.[59] Having balanced the conflicting sets of rights, the court concluded that it was not in the girls' best interests for paternity tests to be ordered, given the potentially disastrous consequences for their family life. As was subsequently argued in the Court of Appeal, they stood to 'lose their psychological father and primary carer. . .a cruel and disastrous blow.'[60]

Ultimately Thorpe LJ decided that 'unpalatable truth can be easier to live with than uncertainty.'[61] The Court re-iterated that 'the incidents of being born a bastard are now almost non-existent'[62] so that there would be no justification in seeking to preserve the formerly rebuttable presumption of 'legitimacy' that attaches to births occurring within marriage. Accordingly, the case was referred back to the High Court, with particular emphasis placed upon the court's inherent discretionary power to order paternity testing (unless this would be contrary to the principle of the best interests of the child) and to the need for separate legal representation for the twins in future hearings.[63]

In *Re D* (*Paternity*) (2006)[64] the adolescent child, raised by his grandparents, declined the taking of blood tests to clarify his parentage, wanting nothing to do with the 'stranger' applicant who was claiming paternity. Hedley J noted that 'the general approach is that it is best for everyone for the truth about a disputed paternity to be known.'[65] Given the age and understanding of the child however, the Court of Appeal sought to have 'particular regard to [his] views'[66] in accordance

---

[58] *Ibid* para 9.

[59] Article 7 of the UNCRC states that 'All children have the right to a legally registered name, officially recognised by the government. Children have the right to a nationality (to belong to a country). Children also have the right to know and, as far as possible, to be cared for by their parents.'

[60] *Re H & A* (*Children*) (2002) *op cit* n 57 para 18; The lower court's aim of keeping the matter secret was called into question at the appeal stage, given that several people outside the family circle were aware of the protracted dispute, and had been invited to give evidence as to the applicant's relationship with the mother (para 20).

[61] *Ibid* para 27.

[62] *S v McC;W v W (1972) AC 24* (where married 'fathers' sought to rebut the legal presumption of paternity) per Hodson LJ at para 29. See also The Family Act 1987; The cases *Re H* (*A Minor*) (*Blood Tests*: *Parental Rights*) (1996) WLR 506 and *Re T* (*A Child*) (*DNA Tests*: *Paternity*) (2001) 3 FCR 577 were distinguished from the instant case on the basis of the high likelihood that the presumed legal fathers (husbands) were infertile or sterile.

[63] s 21 (3) (b) of The Family Law Act 1987.

[64] *Re D* (*Paternity*) (2006) EWHC 3545 (Fam).

[65] At para 22 Citing *Re H and A* (*Children*) (2002) EWCA Civ 383, (2002) 1 FLR 1145.

[66] *Ibid* para 24.

with the statutory welfare checklist. Although 'truth, at the end of the day, is easier to handle than fiction'[67] the child's vulnerable situation (exclusion from school, inclusion on the Child Protection Register for concerns over parental neglect) meant that the immediate ascertainment of genetic truth would not generally further his best interests. The Court did however order storage of a DNA sample so that the child might be able to discover his genetic identity later in life if he so wished.[68] Significantly, the Court stressed that the social workers involved in the case should keep 'these issues in mind in any on-going dealings'[69] that they might have with the child, to acknowledge his longer-term psychological interests; uncertain parentage was unlikely to be forgotten or 'wholly oppress[ed].'[70]

In *Re P* (*A Child*) (2008),[71] the putative father sought contact with his daughter, but only on the condition that they were genetically related. The child perceived this as an unsettling attempt by him to 'disestablish' himself from her, rather than as a bid to strengthen any ties that existed between them. Wilson J distinguished this case from *Re D* (2006) on the basis that emotional and behavioural problems in the earlier case created a 'complex profundity.'[72] Paternity testing there would have carried the risk of upsetting a 'fundamental assumption.. upon which a troubled child's life had been constructed.'[73] The key question in this subsequent case was whether the issue of the girl's welfare had entered 'the equation on a slightly different basis.'[74] Dismissing her mother's application for leave to appeal, the Court also rejected the 'make no order'[75] option with Wilson J noting that, without such a test being carried out, paternal contact questions could not be legally analysed or concluded upon, and the parties would be unable to proceed with their lives 'on a settled basis.'[76] Child welfare was discussed, briefly, with the Court suggesting the involvement of 'an independent professional person' as being beneficial.[77]

The age and understanding of the child in this case were similar to those of the child in *Re D*, but her wishes were effectively disregarded, in favour of establishing genetic truths. The desires of the parents seem to have been a key consideration, with the applicant seeking formalised contact *only* if he was biologically connected

---

[67] *Ibid* para 22.

[68] Under section 20(1) of the Family Law Reform 1969.

[69] *Ibid* para 33.

[70] Per Hedley J para 28.

[71] *Re P* (*A Child*) (2008) EWCA Civ 499.

[72] *Ibid* para 10.

[73] *Ibid.*

[74] *Ibid* para 12.

[75] Permitted under section 1 (5) of the Children Act 1989, where the making of no order would be preferable to actually making an order e.g. Specfic Steps, Contact, etc.

[76] *Ibid* para 14.

[77] *Ibid* para 13.

to the child.[78] Had the reasoning in *Re D* been followed, the welfare needs of the child might have carried more weight; a delay on testing might have been imposed, with the need to collect information on genetic identity perhaps framed as an element of the child's best interests rather than as a pre-requisite component for a contact application.

In *Re L (Identity of Birth Father)* (2008),[79] the Court of Appeal had to consider similar issues. Here the genetic father had not been in contact with his 15 year-old son since his infancy, with the child having grown up in the belief that his step-father was actually his biological father. Although a Contact Order was refused, the question of when exactly the child should be told the truth of his genetic parentage gave rise to some discussion. The child's mother wished to wait a year before revealing this information and did not wish to involve the child's own psychiatrist in the court proceedings. The psychological assessment ultimately led however to the mother being ordered to tell the child as a matter of urgency. Her appeal was not allowed, with the Court holding that the exclusion of previous psychiatric reports meant that judicial decisions on the child's welfare had been based on incomplete evidence. The question therefore had to be re-examined through referral back to the lower court.

The emphasis on the value of obtaining as much psychological welfare-based information as possible, at an early stage in the proceedings, underscores how blood-ties do matter, in family life, if not always in law. Their unwanted presence may threaten family stability while their absence may create uncertainties that the law alone cannot remedy, as reflected in the Court's calls for greater family and child support from social workers and psychologists. Equally, truth has an overwhelming tendency to emerge: in the former case a family friend subsequently announced to the child that the applicant was in fact his 'real father.' In the latter case, papers announcing the application for paternity testing were served minus an envelope through the mother's letterbox and found by the child. Such events illustrate the limitations of the law in trying to ensure that 'welfare-based' veils of secrecy remain permanently in place around the secrets of genetic parentage.

Such jurisprudence also perhaps represents a shift away from the reasoning of *Re F* (1993)[80] where a putative father sought a Contact Order in respect of his presumed biological daughter, whilst also applying to the Court for DNA testing. The Court of Appeal defined the child's welfare almost entirely in terms of her relationship with her mother and the need to protect 'the stability of the family unit'[81] to dismiss the father's claim. As Fortin has noted, Balcombe LJ's response to the child's psychological need for accurate information, and to the issue of not

---

[78] Arguably, his case was helped by the fact that intermittent, brief episodes of familial contact had occurred over the years, and perhaps because the child's mother also stressed that she 'had no doubt' over his paternity *Ibid* para 5.

[79] *Re L* (2008) EWCA Civ 1388.

[80] *Re F* (1993) 1 FLR 598.

[81] Para 604 per Balcombe LJ.

rendering her interests subordinate to those of her parents' wishes, was 'cursory'.[82] Under Hodson LJ's adoption analogy in *S v MC; W v W* (1972) the child's best interests may include a need for identity information but not necessarily the desire to establish some degree of relationship with the genetic parent.[83] As Lind has further argued, disputes over parentage ought not to be settled purely by reference to the contractual terms that exist between a parent and an infertility clinic or through highly convoluted methods of statutory interpretation.[84] Judges should pay greater heed to the child's best interests, and to the requirements of Article 8 of the European Convention with a view to ensuring that the best possible outcome is reached in terms of longer-term child welfare.[85] This would also avoid a return to a 'conceptualisation of paternity that is profoundly genetic and flies in the face of progress that has been made towards the institutionalisation of social parenthood.'[86]

Bainham has however questioned whether the incorporation of human rights principles into private proceedings has significantly modified judicial definitions of family life. The principle of child welfare paramountcy has been frequently re-interpreted, especially in the light of rapidly changing perceptions of what counts as parenthood.[87] He asks which 'values ought to govern the determination of parentage' and suggests that these should reflect 'commitment to truth, individual autonomy and priority for the rights and interests of those primarily affected.'[88] Policies based upon the 'suppression of the truth' might not comply fully with the best interests principle, especially where issues of long-term, psychological welfare arise[89] or 'legislative provisions have made assumptions and the biological facts do not fit the legal fictions.'[90] Again where the veto on disclosure rests with parents, and they are unable to comply with the court's direction or opposed to revealing genetic information, it is difficult to see how courts and social workers can seek to force a resolution that would fully promote the child's 'right' to genetic identity.

---

[82] Fortin (1994), pp. 296–307 at p. 297.

[83] Fortin (1994) at p. 299; the decision in *Re F* was at odds with that of *Re R* (1993) 2 FLR 762, where Butler-Sloss LJ directed a mother to tell her daughter the truth about her genetic father. If she refused to do so, a child psychiatrist was to be engaged to tell the child instead. (para 768).

[84] Lind (2003), pp. 327–340 at p. 333. Here, the question of the child's long-term welfare did appear to influence the Court's thinking on whether to 'afford' legal paternity to the estranged (at the time of implantation) partner of the donor-inseminated mother.

[85] *Ibid* p 330.

[86] *Ibid.*

[87] Bainham (2008), n 7, p. 323.

[88] *Ibid* p. 324.

[89] See for example *W and B v H (Child Abduction: Surrogacy)* (2002) 1 FLR 1008 where the Hague Convention was invoked by a Californian couple to seek the return of twin babies born to an English surrogate mother. See also Lowe et al. (2004), p. 61.

[90] Booth (2004), pp. 270–273 at p. 271.

6.4 Identity Information *'for its own sake'?*

In *Re C*: (*a child*) *v XYZ County Council* & *Anor* (2007)[91] the relinquishing mother of a newborn baby refused to inform the child's father (or her own family) of the birth. The lower court found that local authorities were under a duty to enquire whether 'any of the child's birth family would be suitable carers.'[92] Although only directed to disclose the child's existence to the father if he was identified, social workers made the decision to contact maternal grandparents to seek an interview with them, without explaining why. When they later learned about the child's existence they offered to 'assist in resolving the situation.'[93] At the Court of Appeal stage, Arden LJ stressed that prior to the enactment of the Adoption and Children Act 2002, domestic law did not require local authorities to make any attempt to inform fathers of impending proceedings, especially where the relationship had been fleeting.[94]

The important question seemed to be whether the new legislation created a positive duty of enquiry in respect of gaining paternal consent to adoption.[95] As the father lacked parental responsibility, no duty to seek his consent would arise. Even if he later gained parental responsibility via a court order, he would still be legally regarded as having provided consent to the relinquishment at the same time as the child's mother.[96] The legislation was described as 'child-centred', with the 'lifelong' aspects of adoption meriting fuller consideration.[97] The child's best interests should include 'interest in retaining… identity' given that this was likely to 'be important …in adulthood'. Identity was however only one factor in the 'balance that has to be struck'. The law should not 'privilege the birth family' over prospective adopters[98] and the child's best interests should dictate the nature of any enquiries by social workers. Background information on parentage ought not to be sought simply for its own sake, but to 'further the objective of finding long-term care'.[99] The Court's inherent discretion over child welfare remained extensive but

[91] *Ibid.*

[92] *Ibid* para 1 per Arden LJ.

[93] *Ibid* para 7.

[94] See s.1(4) (f) of the Adoption and Children Act 2002.

[95] *Re C* (*a child*) *v XYZ County Council* (*2007*) EWCA Civ 1206 para 2 per Arden J.

[96] *Ibid* para 9. See also s 52(9) and (10) of the Adoption and Children Act 2002 Under Reg 13 of the Family Procedure (Adoption) Rules 2005; the Court also has discretion over the issue of giving notice to a non-consenting parent and may dispense with consent if they cannot be found (under Regulation 27).

[97] *Re C* (*a child*) *v XYZ County Council* (*2007*) EWCA Civ 1206. See also s.52 Adoption and Children Act 2002.

[98] *Ibid* para 15.

[99] *Ibid* para 3.

the new 'adoption welfare' checklist demanded that placement delays be avoided where possible.[100]

In arriving at this conclusion, the Court of Appeal cited a number of authorities. In the case of *Re H*: *Re G* (2001)[101] the fact that the parties had not lived together was relevant, while in *Z County Council v R* (2001)[102] the mother's profound opposition to paternal involvement meant that the child's best interests precluded informing him of the proceedings. As Lawrence Collins LJ observed, whether a 'potential relationship' exists often depends upon totality of circumstance. Where birth parents were unmarried, the Court might look to the nature of their relationship, especially in respect of paternal commitment before and after birth.[103] Under Strasbourg jurisprudence, child rights were often grounded in 'potentiality; rather than fact'.[104] In lacking legal relatedness, the child also lost the right to establish family life ties with paternal ancestors. Genetic kinfolk are not necessarily 'relatives with whom the child has a relationship which goes beyond the mere status of relative'.[105]

Despite the views of the social workers, the Court stressed that it would not seek to effect a policy of 'natural family preference'[106] in attempting a kin placement of the child. Thorpe LJ referred to the French tradition of anonymous birth[107] with its maternal veto rights, stressing that 'valid social policy considerations' still attach to maintaining such practices. In such situations, it was unlikely that 'one lawful answer alone' would suffice, especially where the mother chose to 'treat ...pregnancy and birth as a private experience.'[108] Describing the situation in this case as 'extraordinary'[109] and the actions of the local authority as 'misjudged',[110] Thorpe LJ suggested that an application for direction would have been more appropriate than initiating Care proceedings, which in turn hindered the 'fast track' process which better reflected the mother's wishes on relinquishment.[111] It is perhaps significant that the court used its inherent discretion to order a fast-tracked adoption, rather than returning the matter to the lower court for

[100] The Court noted also that the four-month old child was '*starting to form bonds*' with her carers and should therefore '*start to live with a family*' (para 5). Arguably, this statement is suggestive of an implication that very young infants are either incapable of forming strong attachments or very resilient in terms of having early attachments broken and replaced.

[101] *Re H*: *Re G (2001)* 1 FLR 646 per Butler-Sloss LJ.

[102] *Z County Council v R (2001)* 1FLR 365 per Holman J.

[103] *Re C op cit* n 97 at para 51 per Lawrence Collins LJ.

[104] *Ibid* citing *Odievre v France* (2003) 1 FLR 621 (distinguishing *Marckx v Belgium* (1979) App no 6833/74).

[105] *Ibid* para 51.

[106] *Ibid* para 43.

[107] *Ibid* para 59.

[108] *Ibid* para 61.

[109] *Ibid* para 63.

[110] *Ibid* para 65.

[111] *Ibid* para 72. See also s 19 of the Adoption and Children Act 2002.

investigations aimed at a kinship care placement. This was done without factoring in any significant 'consideration of the mother's rights to privacy and to autonomous choices'.[112]

Arguably, an opportunity to clarify, or at least add to, the domestic jurisprudence on the welfare paramountcy rights of children was missed here. Although the relinquished child could perhaps refer subsequently to some limited information (gathered 'fortunately'[113] if erroneously by social workers) identification of paternal relatives was banned by the Court, presumably to reflect the mother's wish for privacy. As Douglas has commented, this decision seems to follow the reasoning in *Marckx v Belgium* (1979)[114] (that grandparents *may* have a juridical right to family life under Article 8) but still gives rise to some degree of legal absurdity, where genetic fathers can simply be kept in ignorance of their paternity;

> it must be open to doubt how, where all the relevant players are ignorant of the existence of the child, a grandparent can have a stronger right to know and to form a relationship with the child than the father.[115]

This case also illustrates how easily genetic links may be severed by law. A child's 'right' to know parentage may depend, even in a jurisdiction that allows access to original birth certificates in adulthood, upon the absence of a maternal veto. The statutory obligation to avoid undue delays could, arguably, have been met here by permitting adoptive placement, but allowing social services to gather identifying information. Permanently depriving a child of one half of their genetic ancestry to preserve maternal privacy seems at odds with the need to protect long term welfare. The Court's suggestion that the need to establish a sense of one's 'identity.. [was] only one factor'[116] in gauging best interests, clearly limits the rights of the child in respect of accessing ancestry.

Maternal vetoes were also discussed in the case of *Re L; X County Council v C* (2007).[117] Here the mother refused to provide any identifying information whatsoever on the genetic father. Munby J rejected the guardian ad litem's recommendation that an attempt should be made to obtain paternal consent to adoption and that the mother should be formally required to disclose his surname and last known address. The court accepted that the mother could offer no paternal information despite the fact that their relationship had lasted for several weeks. Understandably, they resisted any temptation to cross-examine her or place her in contempt in a bid to 'extract'[118] additional information, stating that such a strategy would not be appropriate or 'justified merely because we believe, however strongly, that what we

[112] *Ibid* para 81 per Thorpe LJ.

[113] *Ibid* para 79.

[114] *Marckx v Belgium* (1979) ECHR 2.

[115] Douglas (2008), p. 302.

[116] *Re C (a child) v XYZ County Council (2007)* EWCA Civ 1206 para 15.

[117] *Re L* (2007) EWHC 1771 (Fam) (20 July 2007).

[118] *Re C (a child) v XYZ County Council (2007)* EWCA Civ 1206 para 38.

are doing is being done in the best interests of a child.'[119] Holman J's dicta from *Z County Council v R* (2001) was cited, stressing that adoption

> exists to serve many social needs..high among them has been, historically, the desire or need of some mothers to be able to conceal from their own family and friends, the fact of the pregnancy and birth.[120]

The psychological welfare needs of the child were not discussed in any great depth, although the court did acknowledge the importance of being able to trace one's biological origins, declaring that the child should at least have had,

> the emotional security of knowing that all this was investigated, along with any relevant medical, ethnic, and financial/inheritance matters, before the final decision for her adoption was made. Her mother's stance is denying her all these things.[121]

In cases such as these, the concept of 'legal truth'[122] as a means to determining parentage or deciding its parameters, seems to exist as a sort of weighted legal fiction. As Booth has argued, there will always be a 'need to both trace one's antecedents in satisfaction of (the) desire to know where one came from, and find out if one has inherited any diseases.'[123] When faced with overturning parental vetoes however, courts seem to prefer protecting privacy interests, perhaps citing these as 'good enough' reasons for side-lining the child's need for identity and the notion of the 'right to know.'[124] Arguably, in refusing to recognize 'potential relationship(s) between the child and her biological forebears'[125] the law is perhaps seeking to delegate or relinquish its own obligations towards a vulnerable section of the population.

Recently, in *Re G* [2013], the significance of genetic connection was discussed by the English High Court.[126] It was held that, in the wake of legislative change,[127] the sperm donor father was 'as a matter of law...[a] stranger' to the children, but that here, their connection 'was linked in some way to their biological relationship.'[128] Although 'biological kinship ...alone [would] be insufficient to attract the protections of Article 8' of the European Convention, their relationship was likely

---

[119] Para 40.

[120] *Z County Council v R* (2001) 1 FLR 365 at p. 367.

[121] *Re C op cit* at n 97 para 30.

[122] See further Eekelaar (2006).

[123] Booth (2004), pp. 270–273 at p. 272.

[124] See for example Snowden and Mitchell (1981); Cook et al. (1995), pp. 549–559.

[125] Cahn (2000–2001) n 28, p. 4.

[126] *Re G* [2013] EWHC 134 (Fam) Here, a sperm donor father (who had originally been friends with the child's lesbian parents) sought the leave of the Court to make an application for contact with the child.

[127] See The Human Fertilisation and Embryology Act 2008, ss 42, 45, 48. Where birth mothers are in civil partnerships at the time of the child's conception, both they and their civil partner will be named on the child's birth certificate as legal parents. The genetic father will have no legal recognition (or rights) as parent.

[128] *Re G op cit* n 126 at para 116 per Baker LJ.

to fall within the remit of either family or private life.[129] The contact visits were also very significant as was the issue of the child's need for a sense of racial identity and cultural heritage. In terms of acknowledging that blood-ties do matter, it is perhaps significant that the court noted how 'the potential importance of genetic and psychological parenthood is not automatically extinguished by the removal of the status of legal parenthood.'[130]

Skinner has observed that the decision is significant because the court had never before been tasked with considering whether known sperm donors might be afforded leave to apply for a contact order in respect of their genetic child where the legal parents were 'only offering limited identity contact.' The court also had to examine the public policies and legislative thinking (equality for non-heterosexual parents) that underpinned the 2008 statutory changes.[131] Known donors can now seek leave to apply for contact (if not also for a residence order under section 8 of the Children Act 1989, subject to the best interests of the child principle) despite the fact that their parentage is not legally recognised. He advises parents considering sperm donation or co-parenting to seek agreement from the outset as to what role they will play in the child's life as the 'best way to avoid problems later on'[132] but offers no direction in terms of how an affected child might best seek to establish contact with an absent genetic parent, or obtain identifying information where such an initially open agreement might have broken down.

6.5 Surrogacy

A number of recent UK cases on surrogacy[133] have added to the jurisprudence on genetic and legal relatedness, and, to an extent, amended or expanded upon the legal definitions of parenthood. In *A v P* [2011][134] for example the husband (the child's genetic father) passed away before the hearing of the Parental Order.[135] Having used a gestational surrogate mother from India and mixed ova (some from Mrs A and some from an anonymous egg donor) it was unclear whether Mrs A was

[129] *Ibid* para 120.

[130] *Ibid* para 119.

[131] Skinner (2013), p. 60.

[132] *Re G op cit* n 126.

[133] In gestational surrogacy the surrogate has no genetic connection to the child, the eggs or embryos having come either from a donor or from the commissioning parent or couple; 'traditional' surrogacy generally involved genetic connection between the surrogate mother and the baby.

[134] A v P [2011] EWHC 1738 (Fam).

[135] Under s. 54 of the Human Fertilisation and Embryology Act 2008 ("HFEA 2008") the Parental Order transfers legal parenthood from the surrogate mother to the 'commissioning couple'. (See also the Human Fertilisation and Embryology Act 2008 (Parental Orders) (Consequential, Transitional and Saving Provision) Order 2010 which incorporates the 'best interests' requirement of s1 Adoption and Children Act 2002.)

the genetic mother of the child. The High Court (for England and Wales) distinguished this case from an earlier precedent (which had seen the applicant husband's cause extinguished by his death)[136] on the grounds that an application for a parental order was

> essentially declaratory in nature and confers a fundamental status on an applicant and on the child, who is the subject matter of the application. The relevant welfare considerations all relate to the child.[137]

In an earlier case the court had stressed that (under the 2010 Regulations) the welfare of the child was no longer simply the Court's first consideration but had become the paramount one.[138] Looking to the case law on adoption for guidance, it was argued that, subject to the child's interests being met, an adoption order could be made in favour of both parents even where they might now be living separately from one another. In *Re WM* (*Adoption*; *Non-Patrial*) (1997)[139] for example, the court had considered whether the making of the order sought created advantages for the child in terms of becoming a member of the family 'from an emotional and social perspective' and financially.[140] Given the common intention of the applicant parties (commissioning parents and surrogates) as to where parenthood should ultimately vest, and the fact that the child was genetically descended from the deceased applicant there would be advantages to the child in having an order made which would mirror 'the factual circumstances.......that he was the child of his parents as a matter fact and as a matter of law.'[141]

Article 8 of the European Convention also gave rise to a positive state obligation to recognise such de facto relationships.[142] It was further argued that parliament's intentions had clearly not been that 'a child should be denied the legal and social benefits which flow from having his relationship with both parents recognised by law.'[143] Equally, the applicant submitted that the effect of s.54 was such that it would have a 'transformative effect on the legal relationship' between applicants and child, not least in relation to the right to be afforded respect for family life under Article 8. Because the child had lived with both of them and was biologically related to the late Mr A (and possibly also to Mrs A) it would mean that failure to make the parental order would amount to an interference with their family life on the basis that the reality of their relationship would remain unrecognised by law.

[136] D (J) v D(S) [1973] 1 All ER 349 (an application under s.17 of the Matrimonial Causes Act 1965 to vary post nuptial property disposition).

[137] Per Theis LJ DBE, para 12.

[138] *Re L* (*Commercial Surrogacy*) [2011] 1 FLR 1423 per Hedley J at para 9.

[139] *Re WM* (*Adoption*; *Non-Patrial*) [1997] 1 FLR 132 per Johnson J.

[140] *A v C op cit* n 134 at para 20 (under the Matrimonial Causes Act 1973 and The Inheritance (Provision for Family and Dependants) Act 1975.

[141] *Ibid* para 21.

[142] *Marckx v Belgium* (1979) 2 EHRR 330.

[143] Para 23. It was noted that single applicants (and persons under 18 years of age) were precluded by the legislation from commissioning a surrogate.

Given that family life was comprised of 'not only a matter of fact and degree but also the significance of legal relationships'[144] a legal connection between the child and his genetic father should be made by granting the order. Similarly, it was noted that, under *Pini v Romania* [2005] 'protection of the right to family life pre-suppose[d] the factual existence of family life'[145] at which point the State must actively 'facilitate and protect that right.'[146]

The court considered the consequences of not making the order; there would be no legal relationship between the child and his genetic father (who was also the commissioning father) and, significantly, the child would be denied the social and emotional benefits of having that relationship legally recognised. It was important to have a 'legal reality which matches the day to day reality.'[147] In terms of Article 8 of the Children's Convention[148] (which protects the child's right to identity) it was submitted that the notion of identity encompassed legal acknowledgement of the relationships between children and parents. The Supreme Court had previously declared that decision-makers and jurists[149] 'had to have regard to the key principles' of the Children's Convention in terms of compliance with the rights contained in Article 8 of the European Convention and 'in its application to decisions by authorities in this jurisdiction.' The court was perhaps bound by these articles to interpret s 54 (4) so as to allow the child's identity to remain legally linked to his genetic father (and become legally linked to his possibly genetic, but clearly psychological, mother).

In terms of finding an alternative solution, a declaration of parentage under s 55A of the Family Law Act 1986 would have been fundamentally different to the Parental Order, as it would not have established the father as legal parent. 'By its nature a declaration can only declare or state the current legal position, it can't transform the legal position.'[150] The husband of the gestational surrogate mother would therefore have remained the child's legal father. Granting an adoption order instead might have provided a remedy to the breach of Article 8 rights which would

[144] Para 25.

[145] *Pini v Romania* [2005] 2 FLR 596 (at para 143).

[146] Para 25.

[147] Para 26.

[148] Article 8 (1) 'States Parties undertake to respect the right of the child to preserve his or her identity, including nationality, name and family relations as recognised by law without lawful interference.

(2) Where a child is illegally deprived of some or all of the elements of his or her identity, States Parties shall provide appropriate assistance and protections, with a view to re-establishing speedily his or her identity.'

[149] In *ZH (Tanzania) v Secretary of State for the Home Department UKSC 2011 4per Baroness Hale paras 22–25.*

[150] Para 30 (3). In *M v W (declaration of parentage) [2007] 2 FLR 270* it was noted that such a declaration as to genetic parentage would have no affect on an adoption order; it was 'a forever order.' per Hogg J at para 18. Also considered were residence or special guardianship orders but neither of these temporary orders would have been effective to extinguish the legal relationship between the child and the surrogate (legal) parents.

occur if the court did not make the parental order (given its similarly 'transformative effect') but a joint adoption was not possible given the father's death. In granting the parental order, the court appeared to accept the argument that the key issue was one of identity rights and recognition of a relationship 'central to the child'[151] even though the child was clearly too young to remember his father.[152] Most significantly it was observed that although such circumstances were rare, they did throw into sharp relief how 'international' (i.e. foreign) surrogacy arrangements can lead to legal difficulties.[153] As such, parental orders should be sought 'to secure the legal status of the child.'[154]

The domestic legislation reflects the social discomfort that attaches to two aspects of surrogacy processes: non-altruism and single parenthood.[155] Payments by a commissioning couple that go beyond the concept of reasonable expenses suggest exploitation of vulnerable women and an unseemly commercialisation of a process that should be underpinned by a selfless desire to help the childless overcome their infertility or expand their family. That single women were excluded from parliament's remit is also potentially discriminatory and arguably reminiscent of the bias that used to attach to unmarried women seeking to parent a child alone rather than relinquish their baby for the apparent good of all concerned.

In *Re L* (2010)[156] the High Court had to consider whether a commercial surrogacy agreement made in Illinois, United States (where there was no statutory limit in respect of reasonable expenses) might be retrospectively approved as having been within the limit of reasonable expense so as to enable the making of a parental order under s. 54 (8) of the 2008 Act. Referring to *Re S* (*Parental Order*) [2009][157] Hedley J found that 'the policy principles' laid down in that case were still good, and were satisfied here; it was noted however that the notion of 'reasonable expenses' remained 'a somewhat opaque concept.'[158] The use of terms such as 'compensation' suggested that payments went beyond those of reasonable expense. Equally, commissioning parents were prone to receiving inaccurate information, and to being under-informed in terms of immigration advice and re-entering the United Kingdom with the child.

---

[151] Para 30 (8).

[152] In terms of the welfare of the child being the court's paramount consideration, it was noted that the Parental Order Reporter had visited the child and made favourable comments.

[153] *Re X & Y* (*Foreign Surrogacy*) [2009] 1 FLR 733; *Re IJ* (*A Child*) *[2011] EWHC 921.*

[154] Para 37.

[155] The court can however retrospectively sanction any payments that have been made to surrogates that go beyond the reasonable expenses permitted by the statute. Although single applicants are not supposed to be able to commission the services of a surrogate mother, the court did, in the best interests of the child, allow this to occur here, given also that the commissioning mother had been widowed mid-process.

[156] *Re L* [2010] EWHC 3146 (Fam).

[157] *Re S* (*Parental Order*) [2009] EWHC 2977 (Fam).

[158] Ibid para 7.

The main change in the law attached to the child's welfare however; it was no longer merely the 'first consideration' but had become the paramount concern.[159] The balance between public policy considerations and welfare would fall 'decisively in favour of welfare.' Significantly, it was also suggested that if a policy of 'controlling surrogacy arrangements' was to be implemented, then such limitations needed to be effected prior to the cases reaching the courts.[160] That said, it was important for there to be judicial oversight of such applications 'with a view to policing the public policy matters.'[161]

In *Re X and Y* (*Foreign Surrogacy*) [2008][162] the High Court sounded a note of caution in respect of foreign surrogacy agreements. Here the commissioning couple had faced 'enormous delay, stress and expense: the path to parenthood has been less a journey along a primrose path, more a trek through a thorn forest.'[163] The practice of surrogacy was 'ethically controversial' with radically differing approaches being taken depending upon jurisdiction. In some countries (e.g. Italy, Germany, Turkey) it was completely barred while in others commercial contracts were legally and socially sanctioned.[164] The United Kingdom adopted a sort of hybrid approach with paid-for surrogacy being illegal, but the altruism-led model tolerated. Here the applicants went to the Ukraine for gestational surrogacy (anonymous donor eggs fertilised by the applicant father's sperm). Under English law[165] the surrogate and her husband were the legal parents of the twins. It was suggested that their consent would be needed to enable the making of the parental order while the genetic father would require the leave of the court to make relevant applications (e.g. a residence order) under the Children Act 1989. The applicants argued that they were entitled however to apply for a residence order by virtue of the children having been treated as children 'of the family.' Hedley J expressed

> serious doubts as to whether children who are in law 'strangers' can simply become 'children of the family' because for a short time the parents have so acted.[166]

Ukrainian law did not however regard the twins as the legal children of the surrogate mother and her husband. Because of this, they were freed from any duties towards the children and 'had neither the status nor the rights and duties of parents.'[167] Their domestic law would probably have required them to surrender the children to honour the contract for surrogacy, although the children had gained

---

[159] The 2010 HFEA Regulations (S1 2010/986) brought S.54 applications within the remit of s.1 of the Adoption & Children Act 2002 (the 'welfare checklist').

[160] Ibid para 10 'at the border or even before.'

[161] Ibid para 12.

[162] *Re X and Y* (*Foreign Surrogacy*) [2008] EWHC 3030 (Fam).

[163] Per Hedley LJ para 2.

[164] Para 3 (e.g. India, Ukraine and in California, US).

[165] Human Fertilisation and Embryology Act 1990 s 27–30.

[166] Para 7. Further discussion of this point was not forthcoming; such a case had yet to present itself before the court.

[167] Ibid.

no Ukrainian residency or citizenship rights, and would be afforded the protection of accommodation only as an 'act of basic humanity in a state orphanage.'[168] Because the applicant parents' visas were temporary in nature with no long-term right to remain, reside or work, the children had essentially been both orphanised and rendered stateless. Calling citizenship "the right to have rights"[169] it seemed that the twins had been either deprived of their parents by English domestic law or assigned to 'at best, a putative father with no parental responsibility.' Ukrainian law deprived the surrogate parents of all of their rights and duties and removed their legal status as parents.

The children had no free-standing right of entry to the United Kingdom because nationality would not pass to them via the applicants; the applicants similarly had no right to bring the children into the country, although the putative father could apply for leave to do so. DNA tests were needed for immigration purposes, causing delays, anxiety and expense; the parental order did not engender citizenship per se, so that[170] the strict, statutory 6-month time limit on bringing a parental order application could cause problems if immigration difficulties arose. The issue of surrogate mother consent was referred to, with the court noting that there was no power to dispense with withheld consent as in adoption cases.[171] This was so even where, as occurred here, a parent had no legal responsibility for the child and 'however much the welfare of the child [was] prejudiced by such refusal' of consent. An 'absolute veto' thus attached to the birth parent, even where there might be no genetic connection.[172]

In terms of deciding whether the 1990 Act applied abroad to British citizens, it was suggested by the applicants that it did not necessarily 'apply to those of foreign domicile.' They further argued that treating the surrogate's husband as legal father would give rise to several consequences; future surrogacy agreements would most likely involve only unmarried women who might be younger, less experienced, more vulnerable to exploitation or perhaps driven more by financial gain than altruism.[173] It was stressed however that Parliament had clearly intended to 'recognise the particular relevance of marriage in surrogacy arrangements'[174] and that this had no bearing upon issues of domicile. In relation to the money

[168] Para 8.

[169] Para 9.

[170] Para 10.

[171] Para 13. Section 30(5) (of the 1990 Act) provided that "The Court must be satisfied that both the father of the child (including a person who is the father by virtue of Section 28 of this Act), where he is not the husband, and the woman who carried the child have freely, and with full understanding of what is involved, agreed unconditionally to the making of the order."

[172] Ibid.

[173] Para 15 this was 'a fair analysis of fact but whether it assists in statutory construction is more doubtful.' In relation to legal fathers, it was further argued that they might be tempted to seek more money in return for providing consent.

[174] Para 16.

received by the surrogate mother (over €25,000, used to purchase a property) the court noted that the sums involved clearly amounted to more than basic 'expenses reasonably incurred' and therefore contravened English domestic law.[175]

The court opted however to follow the decision of Wall LJ in *RE C* [2002][176] which had posed two key questions. Firstly, the court should ask factually whether the payment was actually "expenses reasonably incurred"; if this was not the case, it should be asked whether the payments could or should be retrospectively authorised. Although the legislation was underpinned by a policy that surrogacy motivated by a desire for commercial gain should be illegal, there was 'a recognition that sometimes there may be reasons to do so.'[177] Ultimately it was decided that 'the welfare of these children require[d] that they be regarded as lifelong members of the applicants' family.' In doing so the court had to however balance 'two competing and potentially irreconcilably conflicting concepts.'[178] Perhaps most significantly it was stressed that the effects of the

> policy consideration ... bear on one wholly unequipped to comprehend it let alone deal with its consequences (i.e. the child concerned) ...that rigour must be mitigated by the application of a consideration of that child's welfare. That approach is both humane and intellectually coherent.[179]

Hedley J also observed that the number of infants available for adoption had fallen and that given the abolition of donor anonymity (adding that this is 'wholly justifiable, of course, from the child's perspective') couples will probably be increasingly inclined to seek out commercial surrogacy arrangements in jurisdictions which permit it.[180] Such issues thus required 'the widest public debate even if the ultimate conclusion be that they continue to be finally resolved on a case by case basis by judicial decision.'[181]

---

[175] Para 19; Section 30(7) of the 1990 Act provided that:

"*The court must be satisfied that no money or other benefit (other than for expenses reasonably incurred) has been given or received by the husband or wife for or in consideration of –*

(a) *the making of the order,*

(b) *any agreement required by subsection (5) above,*

(c) *the handing over of the child to the husband and the wife, or*

(d) *The making of any arrangements with a view to the making of the order, unless authorised by the court.*"

[176] *RE C* (Application by Mr. and Mrs. X under Section 30 of the Human Fertilisation and Embryology Act 1990) [2002] 1FLR 909. (following the decision in Re Adoption Application (Payment for Adoption) [1987] 3WLR 31 which was also endorsed n RE Q (Parental Order) [1996] 1FLR 369).

[177] *Re X and Y (Foreign Surrogacy)* [2008] EWHC 3030 (Fam) *op cit* n 162 para 21.

[178] Ibid para 23.

[179] Ibid para 24.

[180] Ibid para 26.

[181] Ibid para 29.

A recent Irish High Court case is particularly significant insofar as the court looked to the genetic realities underpinning the case to overturn legal parentage.[182] In *M.R & Anor-v-An tArd Chlaraitheoir & Ors* (2013) the gestational surrogate mother registered as the legal mother in accordance with the domestic legislation. The child's genetic mother (whose ova had been used in the conception and who had since their birth been raising the twins with her husband, their genetic and legal father) applied to have herself registered as the legal mother but this was refused by the Superintendent Registrar for Dublin. Despite the provision of DNA evidence, and justified on the basis of the long-standing principle of *mater semper certa est* ('the mother is always certain').[183] The court considered detailed evidence as to the effects of nature (DNA) versus nurture *in utero*,[184] and noted the likely legal consequences of denying the genetic mother any legal recognition; 'significant' succession and taxation issues for example might arise as well as matters falling within the remit of parental responsibility, such as medical treatment.[185] As the applicants argued,

> for over 100 years the courts in this jurisdiction have recognised the importance and the existence of what has been termed....a "blood link" between children and parents.[186]

Referring to the controversial 'Baby Anne' case,[187] they cited the 'natural bond' between child and parent as amounting to a 'primordial constitutional principle' which reflected the law of nature and which saw the child's welfare as best served 'in the heart of its natural family.'[188] Much Irish case law recognised the value of the blood tie,[189] which also tended to be 'clothed' in Constitutional protections.[190] The applicants made an interesting submission in terms of differentiating between the gestational and genetic forms of motherhood, suggesting that a gestational mother

> would, to some degree, protect a child but the natural relationship which exists between a mother and a child is the territory of blood link. This blood link comes from what is described as pure genetics and there cannot be a gestational element in that.[191]

---

[182] *M.R & Anor-v-An tArd Chlaraitheoir & Ors* [2013] IEHC 91.

[183] See s. 63 of the Civil Registration Act (2004) (Ireland).

[184] See in particular paras 7–12.

[185] Para 34. It was also noted that they would be legally entitled to marry their genetic siblings, as cousins. Para 46 (5).

[186] Para 35.

[187] *N v Health Service Executive* [2006] 4 IR 372.

[188] Ibid para 37.

[189] See for example *I O'T v B and Others* [1998] 2 IR 321 (noting how the right to know one's mother still had to be balanced against her right to privacy post-relinquishment); *JPD v MG* [1991] 1 IR 47; *J K v V W and Others* [1992] 2 IR 437; *J McD v PL* [2010] 2 IR 199 per Fennelly J para 304.

[190] Para 46 (1); namely Articles 41 and 42 of the Irish Constitution.

[191] Para 39.

In respect of there being an Irish 'law of blood-ties' it seems that the case law was indeed

> replete with references to the blood link between not just mothers but fathers and their children. . ..Irish case law as a proposition accepts the principle that there is a natural blood bond or blood link between parents and children. Furthermore, genetic links give rise to instinctive understandings between the genetic parents and their children.[192]

In terms of there being a fundamental right to access accurate ancestry, arising by virtue of the blood-tie per se, the court seemed to accept that,

> the right to know the identity of one's natural mother is a basic right flowing from the natural and special relationship which exists between a mother and her child, which relationship is clearly acknowledged.[193]

Equally, it was noted that the Supreme Court had previously held that 'there was an unenumerated constitutional right to know the identity of one's mother' although this ultimately had to be 'balanced in the end against the privacy rights of the mother who had placed or had given the child to the adoption society.' As the gestational mother in this case was the sister of the genetic mother, and no anonymous egg donor was involved, the precedent is perhaps likely to be of limited use in acrimonious situations or commercially underpinned arrangements, where privacy rights, or disputes over parenthood and parentage may well feature as the main question to be determined. That said, as Barron J further observed in *I.O'T v. B and Others* [1998] there was often a need to 'keep the door open' based upon genetics given that 'help from a member of the cognate family might be essential in certain diseases. In *de facto* adoption that is equally important.'[194] The Supreme Court had also previously recognised the nature of the biological relationship between father and child in *J.K v. V.W. and Others* [1992] by holding that

> The blood link between the infant and the father and the possibility for the infant to have the benefit of the guardianship by and the society of its father is one of many factors which may be viewed by the court at relevant to its welfare.[195]

Child welfare was not quite the paramount consideration, but the genetic link was clearly an important factor to be taken into account 'in determining where the custody and where the welfare of a child is to be found.' Granting the application, the court noted that there was also a duty to avoid creating a discriminatory, gendered distinction, whereby biological fathers might be legally recognised solely on the basis of genetic linkss without having to demonstrate any commitment or gestational input.[196] The case clearly seems to have removed the presumption of *mater semper certa est* from Irish law; the law on surrogate maternal parentage

---

[192] Para 39 ibid citing Kenny J in *G v. An Bord Uchtála* [1980] IR 32 at p. 98.

[193] Para 40 citing Hamilton C.J. in *I.O'T v. B and Others* [1998] 2 IR 321 p. 348.

[194] Ibid p. 381.

[195] *J.K v. V.W. and Others* [1992] 2 IR 437 per Finlay C.J. at p. 447.

[196] Para 46(6).

arguably now differs from that in England and Wales given that a Parental Order (underpinned by the child's best interests welfare requirement) would have been needed there to achieve the full 'transfer' of legal parentage to the commissioning mother. Here, the legal status of mother seems to have instead arisen quite naturally by virtue of the blood-tie between egg donor and child. This contrasts very sharply with the stance taken in a recent Canadian case which ostensibly found little if anything in the genetic connection to merit any form of legal protection or to generate a basic entitlement to equality of treatment.

6.6 Fundamental Justice? The *Pratten* Decision

In *Pratten v British Columbia (Attorney General)* [2012][197] the court had to determine whether the provincial legislation which provided a supported means for adoptees to trace their genetic parents was discriminatory (against donor children) and therefore in violation of the Canadian Charter of Rights and Freedoms (and of Part I of the Constitution Act 1982[198]). No equivalent rights provision existed in respect of any donor offspring who might seek information on the identity of their biological father. The lower court had found that a violation of section 15.1 of the Charter had occurred[199] (although none existed in respect of s7) via the information disclosure provisions of the Adoption Act 1996.[200]

---

[197] *Pratten op cit* n 8.

[198] Schedule B, Canada Act 1982 (U.K.) 1982, c. 1s.1. S. 15(1) of the Canadian Charter protects equality rights; it was further argued that a free-standing positive right existed under s. 7 of the Charter in relation to life, liberty and security of the person. See S.1 of The Canadian Charter of Rights and Freedoms which '*guarantees the rights and freedoms set out in it subject only to such reasonable limits prescribed by law as can be demonstrably justified in a free and democratic society.*'

S.7 states that '*Everyone has the right to life, liberty and security of the person and the right not to be deprived thereof except in accordance with the principles of fundamental justice.*'...

s. 15(1) states that '*Every individual is equal before and under the law and has the right to the equal protection and equal benefit of the law without discrimination and, in particular, without discrimination based on race, national or ethnic origin, colour, religion, sex, age or mental or physical disability.*'S.15 (2) qualifies this to the extent that '*Subsection (1) does not preclude any law, program or activity that has as its object the amelioration of conditions of disadvantaged individuals or groups including those that are disadvantaged because of race, national or ethnic origin, colour, religion, sex, age or mental or physical disability.*'

[199] *Pratten v British Columbia (Attorney General)* (2011) BCSC 656 (CanLII), (2011 BCSC 656, 22 B.C.L.R. (5th) 307).

[200] See s.56, s.63, s.65, s.66 (subject to absence of a veto, s.20) S.19 which provides for a post-adoption openness registry while s.22 established a passive reunion registry. S.68 also allows for post-adoption contact and information sharing where 'compelling circumstances' exist which might affect 'anyone's health and safety.'; ss. 71(3) and (4) allow a child or grandchild of an adoptee to apply for assistance in locating birth relatives while ss. 71(5) and (6) allow a birth sibling of an adoptee to apply for assistance in locating the adoptee.

It was noted there that 'donor offspring fear that their health can be compromised, and may be seriously compromised, by the lack of information about their donor' and that it was 'important, psychologically and medically' to access identifying and non-identifying information, given that the psychological and medical needs of donor children were 'substantially the same' as those of adopted persons.[201] Such information was necessary to 'complete their personal identities and to alleviate the stress, anxiety and frustration caused by not knowing.'[202] In particular,

> donor offspring experience sadness, frustration, depression and anxiety – in other words, they suffer psychological and psychosocial difficulties – when they are unable to obtain information. They feel the effects both for themselves and, when they become parents, for their own children.[203]

Despite acknowledging that parents had an 'important and legitimate interest' in decision-making on the timing and level of detail given to their children, the legal and social devices of anonymity and secrecy served to tilt 'the balance heavily in favour of donors and parents, and away from the best interests of donor offspring.'[204] In terms of discriminating between adoptees and donor children, the manner of their conception gave rise to a difference in treatment that in turn caused social disadvantage 'by perpetuating prejudice or stereotyping.'[205] Such a distinction was not a lawful mode of affirmative action aimed at redressing inequality.

The s.7 'positive rights' argument[206] did not however apply here, given the potentially profound implications and serious consequences of 'such a free-standing constitutional right to know one's biological origins.' The court then looked to Ontario (which has a similar system of qualified disclosure to adoptees, based on veto-absence) endorsing its rejection of the concept of a constitutionally-protected right to "know one's past."[207] Although the court found that the impugned provisions violated the Canadian Charter, no direction or suggestion was made as to how any new statutory provision might best be drafted, leaving the matter entirely to the legislators. The Attorney General challenged the declaration of invalidity, arguing that the lower court was mistaken.[208] The Appeal Court

---

[201] Pratten *op cit* n 8 para 17.

[202] Ibid.

[203] Ibid.

[204] Ibid.

[205] Ibid citing the test set out in *R. v. Oakes*, 1986 CanLII 46 (SCC), [1986] 1 S.C.R. 103, the infringement of s. 15(1) was not justified under s. 1 of the *Charter*.

[206] Deciding not to follow the dissent of Arbour MJ in *Gosselin v. Québec (Attorney General)*, 2002 SCC 84 (CanLII), 2002 SCC 84, [2002] 4 S.C.R. 429.

[207] See *Marchand v. Ontario*2006 CanLII 19946 (ON SC), (2006), 81 O.R. (3d) 172 (S.C.J.), aff'd 2007 ONCA 787 (CanLII), 2007 ONCA 787, 88 O.R. (3d) 600, leave to appeal refused [2008] S.C. C.A. No. 37 and *Cheskes v. Ontario (Attorney General)* 2007 CanLII 3837 (ON SC), (2007), 87 O.R. (3d) 581, 288 D.L.R. (4th) 449 (S.C.J.), at paras. 115–116.

[208] Para 27 On the basis that i) 'the benefit sought by Ms. Pratten is prescribed by law as required by s. 15 of the *Charter* ii) Section 15(2) was not a defence to Ms. Pratten's claim; iii) Ms. Pratten established that the challenged provisions of the *Adoption Act* were discriminatory contrary to s. 15(1); and iv) The s. 1 defence was not established.

overturned the lower court, finding that the impugned provisions were constitutionally supported (by s. 15(2) of the Charter) and that there was therefore no need for the other grounds of the appeal to be discussed. The Court outlined the history of adoption legislation and policy in British Columbia,[209] noting that the main purpose of The *Adoption Act*, S.B.C. 1920, c. 2 had been to protect the child welfare of the 'parentless' and stressing how the law had long provided for the 'safekeeping of adoption orders and the use of birth certificates in the adoption process.'[210] Changes to 'social conditions' had led to the 1996 legislative amendments, which, subject to veto-absence, allowed for a qualified right to access information.

Citing the case of *In R v. Kapp* [2008],[211] it was noted that s 15(2) of the Charter did not simply serve to support analyses of potentially discriminatory legislative provisions. It instead operated as 'an independent "saving" provision' aimed at 'protecting "efforts by the state to develop and adopt remedial schemes designed to assist disadvantaged groups".[212] If a program had an 'ameliorative or remedial purpose' and was aimed at helping a 'disadvantaged group, identified by an enumerated or analogous ground' then it was unlikely to fall within the remit of s 15(1). In other words, the provisions of the statute in question were designed to remedy 'the state-sanctioned dissociation of adoptees from their biological parents.[213]

The remarks on adoptees are quite telling; they were described as a group which had been subjected 'historically, if not currently' to a 'negative social characterization.'[214] The court distinguished adoptees from donor children on the basis that adoption involved a change to the legal status of the children concerned, which often served to alienate them from their genetic origins and caused them to 'face barriers in accessing information about their origins.' The impugned sections of the legislation were aimed at reducing these barriers 'while maintaining the essentially socially beneficial stance of the legislation.'[215] Donor children had never had to face such a change of legal status. Citing *Cunningham*, it was further declared that

> governments may not be able to help all members of a disadvantaged group at the same time, and should be permitted to set priorities. If governments are obliged to benefit all disadvantaged people (or all subsets of disadvantaged people) equally, they may be precluded from using targeted programs to achieve specific goals relating to specific groups. The cost of identical treatment for all would be loss of real opportunities to lessen disadvantage and prejudice.[216]

---

[209] Para 30. See also The *Adoption (Amendment) Act*, S.B.C. 1968, c. 4, s. 5.

[210] Ibid *Adoption Act*, R.S.B.C. 1996, c. 5.

[211] *R v Kapp* (2008) 2008 SCC 41 (CanLII), 2008 SCC 41, [2008] 2 S.C.R. 483 on aboriginal salmon fishing rights.

[212] Para 33.

[213] Para 37.

[214] Kapp *op cit* n 211, para 55. In terms of being an adoptee, this had previously been deemed an 'analogous ground' in *Strong v. Marshall Estate*, 2009 NSCA 25 (CanLII), 2009 NSCA 25, 309 D. L.R. (4th) 459 at para. 31. Section 15(2) had also been examined in *Alberta (Aboriginal Affairs and Northern Development) v. Cunningham*, 2011 SCC 37 (CanLII), 2011 SCC 37, [2011] 2 S.C.R. 670.

[215] Ibid para 41.

[216] Ibid.

In terms of s.7 of the Charter giving rise to positive obligations (the 'free-standing right' argument) it was stressed that a departure from the established jurisprudence would be needed. Such a claim would have to show state-deprivation of the right to life, liberty, or security of the person *and* that such deprivation was 'contrary to an identified principle of fundamental justice.' Principles of fundamental justice required 'significant societal consensus' as to their definition, and must be capable of being identified and measured.[217] Further, s.7 of the Charter would have to be re-interpreted to encompass two rights, namely a positive right to life, liberty, and security of the person (which would be subject to qualification only by s.1) and the right to not be deprived of life, liberty, and security of the person, except in accordance with the principles of fundamental justice. The Court declined to do so, rejecting the plaintiff's argument that such a step would be likely to involve only minimal expenditure or resource-usage on the part of the state and declaring that the effects would be wide-reaching, moving beyond the interests of adoptees and donor children.

In sum, although it was desirable for people to be able to access information about their genetic origins, this was not so 'fundamental' a right as to merit constitutionally protected Charter status. The court relied upon *Marchand v. Ontario* (2007) where an adoptee sought information on her birth father's identity in the face of a veto on disclosure.[218] Here it had been held that

> unconditional disclosure of identifying personal information of third parties, even if they are birth parents of the claimant, without regard to the privacy and confidentiality interests of the persons identified and without regard to any serious harm that might result from disclosure, fails to meet the above criteria.[219]

The plaintiff in *Pratten* had also cited in support of her claim the English authority *Rose v. Secretary of State for Health*, [2002][220] where two donor children challenged their lack of access to information and support, arguing an interference with their Article 8 ECHR rights. Here the High Court did find that Article 8 European Convention rights were engaged but declined to offer comment on the issue of whether a violation of them had occurred. The claimants had not been seeking compulsory disclosure of the identity of their sperm donors however, merely a declaration of domestic statute incompatibility. The court noted here the altruism underpinning gamete donation, stating that it was aimed at 'relieving the affliction of infertility and on the clear understanding, if not promise, that their identity would remain undisclosed forever.'[221]A sort of unwritten agreement to

[217] *R. v. Malmo-Levine*; *R. v. Caine*, 2003 SCC 74 (CanLII), 2003 SCC 74, [2003] 3 S.C.R. 571 at para. 113.

[218] *Marchand v. Ontario* (2007) 2007 ONCA 787 (CanLII), 2007 ONCA 787, 288 D.L.R. (4th) 762, aff'g 2006 CanLII 19946 (ON SC), (2006), 81 O.R. (3d) 172 (S.C.J.), [2008] 1 S.C.R. ix, (*sub nom. Infant Number 10968*).

[219] Ibid.

[220] *Rose v. Secretary of State for Health*, 53EWHC 1593 (Admin), [2002] 3 FCR 731.

[221] Para 17.

maintain confidentiality seems to have been framed here as a sort of a reward for this act of kindness, rather than as a legal device to effect a permanent, absolute secrecy aimed at protecting the donor from unwanted future contact or to perhaps shield the new family from outside interference.

The ethical aspect of failing to 'honour that long standing understanding' was that it would be 'manifestly unfair to the donors.' Perhaps most significantly however, in terms of social and public policy concerns, it would serve also to 'drive a coach and horses through the A.I.D. system.'[222] That legislation has since been enacted to remove donor anonymity in the United Kingdom, does not seem to have factored greatly in the British Columbia Court's decision to dismiss the idea that a concept of fundamental justice may be involved in such cases. Equally, the Court chose to ignore the fairly unequivocal message in *Rose* that lessons clearly 'ha[d] been learnt from adoption.'[223] It was further stressed in the 2001 English case that

> A human being is a human being whatever the circumstances of his conception and an A.I. D. child is entitled to establish a picture of his identity as much as anyone else. We live in a much more open society than even 20 years ago. Secrecy nowadays has to be justified where previously it did not. The distinction between identifying and non-identifying information is not relevant at the engagement stage of Article 8, but it is likely to become very relevant when one comes to the important balancing exercise of the other considerations in Article 8(2).[224]

The British Columbia Court distanced itself further from the potential significance of *Rose* (that family life rights do arise in respect of donor anonymity, and may well be violated by a continuing bar on accessing information) by suggesting that Article 8 of the European Convention was not the most relevant provision to the issue of accessing identity. Article 5(1) of the European Convention was deemed to be more relevant here given that it 'partially mirrored' s.7 of the Canadian Charter.[225] In respect of English family law, Article 5 of the European Convention is rarely if ever associated with family or private life rights however, instead dealing with the 'circumstances in which the state may deprive the citizen of his liberty'[226] and thus being more associated for example with the rights of prisoners.[227]

---

[222] Ibid (the term A.I.D. referred to the process of 'artificial insemination by donor').

[223] Para 47.

[224] Ibid.

[225] *Pratten op cit* n 8 para 54. Section 7 of the Canadian Charter states that 'Everyone has the right to life, liberty and security of the person and the right not to be deprived thereof except in accordance with the principles of fundamental justice.'

[226] See for example the Judicial Review in *Duffy (Colin) and Others' Application (No 2)* [2011] NIQB 16.

[227] Article 5 (1) of the ECHR states that;

'Everyone has the right to liberty and security of person. No one shall be deprived of his liberty save in the following cases and in accordance with a procedure prescribed by law;

(a) the lawful detention of a person after conviction by a competent court;

(b) the lawful arrest or detention of a person for non-compliance with the lawful order of a court or in order to secure the fulfilment of any obligation prescribed by law;

Arguably by categorising the right to access genetic identity as a civil or political right, rather than as an element of the right to family life, the Court has afforded a significant degree of recognition to the applicants' arguments that the loss of identity does amount to a fundamental breach of justice, via the loss of a key element of human dignity. Unfortunately the civil or political right that seems to merit the rights-protection of the positive obligation here is not the child's right to access their identity information, but the donor parents' right to genetically relinquish the child and remove all trace of their own existence, vanishing behind the brick wall of anonymity.

It was further declared that the European Convention was merely a Regional Convention with no relevance to Canadian law. That the member states of the Council of Europe might generally accept the need for access to genetic identity information, was not indicative of broad 'support for the proposition that the right "to know one's past" is generally accepted as being of fundamental importance.'[228] Article 8 of The UN Convention on the Rights of the Child (which Canada has ratified) was similarly dismissed on the basis that it did not give rise to any obligation to actually 'provide mechanisms to enable donor offspring to obtain the personal information of third parties who are not their legal parents.'[229] It was suggested also that as adoption matters fell within the legislative remit of British Columbia, the Province was therefore not itself a party to the Children's Convention. Significantly, it was declared that no provisions of the Children's Convention were regarded internationally as giving rise to a right "to know one's past."[230]

The *Pratten* case is significant and disappointing in equal measure. Discrimination has clearly occurred (in respect of the difference made between the rights of adoptees and donor children) but is deemed entirely justified given the legalised prior losses and suffering of adoptees, namely, their lost, original legal status. The best interests of donor children are not a relevant factor here in the sense that the key legal issue seems to be one of determining whether the province's policy of affirmative, redressive discrimination was justified. The fact that donor children are at

(c) the lawful arrest or detention of a person effected for the purpose of bringing him before the competent legal authority on reasonable suspicion of having committed an offence or when it is reasonably considered necessary to prevent his committing an offence or fleeing after having done so;

(d) the detention of a minor by lawful order for the purpose of educational supervision or his lawful detention for the purpose of bringing him before the competent legal authority;

(e) the lawful detention of persons for the prevention of the spreading of infectious diseases, of persons of unsound mind, alcoholics or drug addicts or vagrants;

(f) the lawful arrest or detention of a person to prevent his effecting an unauthorised entry into the country or of a person against whom action is being taken with a view to deportation or extradition

[228] Para 55.

[229] Para 57. The fact that Article 8 of the UNCRC was proposed by an Argentinian delegate (post-Abuelas scandal) was also highlighted. See further Summary Record of the 54th Meeting of the Commission on Human Rights, U.N. Document E/CN.4/1985/SR.54.

[230] See further Concluding Observations (2002) (U.N. Document CRC/C/15/Add.188); Article 8 UNCRC was not mentioned by the Committee.

risk of suffering the same or broadly similar psychological difficulties as adoptees is simply disregarded. It also seems odd that the vulnerabilities of the affected group (adoptees) are firmly grounded in law, loss and their altered legal status.

Had these losses been framed in terms of best interests-led, psychological welfare needs or tied to the socio-cultural stigma of having unknown ancestry, then this perhaps might have aligned the situation of adoptees more to closely to that of donor children. And yet in terms of lost genetic connection, unknown heritage and removed birth right, the two groups face similar if not identical issues. Missing, lost or removed ancestry is still absent, whether it comprises of half or all of one's genetic background. Euphemistically referring to this rights violation (as other jurisdictions have essentially categorised it) as merely a desire to 'know one's past' does not capture the potential extent of the losses and harms that can attach to biological kinlessness and non-ancestry. In terms of birth status enabling access to identity, there is also sanctioned discrimination between triad members; anonymous donors and veto-hidden birth parents are afforded a near-absolute right to privacy whilst their genetic children face real or potential barriers to accessing a right to identity and well-being. Non-disclosure vetoes similarly discriminate between veto-affected adoptees and those who have managed to access their information, either by being born in the right place or at the correct time in history.

Had the Court accepted the arguments of the applicant in *Pratten*, a number of potential consequences might have followed from the decision that the current law was unlawfully discriminatory. The setting up of a donor registry and a ban on anonymity, potentially also retrospectively, might have occurred, although given that many records seem to have been destroyed as a matter of policy, such an outcome would probably have been fairly limited in terms of impact. Arguably the chief significance of a decision in favour of opening records and banning donor anonymity might have been in terms of enabling veto-affected adoptees and donor children to bring challenges to sealed records systems in other provinces on the basis that they too had been victims of a discriminatory system or policy.

In terms of making the child's best interests the paramount consideration, the British Columbia Court essentially opted to avoid this issue entirely and focus instead upon their convoluted statutory framework for gauging the lawfulness or otherwise of discriminatory domestic provisions. In doing so they have turned a blind eye to the emergent Strasbourg consensus on the child's need for genetic identity (as an element of best interests-led welfare paramountcy) and to the opportunity to achieve some measure of reform in respect of their legislation, perhaps along the lines offered by the United Kingdom's system of open birth records and donor reunion registries.

It may be argued that the legally sanctioned withholding of essential information is potentially a charter for discrimination and inequality. As Cahn has observed, 'not providing this information treats these children differently from children in traditional nuclear families.'[231] Where a right of access to accurate genetic

[231] Cahn (2000–2001) n 28, p. 5.

information is the exception rather than the norm, a 'sub-class' of origin-deprived persons arises, possibly stigmatised even further by the various silences that might have enabled their loss of natal kinship.[232] Much of the case law highlights the unique vulnerabilities of genetically relinquished children, and to an extent provides some degree of useful guidance as to how the child's need for ancestral heritage might be protected by reference to the best interests principle, and perhaps freed from the limiting effects of adult-centric balancing exercises aimed largely at protecting traditions of parental privacy.

Haimes' argument is worth revisiting here: the 'unusual social status' of children in 'constructed' families may equate to one which is 'potentially damaging'[233] and not without the risk of marginalisation or stigma for all concerned. As Bainham further argues, a distinct set of

> values ought to govern the determination of parentage. . ..the guiding principles should be a commitment to truth, individual autonomy and priority for the rights and interests of those primarily affected.[234]

Policies that are based upon 'suppression of the truth' might not comply fully with the best interests of the child principle, especially where unique issues of long-term, psychological welfare, well-being or harm might arise[235] or in situations where 'legislative provisions have made assumptions and the biological facts do not fit the legal fictions.'[236] Given that the new assisted reproduction technologies can result in children having up to eight adults involved in their conception, birth or upbringing, a juridical right to genetic truth does not seem an altogether irrational concept.[237] If it is accepted that children are generally a 'category with special interests. . ..[and] a particular nexus of problems'[238] then those who have essentially been genetically relinquished (fully or partially) may be more likely to face a unique set of difficulties on the basis of their lack of authentic 'narrative.'[239] As with a significant number of adoptees, the well-documented need to construct an 'artificial self'[240] to cope with the circumstances of one's past, should not be

[232] On the notion and consequences of illegitimacy (and in particular the doctrine of *filius nullius*) see Herring (2007) n 2, p. 370; See also The Family Law Reform Act 1987 section 19(14) on inheritance of titles; see also *P v B* (*Paternity*: *Damages for Deceit*) [2001] 1 FLR 1041.

[233] Haimes (1988) n 33, p. 47.

[234] Bainham (2008) n 7, p. 324.

[235] See for example *W and B v H* (*Child Abduction*: *Surrogacy*) [2002] 1 FLR 1008 where the Hague Convention was invoked by a Californian couple to seek the return of twin babies born to an English surrogate mother. See also Lowe et al. (2004), p. 61.

[236] Booth (2004), pp. 270–273 at p. 271.

[237] Cahn (2000–2001) n 28, p. 2 (in contrast to the 'four possible parents' scenario that can attach to traditional adoption).

[238] Strathern (2005), p. 3.

[239] Dunbar and Grotevant (2004), pp. 135–161 at p. 135.

[240] Lifton (1990), pp. 85–92 at p. 85. See also Patel (2005), p. 327 at p. 332 on how 'data based on assumptions' might be used to fill identity narrative gaps.

ignored. Where matters such as birthplace or birth date are the deciding factor in being afforded or denied a right to genetic identity, it is difficult to see how issues of inequality do not arise.

6.7 Conclusion

Socio-cultural prejudice surrounding the concept of unknown ancestry or illegitimacy does seem to linger on. To take Kahn's hypothetical example, if a married Jewish woman's eggs were to be fertilized in vitro by an anonymous donor, rather than by her husband, then under a strictly interpreted Halakhic doctrine, 'such a child could be considered a *mamzer*, with all the associated stigma' of an 'adulterous' conception. Talmudic debate also seems to surround the question of determining genetic inheritance of 'Jewishness' and whether it is gestation, parturition or conception that might confer faith status upon donor conceived individuals. A very fear of blood itself may also be inferred in relation to oocyte collection procedures; should a woman's uterus bleed as a result of egg harvesting surgery, she might be classed as 'unclean' and therefore unfit to receive any embryos until seven days have passed and she has undergone a ritual cleansing bath. A child conceived under such circumstances might then be considered a '*ben-niddah*' and suffer social stigma.[241]

It is perhaps not surprising that there is often still a perceived need to shroud the finer details of assisted conception in secrecy. Generally however, the processes associated with assisted reproduction, exist as a type of positive, 'hope technology.'[242] That there has been a more 'medicalised' approach to the need for truth-telling, is perhaps down to different professions having been involved in the process (i.e. doctors rather than social workers, charities or religious orders) so that the advice to recipient parents has possibly been traditionally based more upon the 'health' of the entire family unit than on an individualised promotion of the best interests of the child so conceived.[243] To an extent this mirrors some of the earlier therapeutic approaches to the need for secrecy that were associated with closed adoption practices,[244] although in most jurisdictions initial concerns over *whether* to tell have gradually given way to more reasoned debates over exactly *when* children should be told. Debate still attaches to the issue of exactly *what* information should be made available to adoptees, with domestic laws and policies gradually beginning to focus more attention on *how best* to promote the human rights of genetically relinquished, origin deprived children.

---

[241] Kahn (2004), pp. 362–377 at p. 366. See also Bleich (1991), pp. 82–102.

[242] Franklin (2006) n 18, p. 550.

[243] Haimes (1988) n 33, p. 48 on how 'the professions most closely involved in the two procedures differ as to their power and authority, not only over the resources available to them as family-builders, but also (and perhaps because of) their social status.'

[244] See however Haderka (1987), pp. 72–85 on the need to enshrine a greater degree of secrecy.

This socio-cultural shift in attitude has enabled a move away from absolute models of parental confidentiality (and systems of birth-records secrecy) in many regions, towards an atmosphere of more generalised access and honesty and with the potential for eventual disclosure of information, if not actual reunion.[245] This move towards greater openness was also largely precipitated by social necessity, given that the numbers of 'orphaned' or altruistically relinquished newborns have fallen sharply, and with adoption practice becoming more child-protective in nature and more focussed on child rights in practice. Modern adoption policies have been further influenced by the fact that many older, abused children may still hold distinct memories of their genetic relatives or, as with international or mixed-race adoptees, bear little or limited physical resemblance to their adoptive parents.

It should perhaps be remembered however that what serves to effect a 'negation of the donor' also generally removes the original 'bloodline as ancestral substance'[246] and creates a highly inaccurate, if not dangerous, 'authorship of descent.'[247] Similarly, with childhood now generally recognised in both international and domestic law as a highly vulnerable, rights-bearing 'condition' it is no longer possible to simply sideline the importance of ascertaining genetic parentage in providing some measure of identity narrative.[248] If it is accepted that 'adoption is simply not the equivalent of biological parenthood'[249] then gamete donation should be similarly regarded as an extraordinary process that creates inherent differences. Such an acknowledgement does not diminish or denigrate created parenthood or social kinship, nor does it indict the parenting abilities of gamete recipients; it merely underlines the fact that a number of 'triad' children may be quite vulnerable in terms of achieving psychosocial well-being and having their welfare needs met. In this they may differ significantly from their non-origin deprived peers.[250] As Frith has argued,

> the law recognises that blood can create legal ties, however, as a result of the operation of social convention and morality, not all blood relationships are given legal recognition. . .in utilizing kinship ideals to structure legal determinations, law acts as an informal kinship system.[251]

It is perhaps because of this that there still seems to be a significant 'incongruence'[252] between legal and social parentage. Callus has argued that 'biology,

---

[245] See further Carp (1998).

[246] Grace (2008) n 37, pp. 257–262 at p. 257 on the importance of 'substance' in relation to experiencing or demonstrating kinship.

[247] *Ibid* p. 259.

[248] McWhinnie (1992), p. 37; see also Baran and Pannor (1989).

[249] Theyssen (1998–1999), pp. 711–734 at p. 734. See further O'Donovan (1988), p. 27.

[250] See Grosso and Nagliero (2004), pp. 45–48.

[251] Frith (2001) n 29.

[252] Callus (2012), pp. 347–368. See further Cameron et al. (2010), pp. 95–148 on the various possible consequences of (and models for) abolition of sperm donor anonymity, for example in relation to single sex couples.

intention and functional parenting' may overlap, with the concept of *intentional* parenthood obviously taking centre stage in some cases but not in others. This ad hoc approach is lacking in clarity and coherence; a legal framework making intentional parenthood key could, she argues, yet serve to properly tie legal duties to social realities. Whether such an adult-centric focus on the concept of parental intention would do much to recognise or protect the child's need for identifiable genetic connection is doubtful, however. It may be argued that genetic relinquishment arises as a result of parental 'abandonment' whether in the sense of opting for permanent anonymity, vetoing future informational disclosure or contact, or denying that any moral obligation exists in respect of one's descendants. The issue of having to have actually provided consent to genetic parentage before any moral duties towards one's children might arise should not be confused with the right to decline one's parenthood or parenting responsibilities. The latter often ties directly to and promotes the child's best interests in that it is aimed at preventing dangerous parenting and creating strong ties of welfare-led social parenthood; the former can lead to origin deprivation and a significant level of harm, which directly offends against the principle that the child's welfare is or ought to be the paramount consideration.

Denying the genetic tie either ignores completely the rights and needs of the origin deprived child, or places their best interests firmly at the bottom of a triad rights-hierarchy. Legislative and judicial bars on accessing genetic ancestry serve as a means to legally 'orphanize' children; permanent vetoes on disclosure of information can 'infantilize' adoptees and gamete donor children. This is especially so where there may be a considerable degree of equivocation 'about the importance society attaches to the blood-tie ....an uncertainty reflected in judicial decision-making.'[253] If the need for genetic ancestry were to be framed however by domestic jurists and legislators as a key, equality-based component of the right to identity and as a central element of the best interests discourse, genetic and social parents might find themselves legally obliged to ensure that their children can eventually access biological truths. Such an outcome would accord with the trend towards openness in a number of jurisdictions[254] and perhaps most significantly, support the argument that the definition of a reasonable parent is essentially one who 'places great weight on what will be best for the child'.[255] To do otherwise is to be complicit in justifying 'a lifetime of deceit.'[256]

The chapter which follows looks at the issue of the blood-tie in domestic law contexts, namely the right to engage in some form of contact with one's genetic relatives and the issue of challenging court orders that have 'severed' the genetic connection, for example by dispensing with parental consent to adoption. The issue

---

[253] Fortin (1994) n 82 at p. 307.

[254] See for example Kirton et al. (2000), pp. 6–18; Neil (2009), pp. 5–23.

[255] *Re W* (1971) AC 682. This was followed in *Re D* (1977) AC 602 and withstood the challenge of *Re C* (1993) 2 FLR 260.

[256] Freeman (1996) n 20 at p. 273.

of family contact in particular may hold much relevance as an aspect of the best interests of the child, not least in terms of establishing that there may often be a best-interests led need for genetic truths to be accessible, especially as a means to providing the child with an accurate sense of identity and genetic heritage.

References

Bainham, A. (2008, July). Arguments about parentage. *Cambridge Law Journal, 67*(2), 322–351.

Baran, A., & Pannor, J. (1989). *Lethal secrets*. New York: Warner.

Bestard, J. (2004). Kinship and the new genetics: the changing meaning of biogenetic substance. *Social Anthropology, 12*(3), 253–263.

Bleich, J. D. (1991). In vitro fertilization: Questions of maternal identity and conversion. *Tradition, 25*(4), 82–102.

Blyth, E. D. (2002). How it feels to be a child of donor insemination. *British Medical Journal*, 324.

Blyth, E. D., & Farrand, A. (2004) Anonymity in donor-assisted conception and the UN Convention on the rights of the child. *International Journal of Children's. Rights, 12*, 89–104.

Bonney, H. (2002). The psychopathogenic power of secrecy: child development and family dynamics after heterologous insemination. *Journal of Psychosomatic Obstetrics and Gynecology, 23*(3), 201–208.

Booth, P. (2004). The right to know one's father. *Family Law Journal, 34*, 270–273.

Brandon, J., & Warner, J. (1977). AID and adoption: Some comparisons. *British Journal of Social Work, 7*(3), 335–341.

Braude, P., Johnson, M., & Aiken, R. (1990). The human fertilisation and embryology bill goes to report stage. *British Medical Journal, 300*, 1410–1412.

Brewaeys, A. (2001) Review: Parent–child relationships and child development in donor insemination families. *Human Reproduction Update, 7*, 38–46.

Cahn, N. (2000–2001). Children's interests and information disclosure: Who provided the egg and sperm? Or, mommy, where (and whom) do I come from? *The Georgetown Journal of Gender and Law*, 1–27.

Callus, T. (2012). A new parenthood paradigm for twenty-first century family law in England and Wales? *Legal Studies, 32*(3), 347–368.

Cameron, A., Gruben, V., & Kelly, F. (2010). De-anonymising sperm donors in Canada: Some doubts and directions. *Canadian Journal of Family Law, 26*, 95–148.

Carp, E. W. (1998). *Family matters: Secrecy and disclosure in the history of adoption*. Cambridge: Harvard University Press.

Cook, R., Golombok, S., Bish, A., & Murray, C. (1995). Keeping secrets: A study of parental attitudes toward talking about donor insemination. *American Journal of Orthopsychiatry, 65*, 549–559.

Dennison, M. (2007–2008). Revealing your sources: the case for non-anonymous gamete donation. *Journal of Law and Health, 21*(1), 25.

Douglas, G. (2008). Case reports: Adoption. *Family Law*, 301.

Dunbar, N., & Grotevant, H. D. (2004). Adoption narratives: The construction of adoptive identity during adolescence. In M. W. Pratt, & B. H. Fiese (Eds.), *Family stories and the life course: Across time and generations* (pp. 135–161). Mahwah: Erlbaum.

Eekelaar, J. (2006). *Family law and personal life*. Oxford: Oxford University Press.

Fortin, J. (1994). Re F: The Gooseberry Bush approach. *Modern Law Review, 57*, 296–307.

Franklin, S. (2003). Re-thinking nature-culture: Anthropology and the new genetics. *Anthropological Theory, 3*, 65–85.

Franklin, S. (2006). Origin stories revisited: IVF as an anthropological project. *Culture Medicine and Psychiatry, 30*, 547–555.

Freeman, M. (1996). The new birth right? Identity and the child of the reproductive revolution. *The International Journal of Children's Rights, 4*, 273–297.

Frith, L. (2001). Gamete donation and anonymity: The ethical and legal debate. *Human Reproduction, 16*(5), 818–824.

Grace, V. M. (2008). The psycho-social politics of paternity in the case of male-donated gametes. *Women's Studies International Forum, 31*, 257–262.

Grace, V. M., Daniels, K. R., & Gillet, W. (2008). The donor, the father, and the imaginary constitution of the family: parents' constructions in the case of donor insemination. *Social Science and Medicine, 66*, 301–314.

Grosso, W., & Nagliero, G. (2004). Adoption, fostering and identity. *Journal of Child and Adolescent Mental Health, 16*(1), 45–48.

Haderka, J. (1987). Artificial reproduction in Czechoslovak law. *International Journal of Law and the Family*, 72–85.

Haimes, E. (1988). Secrecy: What can artificial reproduction learn from adoption? *International Journal of Law, Policy and the Family, 2*, 46–61.

Herring, J. (2009, April 6). Revoking adoptions. *New Law Journal.*

Jones, C. (2007). *Why donor insemination requires developments in family law—the need for new definitions of parenthood*. Lewiston: Edwin Mellen Press.

Kahn, S. M. (2004). Eggs and wombs: The origins of jewishness. In R. Parkin, & L. Stone (Eds.), *Kinship and family* (pp. 362–377). Oxford: Blackwell.

Kirton, D., Feast, J., & Howe, D. (2000). Searching, reunion and transracial adoption. *Adoption & Fostering, 24*(3), 6–18.

Leiblum, S. R., & Hamkin, S. E. (1992). To tell or not to tell: attitudes of reproductive endocrinologists concerning disclosure of DI to offspring. *Psychosomatic Obstetrics and Gynaecology, 13*, 267.

Lifton, B. J. (1990). The formation of the adopted self. *Psychotherapy in Private Practice, 8*, 85–92.

Lind, C. (2003). Re R (Paternity of IVF Baby): Unmarried paternity under the Human Fertilisation and Embryology Act 1990. *Canadian Family Law Quarterly*, 327–340.

Lowe, N. V., Everall, M., & Nicholls, M. (2004). *International movement of children: Law, practice and procedure*. London: Jordans Family Law.

McWhinnie, A. (1992). Creating children – the medical and social dilemma of assisted reproduction. *Adoption and Fostering: The Future, 29*, 37.

Mykitiuk, R. (2002). Beyond conception: Legal determinations of filiation in the context of assisted reproductive technologies. *Osgoode Hall Law Journal, 39*(4), 771–781.

Nachtigall, R. (1993). Secrecy: An unresolved issue in the practice of donor insemination. *American Journal of Obstetric Gynaecology, 6*, 1846–1851.

Nachtigall, R. D., Tschann, J. M., Quiroga, S. S., Pitcher, L., & Becker, G. (1997). Stigma, disclosure and family functioning among parents of children conceived through donor insemination. *Fertility and Sterility, 68*, 83.

Neil, E. (2009). Post-adoption contact and openness in adoptive parents' minds: Consequences for children's development. *British Journal of Social Work, 39*, 5–23.

O'Donovan, K. (1988). A right to know one's parentage? *International Journal of Law and the Family, 2*, 27.

Patel, T. (2005). The usefulness of oral life (hi)story to understand and empower: The case of trans-racial adoption. *Qualitative Social Work, 4*, 327.

Rowland, R. (1985). The social and psychological consequences of secrecy in AI by donor programmes. *Social Science & Medicine, 4*, 391–396.

Shenfield, F. (1997). Privacy versus disclosure in gamete donation: A clash of interest, of duties, or an exercise in responsibility? *Journal of Assisted Reproduction and Genetics*, 371–372.

Skinner, K. (2013, December 2). Court examines the rights of biological, but not legal, parents. *LNB News*, 60.

Snowden, R., & Mitchell, G. (1981). *The artificial family: A consideration of artificial insemination by donor*. London: Allen and Unwin.

Strathern, M. (2005). *Kinship, law and the unexpected: Relatives are always a surprise*. New York: Cambridge University Press.

Swanson, H. S. W. (1993–1994). Donor anonymity in artificial insemination: Is it still necessary? *Columbia Journal of Law and Social Problems, 27*, 151.

Theyssen, D. L. (1998–1999). Balancing interests in frozen embryo disputes: Is adoption really a reasonable alternative? *Indiana Law Journal, 74*, 711–734.

Turner, A. J., & Coyle, A. (2000). What does it mean to be a donor offspring? The identity experience of adults conceived by donor insemination and the implication for counselling and therapy. *Human Reproduction, 15*, 2041.

Walker, I., & Broderick, P. (1999). The psychology of assisted reproduction—or psychology assisting its reproduction? *Australian Psychologist, 34*, 38–44.

Case Law

European Court of Human Rights

Gnahoré v France [2000] ECHR 40031/98
Lebbink v The Netherlands (2004) (app no 45582/99)
Marckx v Belgium (1979) (App no 6833/74)
Mikulić v Croatia (2002) (no 53176/99) 54 ECHR
Odievre v France (2003) 1 FLR 621
Pini v Romania [2005] 2 FLR 596
R and H v United Kingdom (2011) 35348/06 [2011] ECHR 844
S H v Austria (2012) 37813/00 EHRLR [2012] 2, 211–214
Y C v United Kingdom (App 4547/10) ECHRR

Canada

Alberta (Aboriginal Affairs and Northern Development) v. Cunningham, 2011 SCC 37 (CanLII), 2011 SCC 37, [2011] 2 S.C.R. 670
Cheskes v. Ontario (Attorney General) 2007 CanLII 3837 (ON SC), (2007), 87 O.R. (3d) 581, 288 D.L.R. (4th) 449 (S.C.J.)
Gosselin v. Québec (Attorney General),2002 SCC 84 (CanLII), 2002 SCC 84, [2002] 4 S.C.R. 429
Marchand v. Ontario (2007) 2007 ONCA 787 (CanLII), 2007 ONCA 787, 288 D.L.R. (4th) 762, aff'g 2006 CanLII 19946 (ON SC), (2006), 81 O.R. (3d) 172 (S.C.J.), [2008] 1 S.C.R. ix, (*sub nom. Infant Number 10968*).
Pratten v British Columbia (Attorney General) (2011) BCSC 656 (CanLII), (2011 BCSC 656, 22 B.C.L.R. (5th) 307)
R v Kapp (2008) 2008 SCC 41 (CanLII), 2008 SCC 41, [2008] 2 S.C.R. 483
R. v. Malmo-Levine; *R. v. Caine*, 2003 SCC 74 (CanLII), 2003 SCC 74, [2003] 3 S.C.R. 571
R. v. Oakes, 1986 CanLII 46 (SCC), [1986] 1 S.C.R. 103
Strong v. Marshall Estate, 2009 NSCA 25 (CanLII), 2009 NSCA 25, 309 D.L.R. (4th) 459

Ireland

I.O'T v. B and Others [1998] 2 IR 321
G v. An Bord Uchtála [1980] IR 32
J McD v PL [2010] 2 IR 199
J.K v. V.W. and Others [1992] 2 IR 437
JPD v MG [1991] 1 IR 47
M.R & Anor -v- An tArd Chlaraitheoir & Ors [2013] IEHC 91
N v Health Service Executive [2006] 4 IR 372

United Kingdom

A v P [2011] EWHC 1738 (Fam)
D (J) v D(S) [1973] 1 All ER 349
Duffy (Colin) and Others' Application (No 2)[2011] NIQB 16
M v W (declaration of parentage) [2007] 2 FLR 270
P v B (Paternity: Damages for Deceit) [2001] 1 FLR 1041
RE C [2002] 1FLR 909.
Re C (1993) 2 FLR 260
Re C (a child) v XYZ County Council (2007) EWCA Civ 1206
Re D (1977) AC 602
Re D (Paternity) (2006) EWHC 3545 (Fam)
Re F (1993) 1 FLR 598
Re G [2013] EWHC 134 (Fam)
Re H & A (Children) (2002) EWCA Civ 383
Re H (A Minor) (Blood Tests: Parental Rights) (1996) WLR 506
Re H: Re G (2001) 1 FLR 646
Re IJ (A Child) [2011] EWHC 921
Re L (2007) EWHC 1771 (Fam) (20 July 2007)
Re L (2008) EWCA Civ 1388
Re L (Commercial Surrogacy) [2011] 1 FLR 1423
Re L [2010] EWHC 3146 (Fam)
Re P (A Child) (2008) EWCA Civ 499
RE Q (Parental Order) [1996] 1FLR 369
Re R (1993) 2 FLR 762
Re R [2003] EWCA Civ 182
Re S (Parental Order) [2009] EWHC 2977 (Fam)
Re T (A Child) (DNA Tests: Paternity) (2001) 3 FCR 577
Re W (1971) AC 682
Re WM (Adoption; Non-Patrial) [1997] 1 FLR 132
Re X & Y (Foreign Surrogacy) [2009] 1 FLR 733
Re X and Y (Foreign Surrogacy) [2008] EWHC 3030 (Fam)
Re: Adoption Application (Payment for Adoption) [1987] 3WLR 31
Rose v. Secretary of State for Health, 53EWHC 1593 (Admin), [2002] 3 FCR 731
S v McC;W v W (1972) AC 24
W and B v H (Child Abduction: Surrogacy) [2002] 1 FLR 1008
Z County Council v R (2001) 1FLR 365
ZH (Tanzania) v Secretary of State for the Home Department UKSC [2011] 4

Chapter 7
'Related' Matters in an Open Records Region: Relinquishment, Contact and Best Interests

7.1 Introduction

> I have been considering whether there is any one noun or adjective which can properly describe this case. I have come to the conclusion that there is not. The four children concerned have been denied the opportunity to argue that they should grow up together with their parents as a family. That is deeply worrying, and, on the face of it, a clear breach of their rights to respect for their family life under Article 8(1).[1]

If a basic right to genetic relatedness is eventually found to exist amongst and emerge from the various strands of jurisprudence that together comprise the 'law' of blood-ties, then access to knowable biological truths is only one aspect of this right. As the Strasbourg court appeared to suggest in *Anayo v Germany*[2] the issues of kin contact, genetic identity and best interests-led decision-making may interconnect and overlap to potentially give rise to positive, justiciable obligations at the level of domestic proceedings involving child welfare, especially in disputes over parenting or parentage. Within the United Kingdom (an open records jurisdiction, where sperm donor anonymity has also been abolished[3]) domestic jurisprudence involving blood-ties has gradually moved towards an acceptance that complete severance of the genetic bond should only occur in exceptional circumstances,

---

[1] *Webster (The Parents) v Norfolk County Council & Ors* (Rev 1) [2009] EWCA Civ 59 (11 February 2009) per Wall LJ at para 2.

[2] *Anayo v Germany* [2010] ECHR 2083 (21 December 2010).

[3] See The Adoption Act 1976 (England and Wales), The Adoption (NI) Order 1987, The Adoption and Children Act 2002 (England and Wales and partially applicable to Northern Ireland e.g. in respect of amending the welfare checklist); The Children Act 1989, The Children (NI) Order 1995; The Adoption and Children Scotland Act 2007, and The Human Fertilisation and Embryology Act 2008. See also however *R v Registrar-General ex p Smith* [1991] 2 QB 393 where an incarcerated adoptee attacked his cell-mate in the belief that he was his birth mother; he was subsequently denied access to his birth certificate.

A. Diver, *A Law of Blood-ties - The 'Right' to Access Genetic Ancestry*,
DOI 10.1007/978-3-319-01071-7_7,

rather than as an automatic, normative aspect of the processes used to create the ties of social kinship.[4]

That said, the context of the child's best interests remains the paramount consideration for courts tasked with determining disputes over child welfare. As Hale LJ observed over a decade ago,

> intervention in the family must be proportionate...the aim should be to reunite the family when the circumstances enable that...cutting off all contact and the relationship between the child and their family is only justified by the overriding necessity of the interests of the child.[5]

The cases in this chapter, drawn from throughout the United Kingdom, tend to be concerned with such key issues as contested relinquishments and objections to adoption or placement,[6] the sharing of parental responsibility, challenges to residence or contact arrangements post-adoption, or matters of 'fair hearing' procedure, such as the use, or lack, of expert evidence on the issue of the child's welfare. In some of the cases referred to the nature of kinship care is discussed, as is the purpose of parenthood; as the Supreme Court recently noted in *Re B (A Child)* (2009) 'it is only as a contributor to the child's welfare that parenthood assumes any significance.'[7]

Similarly, a general atmosphere of aspiring towards information disclosure and familial preservation, does not necessarily always equate to an all-embracing norm of openness.[8] There is no free-standing, over-arching right to be raised by one's genetic parents or to demand the return of a removed child on the basis of a constitutional right, as might be the case in other jurisdictions.[9] In terms of creating and embedding a juridical right to engage in contact with one's genetic relatives, the presence of the blood-tie often carries limited weight. Where parents wish to contest a state decision aimed at removing or severely curtailing their parenthood

---

[4] See further Appell (2010), p. 1, on the equivocal nature of pre-hearing agreements on post-adoption contact arrangements.

[5] Per Hale LJ in *Re C and B (Children) (Care Order: Future Harm)* [2001] 1 FLR 611.

[6] At the time of writing, in England and Wales the Placement Order is used to enable the adoption process; in Northern Ireland the older, but not entirely dissimilar, Freeing Order system is still in use, whilst in Scotland, the Permanence Order is used. In terms of dispensing with parental consent (to adoption) the best interests of the child is the paramount consideration. See *S v L* [2012] UKSC 30 (11 July 2012).

[7] In *Re B (A Child)* [2009] UKSC 5 para 37 per Lord Kerr.

[8] See Smith (2002), pp. 281–302; Bojorge (2002), pp. 266–291.

[9] See for example the Irish Supreme Court decision in *N v Health Service Executive & Ors* [2006] IESC 60 (the 'Baby Anne' case, examined in more detail in the following chapter) saw the return of a relinquished, placed-with-prospective adopters child to her birth parents after several years spent in the care of her psychological parents, not on the basis of welfare paramountcy rights, the best interests principle, her attachments to them or her Article 8 ECHR right to family life, but because of the State's obligation to protect the Constitutional rights of the original (subsequently married) parents. See further however on the recent referendum (10 November 2012) on children's rights, which has amended Article 42.5 of The Irish Constitution http://www.childrensrights.ie/campaign/childrens-rights-and-constitutional-reform (accessed 30.03.13).

(such as a ban on contact, or a court order enabling the adoption of their child) they will usually have to demonstrate that their relationship with the child is comprised of much more than a genetic link, and perhaps also that no harm to the child's welfare will come from these original ties being maintained or reinstated.

Where a closed records region might be considering a move towards the unsealing of anonymous birth records or the abolition of donor anonymity or the birth parent veto, then looking to the laws and policies of an open records region might provide useful guidance, either as a 'what to avoid' comparator or as basic model for positive, child-centric legislative reforms.[10] Where birth records have been opened, other legal and social issues will arise, not least those concerned with achieving a fair, lawful balance between the conflicting rights of triad members (and those outside of it such as siblings, grandparents, donors and surrogates) and ensuring that the best interests of the child become and remains the paramount concern. The changing legal definitions of parenthood and parentage, and the various debates surrounding the extent to which parenthood might or should be shared, continue to give rise to controversies and policy conflicts within this area of UK law.[11] From the perspective of the vulnerable child however, the key issue is fairly straightforward; prevention of harm in the short and long term. Achievement of this in any given case may require that the blood-tie is actively preserved or suspended in the sense that it might be set aside until the child gains adulthood and makes the decision to access his original identity. Equally, adherence to the best interests principle might also demand that the various adults involved in the child's welfare are prepared to sacrifice some elements of their rights parcel e.g. by perhaps sharing the legal status of parent, conceding some aspect of their private life to provide their genetic descendants with essential information and a sense of identity or possibly by relinquishing their right to parent, where the child's best interests have clearly demand such an outcome. As Fair has argued, the most important thing for policy-makers to perhaps remember (and this can be applied equally to children born as a result of gamete donation) is that

> the individual who had no voice in the adoption process, the adoptee, is often the one who bears the psychological . . . .burden of locating his or her roots.[12]

---

[10] See further Fair (2008), p. 1039.

[11] See for example The Norgrove Report (Family Justice Review) (November 2011) available at http://www.justice.gov.uk/downloads/publications/moj/2011/family-justice-review-final-report.pdf (accessed 22.02.13) and The Munro Review of Child Protection (May 2011) available at http://www.official-documents.gov.uk/document/cm80/8062/8062.pdf (accessed 22.02.13).

[12] Fair (2008), p. 1039.

7.2 Opposing 'Relinquishment': Preventing Placement

The English Court of Appeal decision which perhaps illustrated most starkly the difficulties associated with harsh systems of 'all or nothing' blood-tie severance was that of *Webster (The Parents) v Norfolk County Council & Ors (Rev 1)* [2009].[13] Here three children were permanently removed from their birth parents on the assumption that one of them had been physically abused. It was later found that the child's fractures were more likely to have been caused by an iron deficiency, which had itself arisen from a digestive disorder. Wall LJ described the case as 'a disaster, quite apart from any breach of their rights under the ECHR.'[14] The Court stressed however that no blame should attach to the individual professionals who had been involved in the earlier hearings, but did note that the Family Justice System was not without its critics.[15] Because the earlier hearings had pre-dated the introduction of the Adoption and Children Act (2002) the three eldest children had been made the subject of Freeing Orders rather than the new Placement Orders.[16] This meant that the sole issue before the court was whether or not the birth parents were behaving unreasonably in withholding their consent to the children being freed for adoption.

After the birth of their fourth child (whose Care Order proceedings were later dropped by the local authority) the parents sought leave to appeal against the earlier Care, Freeing and Adoption Orders that had been made in respect of their other three removed children, on the basis of fresh medical evidence becoming available.[17] In relation to the adoption of their older children (two of whom had been placed together, with another child being separately cared for) the Court was also asked to consider whether a miscarriage of justice had occurred.

The parents' Counsel argued that the Adoption Orders, if based on error, some procedural irregularity or the denial of natural justice, could become the subject of a successful appeal, especially where this had resulted in a 'denial of justice' for the

---

[13] *Webster op cit* n 1. The child, suffering from scurvy, was thought to have been deliberately neglected having been fed only soya milk as an infant. It later emerged that this had probably been due to a digestive disorder, and that the parents had acted on the advice of the family's GP. The fractures which the child had sustained had probably also occurred as a result of his condition, rather than through deliberate physical abuse. See also Herring (2009). Available http://www.newlawjournal.co.uk/nlj/content/family-revoking-adoptions. Accessed 10.08.11.

[14] *Ibid* at para 3.

[15] *Ibid* at para 7.

[16] Freeing Orders were made under the since-repealed Adoption Act 1976; Placement Orders (and Special Guardianship Orders) were introduced by the 2002 Act, enacted 2005. The youngest child was born in the neighboring jurisdiction of the Republic of Ireland, his parents having fled there before his birth to avoid his being taken into local authority care. Following a satisfactory period of residential assessment, the local authority decided to stay Care Order proceedings and allowed this child to remain in the care of his parents.

[17] The new reports suggested that the injuries were more likely to be due to scurvy and iron deficiency anaemia than deliberate parental abuse.

children.[18] The other side contended that the birth parents had lacked the *locus standi* to appeal, given not only the significant lapse of time[19] but due also to the fact that for the past 3 years the children had, by law, been the children of other parents. They stressed the policy considerations that ought to be taken into account by the Court, arguing that

> If prospective adopters thought that the natural parents could, even in limited circumstances, secure the return of the child after the adoption order was made, that could have a dramatic adverse effect on the number of prospective adopters.[20]

The Guardian ad litem[21] submitted that the Court should seek to 'start at the end of the process' in relation to examining the possible consequences of overturning the Adoption Orders, rather than returning to the beginning to examine afresh the issue of the Freeing Order proceedings. If the adoptions were to be set aside, a form of 'legal chaos' would be likely to surround the status of the children. It would clearly not be in their interests (the word 'best' was, not included in the analysis) to go through the 'emotional turmoil' that would possibly also attach to a complete re-opening of their case should a return to their original family be found necessary.[22]

It was further argued that Article 8 European Convention rights, once engaged, would apply equally to all members of the triad, not just to the birth parents. Ultimately, and from the perspective of the children, given that they had been in foster care for 2 years, the need for 'certainty stability and permanence' had been best met by their being adopted. It followed that 'their Article 8 rights centred exclusively on their new permanent family.'[23] Wall LJ stated however that he was 'unimpressed' with some of the local authority's legal arguments, not least in relation to their contention that the birth parents had lacked any legal status to lodge an appeal.[24] Arguably, he seemed also to acknowledge the significance of the blood-tie, declaring that,

---

[18] *Webster op cit* n 1 para 65.

[19] See *Ladd v Marshall* [1954] 1 WLR 1489 on the rules of admissibility of fresh evidence in respect of re-hearings (per Denning LJ at 1491). See also however the dicta of Waite LJ in *Re S (Discharge of Care Order)* [1995] 2 FLR 639 at 646 which highlighted, 'the willingness of the family jurisdiction to relax (at the appellate stage) the constraints of *Ladd v Marshall* upon the admission of new evidence, does not originate from laxity or benevolence but from recognition that where children are concerned there is liable to be an infinite variety of circumstances whose proper consideration in the best interests of the child is not to be trammelled by the arbitrary imposition of procedural rules. ..the family courts (including the Court of Appeal when it is dealing with applications in the family jurisdiction) will be every bit as alert as courts in other jurisdictions to see to it that no one is allowed to litigate afresh issues that have already been determined.'

[20] *Webster op cit* n 1 para 63.

[21] The Guardian's role is to 'provide a voice' for the child and safeguard their interests in legal proceedings involve Care or Adoption Orders. See for example http://www.nigala.hscni.net/ (accessed 12.03.13).

[22] *Webster op cit* n 1 para 86.

[23] *Ibid* para 89.

[24] *Ibid* para 140.

> Whatever their legal status in relation to A, B and C the simple facts remain that the three children were born to them, and that they have been both removed from their care and adopted by third parties. In my judgment, they would plainly have sufficient interest in the adoption orders – or, to put the matter another way, they would be sufficiently affected by the orders – to warrant an application for permission to appeal against them.[25]

The Court was bound however by the domestic legislation on adoption, which had long served to turn children into 'permanent and full members of a new family' so that they might from then on be treated 'for all purposes as if born to the adopters.'[26] The setting aside of an Adoption Order should only therefore occur in highly exceptional circumstances. Domestic case law similarly provided little in the way of support for the birth parents' cause.

The Court was guided by cases such as *Re B* (*Adoption*: *Jurisdiction to Set Aside*) [1995][27] where the adult adoptee, by then in his mid-thirties, sought to overturn his Adoption Order. He had been raised from infancy in the Jewish faith by his Orthodox adoptive parents. He later learned that his genetic father had however been a Kuwaiti Muslim, while his birth mother had been Catholic. He had been refused permission to live in Israel because of his Arabic heritage but had been similarly unable to settle in Kuwait. His application for the revocation was based upon the argument that a fundamental mistake of fact had formed the basis for the making of the Adoption Order. His application was eventually refused, with Swinton Thomas L J confirming that,

> an adoption order has a quite different standing to almost every other order made by a court... Once an adoption order has been made the adopted child ceases to be the child of his previous parents and becomes the child for all purposes of the adopters as though he were their legitimate child.[28]

Sir Thomas Bingham argued however that Adoption Orders are not simply,

> immune from any challenge. A party to the proceedings can appeal against the order in the usual way...where there has been a failure of natural justice, and a party with a right to be heard on the application for the adoption order has not been notified of the hearing or has not for some other reason been heard, the court has jurisdiction to set aside the order and so make good the failure of natural justice.[29]

The Court also considered the setting aside of the Adoption Order that had occurred in *Re M* (*Minors*) (*Adoption*) [1991].[30] Here, the genetic father had consented to a step-parent adoption, not knowing that his ex-wife was suffering from terminal cancer. Following her death, the adoptive father found it difficult to look after the children, so that they returned to live with the appellant. His appeal was based on the fact that his consent to the process was invalid, having been made

---

[25] *Ibid* para 141.

[26] *Ibid* para 145 See section 67(1) of the Adoption and Children Act 2002.

[27] *Re B* (*Adoption*: *Jurisdiction to set aside*) [1995] Fam 239.

[28] *Ibid* at para 245C.

[29] *Ibid* at para 252 E-F.

[30] *Re M* (*Minors*) (*Adoption*) [1991] 1 FLR 458.

in ignorance of such essential facts. His late appeal was accepted and the Order was set aside on the basis that his lack of knowledge had 'vitiated his consent.'[31] The Court emphasized however the fact that the best interests of the children in this case clearly demanded such an outcome, but also took pains to stress the highly unusual nature of the facts, ultimately noting that it did not create any kind of precedent 'for any other adoption case'.[32] On the issue of procedural irregularity, Wall LJ conceded that the Court of Appeal had previously set aside Adoption Orders, for example in circumstances where birth mothers had not been served with the relevant court papers[33] or where some manner of fraud had occurred.[34] These were extremely unusual circumstances however and therefore unlikely to be relevant to the Webster's application.[35]

Similarly in *Re K (Adoption and Wardship)* [1997][36] where a Bosnian child was adopted by English parents the process was found to have been so 'fatally flawed' that the Adoption Order had to be set aside. Butler-Sloss LJ declared that,

> The law seems to me to be clear that there are cases where a fundamental breach of natural justice will require a court to set an adoption order aside.[37]

Wall LJ noted also that the issue of dispensing with parental consent at the Freeing Order stage appeared to have received only 'cursory' treatment from the lower court.[38] The issue of possibly allowing a kinship care placement with the maternal grandparents had been rejected with very little discussion. Although the Webster's case effectively represented a miscarriage of justice, it did not however warrant revocation of the Freeing or Adoption Orders, nor was there anything in the Strasbourg jurisprudence to suggest that a violation of the parents' Convention rights had occurred.[39] In dismissing their application, the Court stressed the need for 'second opinions' on issues of evidence of physical abuse and neglect in such difficult cases, especially where parents have repeatedly and strongly denied accusations that they were the cause of any non-accidental injuries to their children or where there was scope for conflicting medical diagnoses to be made.[40]

---

[31] *Ibid* at 459 F-G per Glidewell LJ.

[32] *Ibid* per Butler-Sloss LJ at 460B.

[33] See *Re F (R) (An Infant)* (1970) 1 QB 385; see also *Re RA (Minors)* (1974) 4 Fam Law 182.

[34] See *Re F (Infants) (Adoption Order: Validity)* [1977] Fam 165 where the Adoption Orders were deemed valid (if theoretically voidable) as a result of the adopters not being legally married at the time of the hearing.

[35] *Webster* op cit n 1 para 162 per Wall LJ. The Court distinguished the case of *Re K (Non-accidental injuries: Perpetrator: New Evidence)* [2004] EWCA Civ 1181, [2005] 1 FLR 285 on the basis that the adoption hearing had not yet been held.

[36] *Re K (Adoption and Wardship)* [1997] 2 FLR 221.

[37] *Ibid* at 228H per Butler-Sloss LJ.

[38] *Webster op cit* n 1 para 187.

[39] *Ibid* para 175.

[40] *Ibid* para 194.

The concurring opinion of Wilson LJ found that it was now 'far too late for the applicants to bring appeals. . .at each of two levels.' The first level concerned the interests of the adopted children, who by this stage had been living apart from their original parents for over 4 years. During that time no contact at all had been permitted.[41] The second level was based upon the 'vast social importance of not undermining the irrevocability of adoption orders.' Arguably, the fact that the original parents' quiet outrage at the allegations made against them 'was not translated into attempts, with whatever degree of legal difficulty' to convince the courts of their 'innocence' (i.e. by ordering fresh reports) appears to have perhaps worked against them.[42] Whether there is now an onus on birth parents to demand extra, 'second opinion' evidence to help contest Care proceedings, in case they wish to challenge a placement decision at a later stage, remains to be seen. If so, this could create a substantial burden, both in terms of evidence and financial cost, on many parents.[43] The Court did not specifically address the question of exactly when during the proceedings such extra evidence might have to be sought; potentially, the need could arise at the Care Order stage, at the hearing for Placement or Freeing, upon receipt of notice that an Adoption Order is to be applied for or perhaps during pre-hearing meetings between local authorities and parents.

In relation to the issue of post-adoption contact, additional expert evidence might have been particularly useful. Wilson LJ briefly referred to how the children had 'not seen' their parents for over 4 years, which is suggestive of the fact that direct contact had ceased at quite an early stage in the process. It is unclear whether any degree of contact had been permitted between the estranged siblings (only two of whom had been placed together) as the Court did not refer to this issue. Also of significance was the local authority's suggestion that the adoptive parents should be free to decide upon the content, manner and timing of 'telling' the children about their history. This was framed as falling within 'their role and responsibility'.[44] The Court made no comment on the issue, presumably being satisfied with this proposal.

The case perhaps turned mainly upon the fact that 'the children and the adopters all had rights under ECHR Article 8: the current applications undermined them.'[45] The key right, to family life, had simply passed over in its entirety to the adoptive parents at the point when the Adoption Order was made, being completely extinguished in respect of the birth parents. From the earlier point of the Freeing Order however, a key element of the children's Article 8 rights in respect of their natal connections presumably also ceased, given that contact was from then on prohibited. As such, the Convention rights of the adoptive parents replaced rather than outweighed those of the birth parents, with the children's rights being defined

---

[41] *Ibid* para 204 per Wilson LJ.

[42] *Ibid para* 205. Criminal proceedings were not brought against the parents.

[43] Arguably, solicitors acting for non-consenting parents might also perhaps be required to demand extra evidence in such cases, to avoid negligence proceedings.

[44] *Webster op cit* n 1 at para 84.

[45] *Ibid* para 84.

by reference to those of their adoptive parents. A strict, no contact post-adoption approach leaves little room for any form of compromise; in cases such as this, some form of long-term fostering or permanent guardianship might have at least allowed for the possibility of some form of contact to be facilitated.[46] Arguably, the Court did seem to make the suggestion that the blood-tie may to some extent have a rights-bearing status, in terms of affording birth parents the locus standi to appeal against the making of an Adoption Order.

The question of whether legislation dispensing with parental consent to adoption was actually compatible with the European Convention came before the United Kingdom Supreme Court recently, when a birth mother brought a challenge against the provisions of the Adoption and Children (Scotland) Act 2007.[47] In *S v L* [2012][48] a Sheriff's Court had considered the question of whether the child's welfare required maternal consent to the adoption to be dispensed with; the child's mother had argued that her right to respect for family life under Article 8 of the European Convention was not adequately protected by the 2007 Act and that it was thus not compatible with the 'narrower reading' of welfare required by the case law of the European Court of Human Rights.

She sought to rely upon the decision in *Neulinger v Sweden*,[49] that the child's interest was comprised of 'two limbs': in other words, a child's original ties ought to be maintained unless the family has proven to be 'particularly unfit' so that only 'very exceptional circumstances' should sever family ties, with everything being done to preserve personal relations and possibly 'rebuild' the family. Equally, parents cannot cite Article 8 rights in the pursuit of anything that might be harmful to their child's health and development. Referring to subsequent Strasbourg jurisprudence, the Supreme Court found that the best interests of the child were paramount[50] and that the almost identical English legislation on dispensing with consent had already been found to be entirely compatible with the European Convention.[51] In respect of the Grand Chamber's comment in *Neulinger* that family ties ought only to be severed in 'very exceptional circumstances' the Supreme

---

[46] See further Bainham (2007), pp. 520–523.

[47] The court must first consider the issues under s 31(3) (c) (as to whether the where the parent is and is likely to remain unable satisfactorily to discharge their parental rights and responsibilities in respect of the child) before looking at s 31 (3) (d) which provides for situations where a parent is and is likely to remain unable satisfactorily to discharge their parental rights and responsibilities in respect of the child.

[48] *S v L* [2012] UKSC 30 (11 July 2012).

[49] *Neulinger v Sweden* (2010) 54 EHRR 1087.

[50] *R and H v United Kingdom* (2011) 54 EHRR 28; This was further confirmed in *Uyanik v Turkey* 3 May 2012) (app 60328/09) available at http://hudoc.echr.coe.int/sites/eng/pages/search.aspx?i=001-110705 (accessed 15.03.13).

[51] *S v L* [2012] *op cit* n 47 para 40. *Y C v United Kingdom* (App 4547/10) which drew upon previous tests for Convention-compatibility (in Johansen v Norway, Neulinger v Sweden and R and H v UK) re-framing them instead as 'considerations' to be kept in mind. See further McFadden and Thomson (2012), 17, available at http://www.journalonline.co.uk/Magazine/57-9/1011624.aspx#.UV59WKLZHCY (accessed 12.03.13).

Court declared that this was not 'a legal test'[52] but rather an 'observation' on how relatively infrequently such events might occur. The fact that very few children are affected annually by such proceedings (the Court referred to the Scottish population statistics for 2010) was in itself the 'exceptional circumstance' referred to by the Strasbourg Court. Arguably, looking at the issue in terms of numbers is not particularly helpful to a rights-analysis. The majority of children will not enter the care system, nor will most children be denied contact with their natal relatives or access to their birth records; those who do face the loss of their original ties and have perhaps suffered harm as a result will draw little comfort from the thought that their experience was somehow an exceptional or very rare occurrence, in terms of the general population facing such issues. Justifying a potential Convention rights-violation on the basis that it would only affect a small minority group perhaps sets a dangerous precedent in terms of a potentially creating a policy of discrimination or inequality. It is difficult to see how the term 'exceptional circumstance' would not more accurately relate to the individual situation of the family and child facing the loss, so as to ensure that such a harsh severance of family ties is not simply framed as an inevitable, normative feature of created kinship but as a profoundly unusual and undesirable event in terms of affecting rights, going against social and cultural norms of parenthood, having adverse psychological consequence and risking potential harms, rather than simply in respect of frequency of occurrence; it should happen only where the child's welfare clearly demands it.

7.3 Northern Ireland: The 'Need' for Contact Outweighed?

In the absence of court-ordered contact post-adoption, there is often no legal onus upon adoptive parents to actively enable or preserve connections between estranged natal relatives.[53] Until fairly recently, the predominant model within Northern Ireland seemed to be based upon the perception that 'post-adoption contact of any kind may have a limited value.'[54] The need to achieve a high degree of reassurance for prospective adopters often seemed perhaps to be the primary concern for the courts and Social Services Trusts. Arguably, this differed to some extent from the private law presumption which generally sought to promote the preservation of kin contact as largely beneficial.[55] The 'best interests of the child'

[52] *Ibid* para 44.

[53] See for example *Re NI and NS* [2001] NIFam 7 (24 March 2001). Contact seemed until fairly recently to generally be restricted to an indirect form (such as 'letter-box' contact) or only permitted a few times a year, subject to birth kin not behaving in a way that might be disruptive to the placement. See also *Re EFB* [2009] NIFam 7.

[54] *Re A* [2001] NIFam 23 per Gillen LJ.

[55] See for example *Re T (A Minor)* [1998] EWCA Civ 1871 (27 November 1998).

principle often seemed to be subject to a fairly narrow interpretation that tended to leave little room for discussion of the longer-term psychological consequences of losing one's genetic connections.

Pending legislative reform, the role of the judiciary in such cases remains particularly significant within Northern Ireland. Proceedings involving non-consensual relinquishment by birth parents have tended to follow a slightly convoluted statutory template[56]; in dispensing with birth parental consent, the Court firstly frees the child for adoption via the provisions of the Adoption (N I) Order 1987.[57] The best interests of the child are considered under the general Children (NI) Order 1995 (as amended) welfare checklist.[58] A Care Order application is often necessary so that parental responsibility is legally transferred to the Trust;[59] this issue must be dealt with first.[60] This approach to enabling an adoptive placement is aimed at preventing 'compromise (of) the application of the paramountcy principle.'[61] Prior to making the Freeing Order (and before the Care Order is considered) the Court must look at the question of contact post-adoption. This often results in the 'make no order' principle being applied (where it is better for the child for the court to make no order, then this will be the approach taken) so that the issue of contact might be left largely to the discretion of the relevant Social Services Trust or perhaps to the adoptive parents. The dicta of Ormrod LJ in *Re H* [1981] would sometimes be cited, to perhaps provide guidance on the issue of seeking or challenging subsequent decisions on post-adoption contact;

> adoption gives us total security and makes a child part of our family and places us in parental control of the child; long term fostering leaves us exposed to changes of view of the local authority, it leaves us exposed to applications and so on by the natural parent.[62]

Until the incorporation of the European Convention (in October 2000), extensive discussions of human rights principles (in fairness they were persuasive rather than

---

[56] See *Re W (An Infant)* [1971] 2 All ER 49 wherein Lord Hailsham had declared that '*welfare per se is not the test*' in relation to the granting of Freeing Orders; only a finding of parental '*unreasonableness and nothing else*' ought to be considered, given that the key judicial issue '*is not culpability. It is not indifference. It is not failure to discharge parental duties. It is reasonableness and reasonableness in the context of the totality of the circumstances.*' Arguably, the phrase '*totality of the circumstances*' could have elevated the best interests principle to the level of primary consideration. See also *Re F (Adoption: Freeing Order)* 2000 2 FLR 505.

[57] The Order is based upon the English Adoption Act 1976, which was itself inspired by The 1972 Houghton Report. Under Article 16(2) (b) of the Northern Ireland Order, birth parents must be deemed to be '*withholding..agreement unreasonably*' for the Freeing Order to be made.

[58] See Article 3 of The Children (NI) Order 1995. The Adoption Order 1987 Article 9 also states that, '*in deciding any course of action in relation to the adoption of a child, a court or adoption agency shall regard the welfare of the child as the most important consideration.*' See also the expanded adoption welfare checklist s.1 Adoption and Children Act 2002.

[59] See Article 50 of The Children (NI) Order 1995.

[60] See *Re D (Simultaneous Applications For Care And Freeing Order)* [1999] 2 FLR 49.

[61] Per Gillen J in *Re A* [2001] NIFam 23 (24 October 2001) at p. 8.

[62] *Re H* [1981] 3 FLR 386.

binding in nature until the Human Acts Rights 1998 became law) in Freeing Order hearings seemed to happen infrequently.[63] In *Craigavon and Banbridge Community Health and Social Services Trust v JKF* (2000) for example the court applauded the fact that 'very properly, given the facts of this case, Counsel did not invoke to any material degree the European Convention.'[64] Convention rights were not to be 'routinely paraded in cases of this nature.'[65] The need for a wide margin of appreciation was also stressed, with Gillen J citing Kilkelly's suggestion that it was 'a matter of domestic law whether the child's best interests are paramount or merely of 'special importance' in such matters.'[66] The non-binding Children's Convention was referred to however with the Court highlighting 'the fundamental right of every child to belong to a family' and the fact that 'this principle underpins the United Nations Convention.'[67]

Interpretations of the 'best interests of the child' principle also seemed to vary. In *Re A* (2001)[68] for example, it was suggested that in some cases involving post-adoption contact, the 'child's interests are not paramount'.[69] A range of relevant factors were highlighted in this case, including the failure of the birth parents to fully 'appreciate' their social workers' previous efforts at attempting family rehabilitation.[70] Citing the persuasive English authority *Re F* (2000)[71] it was however eventually decided that child welfare ought to be taken into account, despite the fact that it still lacked paramountcy 'at this stage' of the proceedings. In relation to the question of post-adoption contact, 'the crucial test' was the 'best interests' of the child. The court saw no benefit in actively maintaining the ties of genetic kinship, opting instead for complete severance, and directing that the affected parties should be able to 'gradually break their final bonds'[72] over the course of the next few weeks.

In *Re K* (2002)[73] the court similarly decided against actually ordering post-adoption contact. Here the child's maternal grandmother had been refused a Residence Order (and thus parental responsibility) on grounds which included the fact that her 'love for this child (was) simply not enough to secure his care and protection.'[74] The views of the potential adopters were then 'taken carefully into

---

[63] See The Human Rights Act 1998 (in force from October 2000).

[64] *Craigavon and Banbridge Community Health and Social Services Trust v JKF, In. The Matter Of* [2000] NIFam 56 (27 November 2000).

[65] Citing *Daniels v Walker* (2000) 1 WLR 1382 at p. 1387.

[66] See further Kilkelly (1999), p. 298.

[67] DHSS Departmental Guidance (1999) at para 2.1.

[68] *Re A* [2001] NIFam 23.

[69] *Ibid* Per Gillen J.

[70] *Ibid.*

[71] *Re F* [2000] 2 FLR 505.

[72] *Ibid.*

[73] *Re K* [2002] NIFam 13.

[74] *Ibid* at p. 10.

account'[75] with the result that the 'make no order' principle won out. Some discussion of the psychological needs of the child might perhaps have been useful in both of these cases. Although the child protection issues in the latter case would have most likely prevented the child from returning to the inadequate care of the birth parents, the contact issue might well have been decided differently, for example via some form of indirect contact being allowed. The adoptee might then have learned eventually that her birth father had kept 'a plethora of photographs' of her and had 'made numerous toys' for her during his period of incarceration.[76] Similarly, the loss of her other relationships, such as the one that clearly existed between her and her grandmother, might not have been framed simply as an apparently inevitable consequence of being adopted.

In *Re J and S* (2001) the adoptee might have similarly derived some subsequent comfort from learning that her birth mother's refusal to consent to her relinquishment was based on her desire to prevent there being 'any written evidence in writing that she . . .(had) agreed'[77] to it. Her stated motivation was to spare her child from feelings of abandonment in later life. Her fairly prescient concern for her birth daughter's future welfare could only be deemed 'unreasonable' under the legislative framework however, so that the Freeing Order might be made and permit the adoption of her daughter to proceed.

In many such cases, discussion of the significance of (or the need for) genetic relatedness is at best fairly minimal. The Court found in *Re CBCHSST v JKF* [2000] for example that the 'freed for adoption' child had basically never known her birth mother 'as her mother.' As such there was simply 'no existing relationship in being'[78] that might merit consideration. The Court placed a total ban upon all forms of contact between the mother and her child, with the exception of one 'final goodbye visit' and the provision of some photographs, which would apparently provide the child with some measure of a 'life history' as she grew older. The argument underpinning this outcome was that this would avoid creating 'confusion in the child.'[79] Presumably it also provided some measure of reassurance for the adoptive parents and a useful sense of closure for all who had been involved in dealing with the case.

The issue of the need for post-adoption contact between estranged birth siblings arose in *Re NI and NS* [2001][80] where it was held that such forms of kin contact were to be 'encouraged.' The High Court declined however to make any form of Contact Order, preferring instead to confer discretion over the matter upon both the Trust and the adoptive parents. The court cited the Court of Appeal's decision in *Re*

[75] *Ibid* Per Gillen J.

[76] *Ibid.*

[77] *Re J and S* (2001) NIFam 13 (23 May 2001) This also was a factor in the birth mother's refusal to provide consent to relinquishment in *Re CBCHSST v JKF* [2000] NI Fam 76.

[78] Per Gillen J in *Re CBCHSST v JKF op cit* n 64.

[79] *Ibid* para 78.

[80] *Re NI and NS* [2001] NIFam 7 (24 March 2001) per Gillen J.

B-M (2001)[81] to declare that 'the advantage of adoption to the child' justified the over-ruling of the birth parents' wishes on contact. Whether the benefits of being adopted might have also outweighed the need for some form of sibling contact, was not mentioned.

Interestingly, the Northern Ireland High Court went on to refuse the making of a Freeing Order 3 years later in *Re P* [2004], despite there being clear evidence of child abuse having occurred, on the basis that too little was yet known of 'the long term impact of a child learning of his adoption in the context of an opposed application.'[82] Likewise, in *Re E and M* [2001] it was held that 'the making of an adoption order is a very grave matter indeed.'[83] Although domestic violence, alcohol problems, child neglect and emotional abuse were cited as relevant aspects of the case, the Court refused to free the child for adoption, stating that,

> 'Parenthood' also involves ties and relationships of natural love and affection, held by parents towards their children and children towards their parents, even in adversity. To break those ties is a grave matter. To do so against the wishes of the parent who loves that child is more so.[84]

It was stressed that the legislation required 'something less than paramountcy' because of the 'draconian nature of an Adoption Order and also its finality.'[85] With child welfare not occupying a position of paramountcy however, the views and rights of the birth parents were able to be afforded a greater degree of prominence.[86] Referring briefly to Article 8 of the European Convention, the court also found that the parental consent to the adoption process had not been 'withheld unreasonably' under the terms of the domestic statute. With the Freeing Order application refused, a Contact Order was then granted, so that the 'natural ties' which existed between the parties might be renewed whilst their children were under the care of the Social Services Trust.

Although such an outcome could be seen as entirely justified in terms of preserving the parties' blood-ties and protecting their right to family life, it possibly fails to address a number of key child protection issues. The 'best interests' of the child, in terms of being safely or adequately cared for, were not discussed in great depth. This 'wait and see' approach seems to conflict with the reasoning in other, not dissimilar cases where it was felt that 'time [was] of the essence'[87] given that children 'simply cannot wait indefinitely for parents to change.'[88]

---

[81] *Re B-M (a child) (Adoption:Parental Agreement)* 2001 1 ACR 1.

[82] *Re P, (Freeing Without Consent)* [2004] NIFam 6 (11 February 2004).

[83] *Re E and M [2001]* NIFam 2 (02 February 2001).

[84] *Ibid* per Higgins J at pp. 23–25.

[85] *Ibid.*

[86] The Court referred back to the subsequently repealed Adoption Act 1967, which did contain a 'paramountcy' test, noting that this had been omitted from the 1987 legislation. See section 5(1) of the Adoption Act 1967 where the '*welfare of the infant shall be the paramount consideration*'.

[87] *Re C (Freeing for Adoption)* [2002] NIFam 1 (12 January 2002).

[88] Per Gillen J in *Re J (Freeing without consent)* [2002] NIFam 8 (13 March 2002).

The processes surrounding freeing children for adoption in Northern Ireland came in for significant criticism from the House of Lords in *Down Lisburn Health and Social Services Trust v H* (2006).[89] Baroness Hale's Dissenting Opinion highlighted the various disadvantages of the system, namely that it often tended towards the creation of a 'legal limbo' for many children awaiting an adoptive placement, diverted the Court from the real issues, and increased levels of administrative delay.[90] As such, 'adoption practice in Northern Ireland may labour under certain misconceptions.'[91] The children involved in this case were however older; they 'had a history' and had formed 'significant relationships' with members of their birth family.[92] In respect of the issue of post-adoption contact, Baroness Hale reiterated that this might assist the child in finding a sense of identity and self, aid the preservation of 'significant attachments' and prevent such feelings of loss or rejection which 'the child who remembers her birth family may feel if she is completely cut off' from them.[93]

Because Freeing Orders could be granted before suitable prospective adopters were found (i.e. those who are willing to permit some form of post-adoption contact) the Court might not be taking into account all relevant factors in determining whether an eventual adoption would fully represent the best interests of the child. Under the legislation, the question of whether contact-tolerant adoptive parents could be found was relevant to the core issue of whether birth parents were unreasonably withholding their consent.[94] The attitude of both the Trust and the High Court towards the psychological evidence provided was also mentioned here. Professor Triseliotis' report had clearly stressed that the child had existing 'attachments of some significance to her parents' and was therefore benefitting from having episodes of contact with them. If all familial contact were to simply cease there would be a high likelihood that 'psychological scars' would result. Put simply, her connection to them was much stronger than had been initially estimated by the Trust staff.[95]

The Trust's informal policy of not seeking out adoptive parents until after the child had already been freed by the court (and its apparent opposition to the need to

[89] *Down Lisburn Health and Social Services Trust v H* [2006] UKHL 36.

[90] *Ibid* para 24 per Baroness Hale.

[91] *Ibid* para 36. The current legislation was drafted in the wake of The Houghton Report (1972) CMND 5107 HMSO which seemed to suggest that 'permanency' equated either to the rehabilitative reunion with birth kin or to a complete severance from them, in a bid to achieve some form of security for the new 'family for life'.

[92] *Op cit* n 89 per Baroness Hale LJ, para 6.

[93] *Ibid* para 7.

[94] *Ibid* para 8.

[95] *Ibid* para 13–14. The fact that contact-tolerant adopters had not yet been found was also potentially in breach of the Policy Guidance (DHSS) (1999) para 6.9 which recommended that prospective adopters should be found and approved of by Trusts *before* the final Care Order hearing. This would help to avoid delays and serve to connect the central issues of Care and Adoption.

find adoptive parents who would enable post-adoption contact) also contravened the Policy Guidance on the matter, which provided that prospective adopters should at least be prepared to consider the issue of post-adoption contact.[96] The Northern Ireland High Court had eventually decided that adoption was the best option in this case, but then went on to state that if this was proving to be too difficult for the Trust to effect (i.e. after 6 months of making 'all reasonable efforts' to find a suitable placement) then the 'benefits of adoption . . . (would) outweigh the benefits of post-adoption contact.'[97]

The birth parents' objections to the adoption, and their subsequent appeals were based upon the issue of the court's failure to make *any* direction in respect of the issue of interim contact between them and their daughter. Baroness Hale LJ declared in her Dissenting Opinion that under the provisions of the 1987 Order they were essentially acting as 'reasonable parents' would have done in relation to the question of their withholding consent to relinquishment, especially given how *Re W* [1971] had quite clearly defined a reasonable parent as one who 'places great weight on what will be best for the child.'[98] It was also noted that some adoption agencies in Northern Ireland seemed to look to post-adoption contact much less frequently than their counterparts in England and Wales were inclined to do. The fact that expert opinions on the need for contact might be disregarded in favour of a Social Worker Supervisor's report was also not a particularly welcome strategy.[99] It was further suggested that adult members of the triad might benefit from hearing the views of each other on post-adoption contact pre-hearing, so that 'each becomes a person rather than the ogre or the threat they may previously have been.'[100]

It was also noted that the consequences of a Freeing Order would be 'drastic for the child.....even more drastic for the parents'[101] because of the loss of parental responsibility. The suggestion here that birth parents might have 'more to lose' than the removed child is perhaps grounded in the presumption that, as adults, they could plead a more 'significant history' of family life or demonstrate a rights-bearing form of 'remembered relatedness'. Under this line of reasoning, young infants who possess only a genetic connection might stand to 'lose' less than their birth parents in terms of their right to maintain or develop original family connections or genetic identity. The unusual position of the United Kingdom was noted, in that it basically permitted and legalized the complete severance of all family ties via the adoption process, despite being a member of the Council of Europe.[102]

---

[96] *Ibid* para 16.

[97] *Ibid* para 23.

[98] *Re W* [1971] AC 682. This was followed in *Re D* [1977] AC 602 and withstood the challenge posed to it by the case of *Re C* [1993] 2 FLR 260.

[99] *Op cit* n 89 para 32.

[100] *Ibid* para 37.

[101] *Ibid* Per Baroness Hale LJ para 25.

[102] *Ibid* para 34, noting that adoption law in Northern Ireland seemed perhaps to operate outside the spirit of the European Convention.

Baroness Hale was careful to stress that post-adoption contact may not be appropriate in every case, nor should birth parent uncertainties over the enablement of future contact with their relinquished child always simply amount automatically to grounds for objecting to orders freeing the child for adoption. Ultimately, parents 'should try to put aside their own feelings in the interests of the children.'[103] Although directed at the birth parents' failure to consent to relinquishment of the child, this piece of guidance might apply equally as advice to any prospective adoptive parents who would unreasonably seek to automatically prevent all contact between estranged birth kin.

7.4 Freeing and Contact Pre-agreed

In the wake of the comments contained in the Dissenting Opinion, the potential effects of Freeing Orders on birth parents served as a catalyst for legal discourse. In *A Trust v M & P* (2007)[104] Morgan LJ outlined in some detail for example his reasons for the making of an Order for Freeing, and stressed that 'this is an order which will interfere with the family life of the parents and in particular with their Article 8 rights under the Convention.'[105] The prospective adoptive parents here had been on a 'considerable learning curve' in respect of permitting or enabling post-adoption contact yet this was to be 'rooted in an understanding' that such decisions must take into account the best interests of the children affected by the process.[106] The fact that the Trust had at least looked to the possibility of indirect contact (both in the interim period and post-adoption) was welcomed and it was also suggested that direct contact ought not to be completely ruled out as a future option.

Similarly, in *Re L2 and O* (*Post-Adoption Contact*) [2007][107] neither the Court nor the Trust sought to rule out the possibility of direct contact as a potential future outcome. Morgan LJ left the Trust's arrangements for interim indirect contact in place, deeming them to be satisfactory. Some of the birth mother's evidence perhaps provides a useful insight into how adoption might be perceived by those who are most profoundly affected by it. She had stated that she was prepared to accept the reality of the adoptive placement, but that she had real difficulty with the 'concept of adoption' in itself. As she saw it, the process involved a very harsh 'rejection of children by the parent' and then 'the exclusion of the parent from any aspect of the child's life.'[108] Her reluctance to bring about such an outcome perhaps

[103] *Ibid* para 40.

[104] *A Trust v M & P* [2007] NIFam 5.

[105] *Ibid* para 21.

[106] *Ibid* para 18.

[107] *Re L2 and O* (*Post-Adoption Contact*) [2007] NI Fam 12.

[108] *Ibid* para 7.

represents a useful analysis of the realities of the system in many such cases, where for example contact has been completely barred and any chance of eventual reunion perhaps left to be decided by the vagaries of chance and time.[109]

Recent cases have also focused more closely on matters of process and fair hearing. In *South Eastern Health and Social Services Trust v LS and Anor* [2008][110] for example the application for a Freeing Order was adjourned for 6 months on the basis that the birth mother was not withholding her consent unreasonably, given the amount of time that had elapsed since her parenting skills had last been formally assessed by the Trust. The court criticized both the expert witness and the Guardian ad Litem for looking too much to the past and only insufficiently examining and taking into account her present situation, given that the birth mother had since entered into a settled relationship and 'had matured' in regards to her ability to be a parent to her children.[111]

Similarly, in *Re EFB* [2009][112] the birth parents raised an argument in terms of preferring the option of a long-term fostering arrangement which would have allowed for some level of more frequent kin contact to be maintained between them and their child. They offered to provide a form of 'provisional consent' in return for the court making a post-adoption Contact Order at the Freeing Order stage. Neither of these outcomes was accepted by the court as a valid reason for withholding their consent however, under the terms of the current legislation. The court again opted to apply the 'make no Order' principle in respect of issue of post-Adoption contact, allowing for a wide degree of discretion to be given to social workers and adoptive parents as to the forms and frequency of any birth relative contact. The court did however stress that sibling contact would serve here to help 'consolidate a positive genealogical identity' for the adopted child.[113]

The court also noted the psychological evidence provided, namely that the child in question could be at risk of suffering 'significant emotional damage' in later life.[114] Arguably this also pointed to an increased acceptance of the research surrounding the complex, long-term psychological effects of suffering broken attachments in childhood, which some of the earlier cases perhaps lacked. The fact that kinship placement was preferable here did not however mean that the acute needs of the vulnerable child would be compromised simply in a bid to preserve the ties of genetic relatedness. This was especially so given that the benefits of a kinship placement would be longer term.[115] Where relationships with birth kin were abusive in nature, contact was unlikely to be viewed favorably. Perhaps most

[109] See for example the breaches of Articles 3, 6 and 8 ECHR that occurred in recently in *A & S (Children) v Lancashire County Council* [2012] EWHC 1689 (Fam) (21 June 2012).

[110] *South Eastern Health and Social Services Trust v LS* [2008] NI Fam 12 (27 October 2008).

[111] *Ibid* para 23 per Weir J.

[112] *Re EFB* [2009]NIFam 7.

[113] *Ibid* para 13 per Stephens J.

[114] *Ibid* para 49.

[115] *Ibid* para 48.

significantly, the court described the child's difficult relationship with her abusive birth father as having no ties of 'present attachment' but still potentially relevant as a 'relationship in the making.'[116] This observation again seems to contrast with some of the earlier jurisprudence on the issue of the need for biological connection, which suggested that genetic ties will not constitute a 'relationship in being' and therefore merit no judicial recognition or protection whatsoever.[117]

In *J* (*Care Order*) (2008)[118] the court similarly considered in some depth the possibility of a kinship care placement, discussing the individual characteristics of several members of the child's birth family and their capacity, or otherwise, to engage in a caring role.[119] The 'close-knit' nature of this extended family was alluded to but it was the level of their closeness, taken together with lingering child protection concerns over an earlier infant mortality, which was deemed to amount to a severe draw-back in terms of allowing them to act as the substitute carers.[120] The judgment is note-worthy given that the court re-iterated the concerns of the expert witness, who had highlighted the Trust's lack of a discrete kinship care process, adding that such carers 'tend to be older, in poorer health, poorer financially and have complex family relationships.'[121] It was also suggested that some of the procedures employed by the Foster Care Panel might not necessarily fully 'comply with Article 8'. The exclusion for example of the child's grandparents, who were then being considered as possible kinship carers, from much of the pre-hearing proceedings meant that 'this was an inadequate process for ensuring inclusive involvement in the decision-making process.'[122]

Granting the Care Order, Gillen J did refer to the psychologist's recommendation for thrice annual birth family contact and to the need for any prospective adopters to 'take on board the need for ongoing contact with the current primary carer as well as the family'.[123] A clear warning was given to the child's birth relatives however, to be mindful of their future conduct, given that 'post-adoption contact would have to be considered in light of the attitude of the birth family to any adoption that would be ordered by the court.'[124] The issue of contact seems to have been linked here more to the behaviour of the triad adults, and as a privilege rather than an aspect of the child's best interests-led welfare. The decision does emphasize

[116] *Ibid* para 12.

[117] Per Gillen J in *Re CBCHSST v JKF, In The Matter of* [2000] NIFam 76.

[118] *Re J (Care Order)* [2008] NIFam 11 (1 August 2008).

[119] *Ibid* para 54.

[120] *Ibid* para 61. See also para 65, per Gillen J, where he asks '*How then could the M's possibly comply with a protection plan which restricts contact and requires collaborative working with the Trust in circumstances where they might have to align themselves against O and F.?*' (the birth parents).

[121] *Ibid* para 27.

[122] *Ibid* para 46.

[123] *Ibid*.

[124] *Ibid* para 94.

the potential benefits for the child of having some form of post-adoption contact, and suggests that a range of possibilities may well exist in relation to finding workable kinship care alternatives to formal adoption, devoid of contact.

Interestingly, freeing was refused in *Re JJ* [2009][125] on the basis that child welfare paramountcy was 'not the only consideration.' The court had to 'take into account and weigh proportionately'[126] the birth parents' European Convention right to family life, as well as the 'prospect of the development of a secure attachment'[127] between the child and the original family. A request to reduce the current levels of contact was therefore refused. Adoption was also deemed to represent an unlikely option given that 'long-term fostering is not a halfway house to another outcome but a long-term framework for this child.'[128] Similarly in *SEHSST v LS* [2009][129] the court drew upon the principle of 'good enough parenting' to refuse the Freeing Order. Here, 3 years had elapsed since the initial Care Order had been made, with no prospective adopters yet being found. Given the need to apply the 'likelihood test'[130] the court conceded that its remit to decide upon the vulnerable child's best interests was limited by the presence of the Care Order. In other words the child's future welfare was now in law 'a Trust matter'. It was held that a 'compelling reason' would have to be shown before a Freeing Order could be granted, with the court noting that 'in the ordinary way the rearing of a child by his or her biological parent can be expected to be in the child's best interests.'[131]

That said, the promotion of new family autonomy and security still remains an important aim. In *Re EFB* [2009], for example the Court had not wished to 'overload the adoptive family with too many arrangements'[132] in respect of having to actively facilitate post-adoption contact. There was similar reluctance to require of them anything which might have served to undermine 'the control of the adoptive parents.'[133] In the absence of court-ordered contact arrangements, it is difficult to see how Trusts might be able to remedy a situation where, for example, previously content adoptive parents subsequently decide that they can no longer facilitate or tolerate contact with birth relatives. The court stressed here that adoption emphasized stability, commitment and security for the child, by engendering a 'greater sense of belonging', allowing for 'less intrusion' from Social Services, and permitting 'less chance of the child suffering from 'drift' or feelings

[125] *Re JJ* [2009] NIFam 2.

[126] *Ibid* para 15 per Morgan LJ.

[127] *Ibid.*

[128] *Ibid* para 24.

[129] *SEHSST* v LS [2009] NI Fam 14.

[130] *Ibid* per Weir J [at para 21citing Lord Nicholls of Birkenhead *In Re G (Children) (FC)* [2006] UKHL 43].

[131] *Ibid* at para 2.

[132] *Re EFB op cit* at n 112.

[133] *Ibid.*

of 'impermanence.'[134] In comparison with long-term fostering, adoption may well bring 'a different commitment to the task of parenting'[135] and promote 'greater closeness between parent and child.'[136]

Because much of the initial task of decision-making in respect of post-adoption contact now seems to fall primarily to social workers, via pre-hearing meetings aimed at finding workable agreements on the issue, an obligation to comply with the 'fair hearing' provisions of Article 6 of the European Convention exists in respect of involving birth relatives. Equally, if the courts have essentially delegated much of the management of post-adoption contact issues to Trust social workers or parents, then they perhaps bear a fairly substantial burden, not least in respect of having to ensure that the human rights of all concerned are promoted and protected against unlawful interference.[137] Some of the procedural aspects of such meetings, such as whether funded legal representation is automatically available for meetings aimed at negotiating the terms of the agreement for contact, must be seen to comply with the European Convention, but should perhaps also aim to reflect the principles of the Children's Convention, for example in relation to all parties having access to fair and adequate assessment processes, with up to date, independent expert evidence being made available. The consequences of giving informed consent to relinquishment or of agreeing to limited or nil contact should also presumably have been fully explained to all who will be most affected by the agreement.[138]

---

[134] *Re EFB op cit* n 112 para 36 per Stephens J.

[135] *Ibid.*

[136] *Ibid.*

[137] Article 6 of the ECHR protects the right to '*fair and public hearing within a reasonable time by an independent and impartial tribunal established by law.*' Available (see the Schedule to the Human Rights Act 1998) http://www.opsi.gov.uk/acts1998/ukpga_19980042_en_3. Accessed 20.10.09. See also Policy Document (2010) http://www.baaf.org.uk/webfm_send/2565 (2010) (accessed 31.10.11) and the (2006) Consultation Document on the need for legislative reform in Northern Ireland http://www.dhsspsni.gov.uk/adopting-the-future-consultation-report-final-2.pdf (accessed 10.11.11).

[138] Whether such meetings might to some extent warrant an increased formalization (to ensure for example the availability of right of appeal or judicial oversight if either set of parents subsequently change their mind on the issue of contact) is perhaps a question that requires more empirical research into the various processes currently used by Social Services Trusts. In respect of how pre-proceedings meetings are currently conducted in England see the recent evidence of Masson (2013) to the Education Committee – '*Children first*: *The Child Protection System in England.* Available http://www.publications.parliament.uk/pa/cm201213/cmselect/cmeduc/137/137vw79.htm (accessed 28.02.13).

7.5 Contact as Best Interests?

Although Freeing Orders have been replaced by Placement Orders in England and Wales[139] the issue of the decision-making remit of local authorities has recently arisen in a number of cases there. In the English case *Re K (Care Proceedings: Care Plan)* [2007][140] for example, the question of indirect contact arose in the wake of a bar on direct contact as part of the child's care plan, which in turn led to the court's refusal of the Care Order. The matter was eventually resolved on appeal, with the Care Order being granted and questions over contact being returned to the High Court for re-hearing. As Bridge later commented, 'what is clear is that in the final analysis the court.. ..determines questions of contact whatever the care plan may or may not say on the subject'.[141]

In *SB v A County Council; Re P* [2008][142] the Court of Appeal had to determine whether the statutory welfare test was to be regarded as a 'simple welfare test' or as an 'enhanced test'. Given that the family life rights of parents were enshrined in Article 8 of the European Convention, the legislation seemed likely to demand more than a simple approach to the concept of child welfare, with increased emphasis being placed upon the lifelong, longer term aspects of adoption. A process of 'dual planning' (simultaneous searches for adopters or foster carers, who would be willing to enable contact between siblings) would be unlikely to offend against the legislative provisions.[143] The Court stressed that decisions on post-adoption contact ought however to be taken by the judiciary rather than by the prospective adopters or the adoption agencies. An order for contact could therefore be granted at the same time as the adoption application hearing.[144] The Court noted with regret that the local authority social workers had unilaterally taken the decision to terminate contact between the mother and one of the children in this case. It was held that, where possible, siblings should be permitted to remain in contact with one another post-adoption or during fostering.

In *K (Children) v Sheffield City Council* [2011][145] matters of procedural importance were again highlighted as the child's parents sought to oppose the making of a Special Guardianship Order in favour of the maternal grandmother and argued for another assessment of their own parenting skills to be made. They expressed particular concern as to whether the child's ethnic and religious heritage (she was Muslim) would be adequately protected and provided for within the grandmother's secular household, and also argued that they had hoped eventually (post-therapy and with assistance) to be able to have the child returned to their care. The Court of

[139] S. 52 (1) (b) The Adoption and Children Act 2002.

[140] *Re K (Care Proceedings: Care Plan)* [2007] EWHC 393 (Fam).

[141] Bridge (2008), 114.

[142] *SB v A County Council; Re P*[2008] EWCA Civ 535.

[143] *Ibid* para 154 per Wall LJ.

[144] *Ibid* para 153.

[145] *K (Children) v Sheffield City Council* [2011] EWCA Civ 635 (25 May 2011).

Appeal found that the correct balance had been achieved by the lower court in respect of addressing the

> need to keep [the child] in her own family alongside considerations as to culture and religion...to adjourn for a further report ... would have been to generate further litigation.[146]

The lower court had also opted to make a s. 91(14) order (under the Children Act 1989) so as to prevent the parents from bringing any further court applications for a period of 2 years and 4 months, considering this to be entirely in the child's best interests, as it provided for a 'period of calm' which would be free from protracted or unnecessary litigation. It was however stressed that the making of such an order was an 'exceptional' occurrence; the child's welfare was of paramount importance in such cases as this one and therefore clearly required this as an outcome. As Black LJ noted 'this case was not ...a run of the mill case but an unusual one.'[147]

In *Re H* [2012][148] Thorpe LJ similarly noted that 'to be brought up by the birth mother is generally of cardinal importance to the well-being of children.'[149] As such, courts should be careful to avoid effecting a 'complete absence of recognition of the right to family life, which is the entitlement of not only the mother but of the children.' Equally, 'the principle that siblings should not be separated...[was] a generally recognised principle in family law.'[150] Article 8 Convention rights should be considered to a meaningful extent as should the expert evidence of witnesses. The Court also outlined the need for pre-hearing agreement procedures to reflect the Article 6 Convention requirements of fair hearing;

> Placement orders do not follow automatically or consequentially from the making of care orders. They are orders of the greatest significance and, insofar as they require overriding parental objection, albeit on welfare grounds, they must surely be preceded by preparations and processes that duly respect the Article 6 rights of the parents.[151]

In *Re K (Children)* [2011][152] the Court found that the removal of direct parental contact was one of the most 'extreme ...interventions that a court can make in ordinary private family life.'[153] The impact upon parent and child was likely to be 'significant' given that such contact generally tended to be in the best interests of the child. Referring to an earlier judgment, it was further observed that,

---

[146] *Ibid* para 43.

[147] *Ibid* para 53.

[148] *H (Children), Re* [2012] EWCA Civ 743 (16 May 2012).

[149] *Ibid* para 17.

[150] *Ibid.*

[151] Para 27. See further *Re B* (*a child*) (*residence order*) [2010] 1 All ER 223 where the Supreme Court declared that framing issues in terms of the right to be brought up by a biological parent rather than in terms of the best interests of the child served to alter the focus away from the child's welfare in decisions on residence.

[152] *K (Children), Re* [2011] EWCA Civ 1064 (20 July 2011).

[153] *Ibid* para 35.

> There is a positive obligation on the State . . .to take measures to maintain and to reconstitute the relationship between parent and child, in short, to maintain or restore contact. The judge has a positive duty to attempt to promote contact. The judge must grapple with all the available alternatives before abandoning hope of achieving some contact. He must be careful not to come to a premature decision, for contact is to be stopped only as a last resort and only once it has become clear that the child will not benefit from continuing the attempt.[154]

That said, the father in this case had a history of child sex offences; the question of balancing the conflicting rights and interests was relevant given that 'the more finely balanced a decision is in a case regarding children the more impossible it will be to challenge on appeal.'[155] The Court of Appeal could not simply substitute its own discretion for that of the lower court, so that the father's appeal against the ban on him having direct contact did not succeed here. Had the presence of legal or genetic relatedness *per se* been sufficient here to give rise to some sort of absolute parental right to engage in contact, the best interests of the child principle would clearly have been completely sidelined given the obvious risk of harm to the child. The main significance of the case perhaps lies however in its endorsement of the notion of a positive state obligation attaching to the preservation of familial contact as being generally in the best interests of children.

The case of *Coventry City Council v PGO & Ors* [2011][156] is of interest in relation to determining the point at which placement occurs. Here foster parents decided suddenly at a late stage in the process to seek an injunction preventing the adoptive placement of the two infants who were in their care. The two main issues for the Court of Appeal were whether the lower court had had the jurisdiction to do this and whether a child was legally 'placed' for adoption at the point when an adoption agency or local authority had ratified a match with prospective adopters, or whether legal placement only occurred when the child took up actual residence with the prospective adoptive parents. Despite the birth mother's expressed support for the idea of their remaining with their foster parents, the local authority had expressed concern over their suitability as adopters. This was based upon a number of issues including their financial situation, the foster father's previous history of offences committed as a young man, and his lack of contact with his own children from a former relationship, despite the fact that they had by then successfully foster-parented twenty three children.[157]

In terms of the injunction being granted, it was held that the judge should have asked whether the local authority's decision to remove the children from the foster parents for adoptive placement was irrational, disproportionate or otherwise unlawful, or in breach of any of the Article 8 ECHR rights of any member of this unusual 'triad' (i.e. the adopters, foster parents and children). None of these issues were

---

[154] *Ibid* para 14, Citing Mumby LJ in *Re C (a child)* [2011] EWCA Civ 521 (at para 47).

[155] *Ibid* para 35.

[156] *Coventry City Council v PGO & Ors* [2011] EWCA Civ 729 [2011] All ER (D) 159 (available at http://www.bailii.org/ew/cases/EWCA/Civ/2011/729.html accessed 11.03.13).

[157] *Ibid* para 48.

deemed to be relevant here; the welfare of the children was meant to be considered in respect of how the granting of an injunction preventing their placement might have affected it.

Ultimately the disappointed prospective adopters withdrew their candidacy to adopt so that the children were not removed from their foster home.[158] The Court expressed considerable sympathy for the disappointed adopters (they had for example spent a considerable amount of money and time preparing for the children's arrival) and went on to make several interesting observations on the differences between foster and adoptive parenting (*per curiam*):

> Foster parents, particularly short-term foster parents, and adopters have entirely different roles; and the appraisals which local authorities conduct before approving persons as adopters and as foster carers respectively are of an entirely different nature, extent and perhaps even intensity.[159]

As such, foster parents should apparently be and remain prepared for the eventuality that they are not necessarily likely to retain their role as psychological parent by becoming eligible to apply to be the child's adopter. There was also a clear

> need for them to accept that in all likelihood the child will be moving away from their home and thus to fashion a relationship with him on that limited basis.[160]

It is difficult to see how the issue of the children's emotional attachments might fit within this guidance, especially where, as was the case here, they had been cared for since birth (and very young infancy, respectively) by their foster parents. One could wonder also as to how foster parents might be expected to parent the children differently, so as to prevent strong bonds of affection or attachment (associated with good parenting) from forming, just in case suitable prospective adoptive parents might eventually found. Arguably, a little fuller discussion of the children's Article 8 rights and of how their best interests (in terms of suffering broken psychological attachments) might have been protected in such situations might have been useful here. Some arrangement that might have facilitated contact between the children and their former psychological parents might have been appropriate as an alternative to the 'all or nothing' exclusivity and finality of an Adoption Order had the system (and the prospective adopters) provided scope for such a solution.

The protection of the best interests of the child was the key issue in *Re Q (a child)* [2011].[161] Here the father of the relinquished child, on learning of her placement applied for a Residence Order and sought to appeal the Adoption Order that had by then been made. He had since married and had another child with a woman who was not the child's mother. As an unmarried father, the lower

[158] *Ibid* para 52.

[159] *Ibid* para 53 per Wilson J.

[160] *Ibid*.

[161] *Q (A Child)* [2011] EWCA Civ 1610 (21 December 2011).

court had been under no statutory obligation to inform him of the birth or the placement, or to obtain or dispense with his consent to her being adopted. The child's mother had consented to the proceedings to keep both herself and the child safe from any possible 'honor killing' by the child's maternal grandfather. The father's appeal included the argument that (citing *Anayo v Germany*) his Article 8 ECHR rights had been engaged, and potentially infringed, in terms of family life or private life.[162] As a result, the adoption of the child had therefore proceeded against the wishes of her birth father, and should have been an outcome 'permissible only in "exceptional" circumstances of "overriding necessity"' which, it was argued, did not exist in respect of this case.[163]

The Court noted firstly that the domestic legislation on dispensing with consent complied with the Strasbourg jurisprudence on 'necessity' in such instances ('the word "requires" in section 52(1)(b) of the 2002 Act accurately captures the essence of the Strasburg jurisprudence on necessity.'[164]) It was then found that 'no enhanced welfare test [was] to be applied in cases of adoption.'[165] The lower court had quite correctly relied upon the question of whether the child's welfare 'throughout her life' would require her to be adopted. This was exactly the same test which would have been applied even if her birth father had somehow obtained parental responsibility for her, so that his Article 8 Convention rights had actually become engaged.[166] Perhaps most significantly the Court observed that cases such as *Anayo* served mainly to,

> raise important and difficult questions best left for determination when the need arises and in the light of fuller argument than, understandably, we have had in this case.[167]

The Court also confirmed that the lower court had clearly taken into account and addressed the child's 'issues of attachment and relationships with particular concern;' the differing views of the guardian and social worker had been discussed and evaluated and an adjournment aimed at allowing a child and adolescent psychiatrist to furnish a further report would have most likely led to unacceptable delays and been superfluous given the 'expertise of the professionals appearing before her [and] their united opposition to this proposal.'[168]

Several of the court's observations were perhaps particularly telling; it was noted that the adoptive parents' seemed fully committed to keeping the adopted child informed of the details of her background. Equally, 'the prolonged agony of parting

[162] *Ibid* para 57.

[163] *Johansen v Norway* (1996) 23 EHRR 33 and *Re C and B* (*Care Order*: *Future Harm*) [2001] 1 FLR 611 were argued in support of this point.

[164] *Ibid* para 58 (*citing Re P* (*Placement Orders*: *Parental Consent*)[2008] EWCA Civ 535, [2008] 2 FLR 625, paras 118–127).

[165] *Ibid* 'We do not understand what Arden LJ said in *C* (*A Child*) *v XYZ County Council and EC* [2007] EWCA Civ 1206, para [40], as qualifying any of this well-established learning.'

[166] *Ibid* para 60.

[167] *Ibid* para 61.

[168] *Ibid* para 66.

would cause Q immense distress.'[169] In respect of moving the child to the care of her father and his new family, it was felt that despite the 'care and warmth' that was being offered by them to 'their own child' (and in spite of 'the importance of natural ties') it was suggested that they

> would feel differently about Q from their own child, that they were unrealistic in their approach to the integration of Q into their family and that there was a significant risk that Q would not make an adequate transfer of attachment.[170]

The English High Court case of *Re A & S* [2012][171] perhaps illustrates most sharply the difficulties that can be associated with the processes of adoption. Here, the children were freed for adoption as infants but never actually placed with any prospective adoptive parents; the Order meant that they had ceased to be members of their original family and thus lost all contact with them, including any with their siblings. This was so despite thirty five LAC ('Looked After Children') reviews having taken place over the years and several requests for contact with their original family being made. The Freeing Orders stayed in place for 11 years, with the children being moved between numerous foster parents, some of whom were extremely abusive to the children. Had the children remained under or been returned to a Care Order instead, the local authority would been under a statutory obligation to both permit and promote a reasonable level of contact with their mother and their siblings, and to allow them access to the services and oversight of the Guardian ad litem and legal representation.[172] As Jackson J observed,

> One lesson is that the legal status of children in care has welfare consequences. If the court has authorised a local authority to look for an adoptive home under a freeing order or a placement order, all energies will be directed towards achieving that outcome, with the maintenance of family ties a much lower priority. That is, in the current jargon, the direction of travel. But if adoption is not possible, the direction changes. The child will not be losing membership of his birth family and the question of what family connections (if any) are in his interests needs to be reviewed.[173]

The details of their childhood make for an upsetting catalogue of harmful episodes of substitute 'parenting' by a chain of 'carers' and underscore the seriousness of the potential for negligence and errors of judgement that occurred. As the Guardian ad litem subsequently noted 'the "havoc" in the children's lives [meant that] permanence …amounted in reality to "permanently looked after disruption".'[174] In terms of the boys' European Convention rights under Articles 3, 6 and 8, the court declared incompatibility; the local authority and the Independent Review Officer ('IRO') both accepted that several breaches of their rights had

---

[169] *Ibid* paras 29 and 30.

[170] *Ibid.*

[171] *A & S (Children) v Lancashire County Council* [2012] EWHC 1689 (Fam) (21 June 2012).

[172] See The Children Act 1989 s 34(1) and Schedule 2, para 15 (1).

[173] *Ibid* para 11.

[174] *Ibid* para 9.

occurred. The complete bar on family contact[175] had for example served to offend against Article 8, as had the failure of the local authority to provide the children with any opportunity of finding an adoptive placement that might have allowed them to have a settled and secure home life. In terms of the various procedures associated with their 'freeing', Articles 6 and 8 were clearly engaged and had been infringed, while Article 3 (which prohibits torture, inhuman or degrading treatment) was relevant in terms of the boys' abusive treatment at the hands of some of their foster parents.

The court expressed concern that the apology offered by social services had been limited and that, for example, no explanation had ever been given as to why social workers had repeatedly informed the IROs that legal proceedings to revoke the freeing orders had been initiated when this was not the case.[176] Equally, no meaningful answer had been given to the central questions of the case, with the local authority instead pleading "competing demands and priorities" on the part of the boys' previous social worker (who had dealt with the case for several years) and then arguing that 'too many crises' had arisen to allow for any progress on the issue of revoking the freeing order.[177] The argument from their current social worker (that their incorrect legal status 'did not have any welfare consequences') was not accepted by the court and clearly showed a profound lack of awareness of how a child's human rights, basic needs and best interests entitlements are often irrevocably tied to their legal status, not least in terms of preventing the harms associated with being a 'statutory orphan.' As the court stressed, such children are particularly vulnerable to a wide range of harms,

> This not simply a matter of legal technicality, but has . . .welfare consequences. . .there are also real problems with sole corporate parental responsibility existing for anything other than a short transitional period. Anyone, let alone a child, who relies entirely on professionals to make decisions for him depends on effective systems and professional commitment. Foster carers, employed by the local authority, are responsible for the day to day care of children. They are not well placed to hold local authorities to account, even if, as here, some do. The situation of a child in a children's home is even more exposed.[178]

The case serves as a harsh lesson and as a reminder of how a child's legal status may easily dictate how key issues of child welfare (such as the need for strong attachments, good parenting, family contact, protection of basic human rights) might be determined simply on the basis of absent paperwork and the passage of time, rather than by reference to the needs of the child and the duty of decision-makers to avoid causing them harm.

---

[175] *Ibid* see paras 119–121.

[176] *Ibid* para 128.

[177] *Ibid* para 130.

[178] *Ibid* para 150. Had one of the boys not independently sought out legal advice, it is possible that the situation would simply never have been discovered or addressed. Para 154.

7.6 Conclusion

From the perspective of triad members, adoption still remains 'the most extreme interference in family life that the state can lawfully undertake.'[179] The 'classic' model of blood-tie severance based upon a ban on contact (if not also on identifying information) meant that family courts often had to choose between 'the parent maintaining responsibility on the one hand, and, on the other, losing parental responsibility altogether, upon the child being adopted.'[180] The process has moved however beyond its original purpose of enhancing the 'opportunities for identity' (and increased social standing) of vulnerable, usually illegitimate, 'freely relinquished' children. The desire to achieve a useful level of 'opaqueness, by putting barriers between birth families and adopters'[181] has given way to a culture of more general openness, based upon the argument that the child's welfare may demand retention of some measure of connection with their original relatives. The need 'to reduce ... feelings of being rejected'[182] has also been acknowledged by policy-makers; as Gilmore has recently also observed in relation to child protection, the generally accepted presumption on contact is now that it is largely 'beneficial to the child's welfare.'[183] That said, disputes over contact and child welfare are tending increasingly to be 'decided by facts not presumptions.'[184]

Humphrys' argument is worth recalling however, namely that contact between estranged genetic relatives, even in the context of some private law disputes, can serve to enable 'a vulnerable point for the continuation of abuse.'[185] As Cullen has also suggested in respect of the legislative reforms on adoption, 'the processes are altered but the fundamental concept of adoption is not.'[186] There will always be a need to protect abandoned and abused children from inadequate and dangerous parenting. As such, it seems likely that 'the model of control that is intrinsic to child protection intervention [will] not sit easily with a model of partnership and empowerment.'[187] Intervention should not however 'automatically involve the severance of all emotional ties' but be aimed instead at the duty to 'produce a more unbroken continuum between the options for permanence.'[188] Outcomes should be 'rooted in

[179] Langdale and Weston (2008), pp. 769–772 at p. 769.

[180] S Mahmood (2004), pp. 449–452.

[181] Ross (2001).

[182] Thomas et al. (1999).

[183] Gilmore (2008), pp. 285–310 at p. 285.

[184] *Ibid* p 300. Gilmore seems to argue that contact ought not to be regarded as a right, but instead merits some level of legal 'respect' as an element of family life; this might avoid any judicial obligation to assume that it is beneficial.

[185] Humphrys (2003), p. 237.

[186] Cullen (2005), pp. 475–486 at p. 484.

[187] Calder and Hackett (2003).

[188] Cullen (2005). That said, where a child has suffered harm at the hands of their original family, this argument is difficult to justify. See further *Re P (A Child)* [2008] EWCA Civ 535 per Wall J;

child-focused, empowering practice'[189] and reports by expert witnesses should not be used to simply

> reinforce *a priori* judicial assumptions about the benefits of contact, with judges tending to simplify expert opinion and to discount or reject that which runs counter to their own *a priori* assumptions.[190]

As Bailey-Harris has argued, confusion may attach to the question of how the 'psychiatric underpinnings of a child-focused approached are now emphasised' within family law cases.[191] The importance of updating regional policies cannot be over-stated, especially where there may be delays over legislative reform and differing strategy approaches between local authorities[192]and particularly in relation to ensuring that front-line decision-makers are fully aware of the human rights aspects of the cases they must deal with. Although hard cases seldom make for good law, they do tend to offer sharp lessons and useful guidance in terms of seeing how domestic jurists have interpreted and implemented such rights as respect for family life and fair hearing to comply with best-interests-led, child welfare paramountcy principles. Often, the facts of the case will require judges to display a King Solomon-like wisdom in deciding upon, if not predicting, what the best future will be for the children involved in intractable disputes over exclusive parentage and parenting. What appears to be a successful case outcome for those who support an unqualified, absolute right to genetic connection often amounts instead to a Pyrrhic victory, where for example children have simply been returned to abusive birth relatives or lose the attachments and protective care of the social parents on the basis of such firmly adult-centric issues as flawed consent, a need to preserve communal heritage or Constitutional principle. The chapter which follows examines such 'blood-tie paramount' jurisprudence from several jurisdictions.

References

Appell, A. (2010). Reflections on the movement toward a more child-centered adoption. *Western New England Law Review, 32*, 1.

Bailey-Harris, R. (2001). Contact-challenging conventional wisdom? *CFam, 13*(4), 361–370.

Bainham, A. (2007). Permanence for children: Special guardianship or adoption? *Cambridge Law Journal, 66*(3), 520–523.

Re R (*Adoption*: *Contact*) [2005] EWCA Civ 1128 at para 49; *Re T* (*Adopted Children*: *Contact*) [1995] 2 FLR 251 on the 'dual planning' approach to adoption placement hearings, and the onus now placed upon adoption agencies to select contact-tolerant prospective adopters.

[189] Talbot and Kidd (2004).

[190] Bailey-Harris (2001), p. 361.

[191] *Ibid.*

[192] See further Kelly and McSherry (2003); Kelly and McSherry (2002), pp. 297–309; '*Adopting the Future*' Consultation Report Responses (2006) DHSSPS available at http://www.dhsspsni.gov.uk/adopting-the-future-consultation-report-final-2.pdf (accessed 05.10.11).

Bojorge, C. (2002). Intercountry adoptions: In the best interests of the child? *Law and Justice Journal, 2*(2):266–291.

Bridge, C. (2008). Case reports: Care proceedings. *Family Law* 114.

Calder, M. C., & Hackett, S. (2003). *Assessment in child care – Using and developing frameworks for practice*. Lyme Regis: Russell House Publishing.

Cullen, D. (2005). Adoption – A (fairly) new approach. *Child and Family Law Quarterly, 17*(4), 475–486.

DHSS. (1999). *Permanency planning for children: Adoption – Achieving the right balance*. DHSS: Belfast.

DHSS. (2006). *Adopting the future*. Consultation report responses (2006). Available at http://www.dhsspsni.gov.uk/adopting-the-future-consultation-report-final-2.pdf.

Fair, L. M. (2008). Shame on U.S: The need for uniform open adoption records legislation in the United States. *Santa Clara Law Review, 48*, 1039.

Gilmore, S. (2008). Disputing contact: Challenging some assumptions. *Child and Family Law Quarterly, 20*, 285–310.

Herring, J. (2009). Revoking adoptions. *New Law Journal* 377.

Houghton Report (1972) CMND 5107 HMSO.

Humphrys, C. (2003). Focusing on safety-domestic violence and the role of child contact centres. *CFam, 15*(3), 237.

Kelly, G., & McSherry, D. (2002). Adoption from care in Northern Ireland: Problems in the process. *Child and Family Social Work, 7*, 297–309.

Kelly, G., & McSherry, D. (2003) *Review of the freeing order processes in Northern Ireland*. Belfast: DHSS Report

Kilkelly, U. (1999). *The child and the ECHR*. Aldershot: Ashgate.

Langdale, R., & Weston, J. (2008) Guidance on the Adoption and Children Act 2002. *Re P. R. Langdale and J. Weston. Family Law 38*, 769–772

Mahmood, S. (2004). In Practice-Adoption and Children Act 2002: Where are we so far? *Family Law Journal, 34*, 449–452.

Masson, J. (2013) Report to the education committee – '*Children first: The Child Protection System in England*'. Available http://www.publications.parliament.uk/pa/cm201213/cmselect/cmeduc/137/137vw79.htm. Accessed 28 Feb 2013.

McFadden, C., & Thomson, C. (2012) Welfare still rules. *The Journal of the Law Society of Scotland* 17.

Munro Review of Child Protection (May 2011) available at http://www.official-documents.gov.uk/document/cm80/8062/8062.pdf. Accessed 22 Feb 2013.

Norgrove Report (Family Justice Review) (November 2011) available at http://www.justice.gov.uk/downloads/publications/moj/2011/family-justice-review-final-report.pdf. Accessed 22 Feb 2013.

Ross, M. (2001). Adoption in the 21st century: Still image against a moving picture? In J. Scoular (Ed.), *Family dynamics-contemporary issues in family law*. Butterworths: Edinburgh.

Smith, C. (2002). Adoptive parenthood as a 'Legal Fiction' – Its consequences for direct post-adoption contact. *Child and Family Law Quarterly, 14*(3), 281–302.

Talbot, C., & Kidd, P. (2004). Special guardianship orders-issues in respect of family assessment. *Family Law Journal 34*(273).

Thomas, C., Beckford, V., Lowe, N., & Murch, M. (1999). *Adopted children speaking*. London: BAAF.

Cases

United Kingdom
Daniels v Walker (2000) 1 WLR 1382
Down Lisburn Health and Social Services Trust v H [2006] UKHL 36
K (Children)[2011] EWCA Civ 1064 (20 July 2011)
Ladd v Marshall [1954] 1 WLR 1489
R v Registrar-General ex p Smith [1991] 2 QB 393
Re A & S (Children) v Lancashire County Council [2012] EWHC 1689 (Fam) (21 June 2012)
Re A [2001] NIFam 23
Re A Trust v M & P [2007] NI Fam 5
Re B (a child) (residence order) [2010] 1 All ER 223
Re B (A Child) [2009] UKSC 5
Re B (Adoption: Jurisdiction to set aside) [1995] Fam 239
Re B-M (a child) (Adoption:Parental Agreement) [2001] 1 ACR 1
Re C (A Child) v XYZ County Council and EC [2007] EWCA Civ 1206
Re C (Freeing for Adoption) [2002] NIFam 1 (12 January 2002)
Re C [1993] 2 FLR 260
Re C and B (Children) (Care Order: Future Harm) [2001] 1 FLR 611
Re CBCHSST v JKF [2000] NI Fam 56 (27 November 2000)
Re Coventry City Council v PGO & Ors [2011] EWCA Civ 729 [2011] All ER (D) 159
Re D (Simultaneous Applications For Care And Freeing Order) [1999] 2 FLR 49
Re D [1977] AC 602
Re E and M [2001] NIFam 2 (02 February 2001)
Re EFB [2009]NIFam 7
Re F (Adoption: Freeing Order) [2000] 2 FLR 505
Re F (Infants) (Adoption Order: Validity) [1977] Fam 165
Re F (R) (An Infant) (1970) 1 QB 385
Re F [2000] 2 FLR 505
Re G (Children) (FC) [2006] UKHL 43
Re H (Children) [2012] EWCA Civ 743 (16 May 2012)
Re H [1981] 3 FLR 386
Re J (Care Order) [2008] NIFam 11 (1 August 2008)
Re J (Freeing without consent) [2002] NIFam 8 (13 March 2002)
Re J and S (2001) NIFam 13 (23 May 2001)
Re JJ [2009] NIFam 2
Re K (Adoption and Wardship) [1997] 2 FLR 221
Re K (Care Proceedings: Care Plan) [2007] EWHC 393 (Fam)
Re K (Children) v Sheffield City Council [2011] EWCA Civ 635 (25 May 2011)
Re K (Non-accidental injuries: Perpetrator: New Evidence) [2004] EWCA Civ 1181, [2005] 1 FLR 285
Re K [2002] NIFam 13
Re L2 and O (Post-Adoption Contact) [2007] NI Fam 12
Re M (Minors) (Adoption) [1991] 1 FLR 458
Re NI and NS [2001] NIFam 7 (24 March 2001)
Re P (Placement Orders: Parental Consent)[2008] EWCA Civ 535, [2008] 2 FLR 625
Re P, (Freeing Without Consent) [2004] NIFam 6 (11 February 2004)
Re Q (A Child) [2011] EWCA Civ 1610 (21 December 2011)
Re R (Adoption: Contact) [2005] EWCA Civ 1128
Re RA (Minors) (1974) 4 Fam Law 182
Re S (Discharge of Care Order) [1995] 2 FLR 639
Re SEHSST v LS [2008] NI Fam 12 (27 October 2008)
Re SEHSST v LS [2009] NI Fam 14

Re T (A Minor) [1998] EWCA Civ 1871 (27 November 1998)
Re T (Adopted Children: Contact) [1995] 2 FLR 251
Re W (An Infant) [1971] 2 All ER 49
Re C (a child) [2011] EWCA Civ 521
Webster (The Parents) v Norfolk County Council & Ors (Rev 1) [2009] EWCA Civ 59

Ireland
N v Health Service Executive & Ors [2006] IESC 60 (Ireland)

Strasbourg
Anayo v Germany [2010] ECHR 2083 (21 December 2010)
Johansen v Norway (1996) 23 EHRR 33
Neulinger v Sweden (2010) 54 EHRR 1087
R and H v United Kingdom (2011) 54 EHRR 28
S v L [2012] UKSC 30 (11 July 2012)
Uyanik v Turkey (3 May 2012) (app 60328/09)
Y C v United Kingdom [2012] (App 4547/10)

Legislation
The Adoption Act 1976 (England and Wales)
The Adoption (NI) Order 1987
The Adoption and Children Act 2002 (England and Wales)
The Children Act 1989
The Children (NI) Order 1995
The Adoption and Children Scotland Act 2007
The Human Fertilisation and Embryology Act 2008

Chapter 8
Blood-Tie Preservation as Paramount: Best Interests of the Child Outweighed?

8.1 Introduction

Concealment is not as easy when the adoption takes place across racial lines.[1]

In terms of the need to preserve blood-ties via family reunification, the various strands of case law seem to fall into three, occasionally overlapping categories; a Constitutionally protected right to not be deprived of parenthood,[2] consent to relinquishment that is flawed, absent or subsequently revoked,[3] and a statutory duty to protect tribal cultural heritage.[4] The arguments in these cases have all served to either overturn an adoptive placement or prevent the termination of parental rights. The focus is largely adult-centric, with the child's best interests generally not appearing as the paramount concern. Making the preservation of

[1] McGillivray (1983–1985), pp. 437–467 at p. 449.

[2] See for example United States case law such as *Thurnwald v A* E (2007) UT 38, 163 P 3d 623; *Sanchez v LDS Social Services* (1984) 680 P 2d, or *In Re Custody of MCC (A Minor) v Aisha Umer* (June 27, 2008) 06 D 79411 (No 1-08-0203) and the Irish case *N v Health Service Executive & Ors* [2006] IESC 60 (the 'Baby Anne' case).

[3] See for example English authorities *Re F (R) (An Infant)* [1970] 1 QB 385; *Re RA (Minors)* (1974) 4 Fam Law 182; *Re F (Infants) (Adoption Order: Validity)* [1977] Fam 165 on 'procedural irregularities' e.g. failure to serve notice of proceedings upon birth parents or parental ignorance of proceedings. See also *Webster (The Parents) v Norfolk County Council & Ors* (Rev 1) [2009] EWCA Civ 59 however for the Court of Appeal's discussion of 'natural justice' in relation to the question of allowing birth parents an opportunity to object.

[4] See for example *Re Bridget R et al (Minors)* (1995) 25 USCA 1901 BO93520 (the 'Rost twins' case); see also *Nelson v Children's Aid Society of Eastern Manitoba* [1975] 5 WWR 45, 56 DIR (3d) 567, 21RFL 222 (Man CA) [1974] 5 WWR 449, 46 DIR (3d) 633 (Man QB) on whether the child's 'Indian status' heritage would be extinguished by adoption. See also *Adoptive Couple v Baby Girl* (2013) No. 12-399 (The 'Baby Veronica' case) Opening day oral transcript hearing available at http://www.supremecourt.gov/oral_arguments/argument_transcripts/12-399.pdf (accessed 16.04.13) which, at the time of writing, is being heard by the US Supreme Court, on appeal from the Supreme Court of South Carolina (*Adoptive Couple* v. *Baby Girl*, 398 S.C. 625; 731 S.E.2d 550 (S.C.) (2012).

A. Diver, *A Law of Blood-ties - The 'Right' to Access Genetic Ancestry*,
DOI 10.1007/978-3-319-01071-7_8,

genetic connection a paramount consideration in cases involving parentage and parenthood disputes can create a precarious template, not least in respect of the need to ensure that child welfare is being promoted and protected. The presence of the blood-tie offers no guarantee that the child's best interests will be met; as in those cases where contact and information have been vetoed in accordance with parental wishes for privacy, so too have the child's welfare needs been both defined and delineated in terms of the 'weightiness' of those adult interests which may be in direct conflict with the child's best interests. As such, it seems fair to state that 'the law has been slow to adapt adoption to meet the demands of the majority of the people most affected'[5] by it.[6]

This chapter will describe the various legal frameworks and precedents which exist to support 'blood-tie paramountcy' and evaluate the extent to which they might allow for the best interests of the child to be afforded greater recognition. It will make the suggestion that rights-balancing exercises and repatriative approaches to redressing or preventing the harms of origin deprivation have progressively tended to highlight the civil and political nature of this difficult area of child and family law. As international adoptions continue to receive much media attention, issues such as the loss or removal of the adoptee's original identity in domestic adoptive placements seem likely to also come increasingly under the spotlight, especially in relation to hearings on issues such as contact, contested relinquishments and information disclosure. Irrespective of context, there is a clear need for decision-makers, legislators and overseeing jurists to be mindful of the fact that 'children, too, have fundamental rights – including the fundamental right to be protected from neglect and to have a placement that is stable (and) permanent'.[7]

8.2 *Habeas Corpus*: Ireland's '*Baby Anne*' Case

Belated revocations of consent to adoptive placement do not generally sit well with the wider, underpinning principle of child welfare paramountcy. As Baldassi has observed, 'most traditions look poorly on the idea of biological family changing their mind and attempting to reclaim the child.'[8] In domestic disputes over parentage or parenthood, judicial attention may however have to focus primarily upon such issues as the legality or otherwise of parental consent to the process of relinquishment. In this sense, a sort of flawed 'contract' may have arisen, either between the triad adults, or in relation to the birth parents and social services; a lack

[5] Baldassi (2004–2005), pp. 212–265 at p. 219.

[6] See however Cahn (2002) 163 who has noted that post-adoption contact agreements may be protected by statute, at least to an extent, within some US regions.

[7] *In re Jasmon O.* (1994) 8 Cal.4th 398, 419 (citing *In re Marilyn* (1993) H., 5 Cal.4th) at p. 306.

[8] Baldassi (2004–2005), p. 75.

of informed, meaningful consent may be inferred, which might leave the court no other option but to allow the overturning of the adoptive placement. Equally, a constitutionally protected right to parenthood might be declared, so that again judiciaries are limited by law in respect of finding workable compromise solutions based upon the child's best interests-led welfare 'rights'.

In *N & Anor v Health Service Executive* (2006)[9] (the 'Baby Anne' case) for example the Irish Supreme Court looked to Constitutional provisions for guidance when faced with birth parental demands for the return of their previously relinquished child, who had by then spent 2 years with her prospective adoptive parents. The birth mother sought to argue a claim of *habeas corpus* in respect of the child being held 'unlawfully' by her substitute carers. The proposed adoption was initially to have been relatively 'open' in nature, with the original parents set to have annual contact visits with the child. The mother belatedly changed her mind over the relinquishment however and subsequently married the child's genetic father, pursuant to legal advice that this would be likely to strengthen their legal bid to reclaim their daughter from the prospective adopters.

The other side argued against returning the child to the original parents on the basis of the best interests of the child, via the welfare paramountcy principle. Because a phased transfer of custody had proven impossible (due to a complete breakdown in trust between the two sets of parents) the Irish High Court had initially suggested that the child should be made a Ward of Court but remain in the care of her prospective adopters. It was at this stage of the proceedings that the High Court declared that the birth parents had clearly both 'abandoned' the child and failed in their fundamental parental duties towards her. On appeal, the degree of weight that the High Court had placed upon the issue of the child's attachments was questioned. It was argued that insufficient emphasis had been placed upon the inalienable rights of married parents both under the Irish Constitution (Articles 40–42[10]) and in respect of those arising by virtue of Article 8 of the European Convention on Human Rights. By still keeping the child in their custody, the prospective adoptive parents were said to be infringing the birth parents' Constitutional rights to exist as a family. The child's psychological and emotional attachments to her social parents was further deemed to be insufficiently compelling to defeat the original parents' right to retain, or rather, regain, their custody of her.

The birth parents had also argued the Constitutional rights of the child, the legal presumption being that a child's welfare needs would be best met within the framework of the original family unit. This was an entirely rebuttable presumption however, albeit likely to be denied only in such exceptional circumstances as the

---

[9] *N v Health Service Executive & Ors* [2006] IESC 60.

[10] Article 40.4.(1) and (2). provide that '*No person shall be deprived of his personal liberty save in accordance with the law*'. Articles 41 and 42 provide that '*The State recognizes the Family as the natural, primary and fundamental unit group of Society…possessing inalienable and imprescriptible rights*'.

total failure of the parental duties of care.[11] The Supreme Court, relying upon an earlier judgment,[12] stressed the history of adoption law and policy in Ireland, stating that up until 1988,

> only illegitimate or orphaned children were eligible for adoption. The children of married parents who formed part of families whose rights arose under Articles 41 and 41 of the Constitution could never be adopted.[13]

The Court sought to distinguish the 'Baby Anne' case however, on the grounds that in the earlier case the mentally ill birthmother had remained unmarried and had completely 'failed in her duty as mother by having 'abandoned' her disabled child to twelve years of foster care, with relatively little contact taking place throughout that time.'[14] Turning instead to the earlier, and more factually relevant authority of *In Re JH* (1985)[15] the Court looked particularly to the matter of the child having formed strong emotional attachments to her social parents, the prospective adopters. Lynch J's High Court dicta on the dangers of long-term psychological harm, and the need to give 'considerable weight' to such a risk, were also referred to here, with the Supreme Court endorsing the lower court's call to regard this case 'through the eyes of the child.'[16] Using this formula the High Court had been able to previously conclude that the child would have been likely to gain very little from being returned to the care of her birth parents; a right of contact was however granted to them.

On Appeal, the Irish Supreme Court overturned the High Court's decision however in favour of returning the child to the care of her birth parents. Finlay CJ stressed that the child's birth status had been retrospectively legitimated by the subsequent marriage of her birth parents; she therefore gained the full Constitutional rights protection of a legal 'family member.' This in turn meant that 'the balance of welfare'[17] was not necessarily the sole factor in the determination of such a dispute. Only a finding of 'exceptional circumstance' would therefore enable the Court to deprive the married birth parents of their near-absolute right to parenthood and to custody of the child. Following its remittal to the High Court, Lynch J continued to express serious reluctance over this outcome. Fresh expert evidence on the issue of the child's psychological attachments was submitted and the Court's misgivings over the effects of a custody transfer, and the ability of the birth parents to care for the child, were expressed. These were not sufficient however to rebut the legally enshrined presumption that child welfare would be

[11] *N v Health Service Executive* [2006] *op cit* n 9 p. 15.

[12] *Northern Area Health Board v An Bord Uchtála* [2002] 4 IR 252.

[13] *N v Health Service Executive* [2006] *op cit* n 9 p. 1 per McGuinness J.

[14] She therefore could claim a right to custody under Article 40, but had no entitlement to the '*protections and presumptions which benefit the married family under Articles 41 and 42*'. *Northern Area Health Board v An Bord Uchtála* [2002] op cit n 12 Per Murray CJ.

[15] *In Re JH* [1985] IR 375.

[16] *Ibid* pp. 388–389.

[17] s.3 Guardianship of Infants Act 1964.

best protected 'within the family.' The child was returned to her birth parents, despite having spent several years in the care of her prospective adoptive parents.[18]

Bound by the reasoning in *In Re JH* the Irish Supreme Court had no other option but to find that no 'failure of duty' had occurred on the part of the child's birth parents, solely by virtue of their having placed her for adoption.[19] A contrary finding would have posed a considerable threat to the national system of substitutive care and indeed to the practice of adoption generally, and would also have diminished the actions of any birth parents seeking to effect relinquishment in an altruistic bid to protect the child's welfare. The fact that this adoption was originally intended to be 'open' in nature was also taken by the Court to mean that her original parents had shown no intention towards a total 'abandonment' of her or of their right to parenthood.

In relation to the question of an over-riding 'compelling interest', the Court endorsed the expert opinion that a sudden change of carer would be likely to result in 'severe psychological damage to the child', probably 'ending in despairing indifference towards her carers'. Even if the transition were to be well managed, it was noted as entirely possible that the child could still suffer some form of 'lasting harm'.[20] Bound by the decision in *In Re JH* however, the Court felt itself unable to make a finding that such a compelling interest had arisen here. The fact that the birth parents had since married also served to direct the main focus of the Court away from the issue of the child's best interests, moving them more towards gauging whether illegality now attached to the prospective adopters' continuing custody.[21]

It is difficult to see how such an approach fully reflects the spirit of the Children's Convention, with its chief emphasis being on child welfare primacy. Arguably, subsuming a particular child's welfare needs into the more generalized, adult-centric Constitutional rights of 'the family' (which in this context, meant only the genetic parents) mirrors much of the U.S. jurisprudence on revoked adoptions, not least by defining the child's best interests through reference to parental rights, interests and preferences and then outweighing them to preserve familial autonomy.[22] It was only with 'reluctance and some regret'[23] that the Court eventually allowed the appeal of the birth parents and ordered the return of the 2-year old child to them. Echoing the concerns of Lynch J in *In Re J H* however the Court expressed serious apprehension over the likely effects of this transfer of custody and indicated profound disappointment that the matter could no longer be viewed through the 'eyes of the child' in question. McGuinness LJ stressed the need (in 'her personal view') to bear in mind 'the terms of such international instruments as the United

[18] *In Re JH* [1985] *op cit* n 15.

[19] *N v Health Service Executive* [2006] *op cit* n 9 p. 22.

[20] *Ibid* p. 23.

[21] *Ibid* p. 24.

[22] See further Caballero (2005–2006), pp. 291–313.

[23] *N v Health Service Executive* [2006] *op cit* n 9 p 30.

Nations Convention on the Rights of the Child' and 'EU Regulation 2201/Nov 2003 (Brussels 2)'. The vulnerable position of the child under the Irish Constitution was also criticized. Hardiman LJ referred to the reasoning underpinning the earlier, persuasive English authority of *Re G (Children)* (2006)[24] which had stressed that

> in the ordinary way the rearing of a child by his biological parents can be expected to be in the child's best interests, both in the short terms and also, and importantly, in the long term.....a child should not be removed from the primary care of his biological parents without compelling reasons. Where such a reason exists the judge should spell this out explicitly.[25]

Although such a formula endorses the significance of the genetic tie, it seems to be grounded upon the dangerous assumption that all biological parents will always act in the best interests of their child. Such a template may clearly also be called into question where a child has been, essentially since her birth, in the long-term care of social parents. The Court suggested here further that a 'child centred' approach to such situations may simply not be possible where the child in question is very young. It is difficult to see however how issues such as psychological attachment can be so completely dispensed with in favour of protecting the absent parents' Constitutional rights.[26] As one expert witness noted during the proceedings,

> attachment of children to parents is seen as the most fundamental requirement of emotional development....secure, uninterrupted attachment to a new family ....would give them a good start and a sense of security and belonging for life. Disruption of primary attachment can be extremely painful and damaging to the child's emotional development and can lead to serious emotional and behavioural problems.[27]

The Court was not at liberty to apply the succinct, child welfare-focused test of the High Court in *Re JH*, because this had itself been over-ruled on appeal to the Supreme Court.[28] Had it been endorsed, such a test might have allowed the Court to rely more upon the psychological evidence presented to it and the eventual outcome of the proceedings may have differed significantly. The simple formula for determining where the child's future care should lie, could have perhaps settled the best interests dilemma here: '*is anything worthwhile to be gained for the child by transferring her from the adopting parents to the parents?*' Such a question could have, arguably, permitted fuller consideration of child welfare paramountcy and best interests matters, in isolation from the constraining influences of other adult-centric factors which the Court had been obliged to consider.

The fact that the child's parents had married, for example, meant that she now bore their surname and had been 'legitimated' in Irish law; this in turn made her a family member, which then entitled her genetic parents to apparently inalienable Constitutional protections. This further enabled the presumption that child welfare

[24] *Re G (Children) (residence: same sex partner)* [2006] 4 All ER 241.

[25] *Ibid* per Lord Nicholls of Birkenhead p. 243.

[26] *Ibid* per Hardiman LJ.

[27] See Professor Iwaniec's Report (for the Respondents) as cited by Fennelly J.

[28] *In Re JH* [1985] *op cit* at n 15 per Finlay LJ p. 394.

is best served within the environment of the 'natural family' unless exceptional grounds existed which would preclude this. Had the parents not married, or if only one of them had alone sought to reinstate their original ties of kinship, it seems likely that the Court would have perhaps arrived at the opposite conclusion based upon a lack of Constitutional protections. Likewise, had the Irish Supreme Court been at liberty to adopt the reasoning of the English Court of Appeal in *Webster*,[29] the eventual outcome may have been entirely different. Although the facts differ (the children in Webster had been through the adoption hearings, and had been non-consensually removed on suspicion of parental abuse, and, by the time of the appeal, had spent 4 years with their adoptive parents) the core issue of the remit of the best interests of the child principle is not entirely dissimilar.

In both scenarios substitute parenting was meant to achieve protective permanency, yet the genetic parents had either established some form of relationship through caring for their children during infancy or by maintaining some degree of contact post-placement. Either of these activities could arguably have amounted to a justiciable form of 'family life' under Article 8 of the European Convention, and thus perhaps have warranted some fuller degree of protection. The key differences between the two cases are almost entirely adult-focused, turning upon the nature of the children's relinquishments and the timing of the act. The definitional remits of concepts such as legal 'relatedness', family life and best interests are relevant however to both cases.

The Websters did not consent to the removal of their children (and serious questions later attached to the issue of whether they had deliberately abused or neglected them) yet they lost all rights to parental responsibility and to family life, including contact with the children, on the basis that too much time had elapsed since the adoption had occurred. The birth parents in the Baby Anne case had initially relinquished their child consensually at birth, with slightly less time having elapsed since the child was placed in the care of substitute parents. The 'paper magic' (of the written-down Constitutional protections in the Irish case and of the adoption order having been granted in the English one) is quite evident; the outcomes differ largely on the basis of these legal documents being present. In both of these cases the courts were perhaps influenced by wider policy concerns, possibly in the belief that an 'uncertainty of jurisprudence' might 'cause greater trauma'[30] in the long run for jurists subsequently faced with similarly difficult cases.[31]

In the Webster case for example the need to underscore the 'vast social importance of not undermining the irrevocability of adoption orders'[32] was cited.

---

[29] *Webster v Norfolk County Council* [2009] EWCA Civ 59.

[30] *N v Health Service Executive op cit* at n 9.

[31] Media reaction has been fairly unequivocal in its criticism of the decision; see for example www.independent.ie/ntional-news/baby-ann-returned-to-parents-693265.html; or www.timesonline.co.uk/tol/news/world/ireland/article641483.ece (accessed 16.04.11).

[32] *Webster v Norfolk op cit* at n 29 para 204.

Arguably, in both cases, had the best interests of the child principle been more fully enshrined in domestic law as the over-riding, paramount consideration, the Courts might have perhaps felt more inclined or able to make the suggestion of some measure of contact post-adoption (or post-reunion) between estranged kinfolk on the basis that such a result might offer the best means of ensuring a child welfare-led outcome.

8.3 Lack of Parental Consent to Relinquishment

Several cases involving challenges to relinquishment focus on the question of whether parental consent was in some way flawed. The U S case of *In Re Haley A. (Adoption)* (1996)[33] for example seemed to turn mainly upon the issue of whether parental consent to the child's relinquishment had actually occurred. Here the birth mother sought the return of her child, having initially consented to the adoptive placement process going ahead. She subsequently refused to sign the consent papers, which meant that the statutory requirement of 'six months' abandonment had not been met. The court took into consideration the evidence that had been provided by the psychologist's report, which had made the recommendation that the prospective adoptive parents' home would have provided an environment which, in sharp contrast to the birth mother's circumstances, was likely to create a residence for the child that would be 'stable, supportive and nurturing, having no record of violence, threats of violence, alcoholism or chaos.'[34]

It was also noted that the placed child had clearly formed a strong bond of attachment with her substitute carers, already regarding them as her parents. Moving her out of the adoptive placement could serve to make her 'fearful and anxious as a short term effect.'[35] The earlier case of *In Re Bridget R.* (1995)[36] was distinguished here on its facts by virtue of the parents in that case having actually signed the consent papers. Here, the birth mother had clearly exercised her legal right to change her mind within the statutory 6 month period. The Court stressed that 'the ensuing litigation ha[d] consumed another two years during which this child has lived with the only parents she has known as such'[37] but found that they had no other option in law but to order the return of the child. They displayed strong reluctance to do so, stating that,

> Were this fact to determine the outcome of the case, however, we would not only condone the contravention of appellant's rights but encourage adoptive parents to refuse to honor legitimate demands for return of a child in the hope that sufficient time would pass to convince courts to retain the status quo.[38]

---

[33] *Haley A Mark et al v Elizabeth L.* (1996) Contra Costa County Superior Court No. 9400055.

[34] *Ibid* p. 11.

[35] *Ibid.*

[36] *Re Bridget (1995) op cit* n 4.

[37] *Ibid* per Kline J at p. 25.

[38] *Ibid.*

Conceding that lengthy litigation would, through the passage of time, eventually work to deprive the birth mother of any legal claim to parenthood, the Court held that they were bound by the statutory provisions.

> While we recognize, and deeply regret, the havoc that will be wreaked upon the life of a two and a half year old child as a result of this decision, the directive of section 8804 must be followed and the child returned to the appellant.[39]

The suggestion that the transition between homes and parents should be effected 'sensitively' does indicate awareness of the need to consider the child's best interests. That much of the expert psychological opinion had to be disregarded however suggests that the child welfare paramountcy principle was not the central consideration under the domestic legislation, but that the 'contractual' issue of consent was the key matter. Clayton has made the observation that the Supreme Court appears to have embraced the view that 'biological ties are not enough, alone, to merit parental rights'.[40] The rights of unmarried fathers for example to object to adoption proceedings have traditionally been subject to the test of whether the requisite degree of 'commitment' to the child had been demonstrably present.[41] Cases such as *In re BGC* (1992)[42] (the 'Baby Jessica' case) have however seen putative fathers claiming custody, on the basis that they were not given any opportunity to object to the birth mother's relinquishment of the child for adoption. Here, the mother had provided adoption relinquishment consents after initially making a false identification of another man as the father of the child. She later changed her mind; the child's birth father eventually convinced the court to overturn the placement several years after the child had been placed for adoption.

A similar outcome was seen *In re Kirchner* (1995)[43] (the 'Baby Richard' case) where the unmarried father had been told that the baby had been stillborn so as to prevent him from making an objection to the adoption relinquishment. On finding out the truth, he quickly married the child's mother and together they brought court

[39] *Ibid.*

[40] Clayton (2007–2008), pp. 249–266 at p. 249. See also *Lehr v Robertson* (1983) 463 U S 248.

[41] See for example *Thomas v Children's Aid Society of Utah* (1961) 364 P (2d) 1029; see also however *Stanley v Illinois* (1972) 405 US 645 (which held that the presumption that unmarried fathers were automatically unfit to parent was unconstitutional); *Ellis v Social Services Department of the Church of Jesus Christ of Latter Day Saints* (1980) 615 P 2d (Utah) (which created an exception for putative fathers who could show that they had not been afforded reasonable opportunity to object, for example where the birth mother had not informed him of the relinquishment); *Wells v Children's Aid Society of Utah* (1984) 681 P 2d (Utah) (where the Court found that public policy demanded that prospective adoptive parents be protected and encouraged to come forward); *Sanchez v LDS Social Services* (1984) 680 P 2d (where the father had demonstrated profound commitment and found out about the birth mother's relinquishment only days after the event, as cited by Clayton *ibid*).

[42] In *Re BGC* (1992) 496 NW 2d 239 (Iowa).

[43] *In re Kirchner* (1995) 649 N E 2d 324 (Illinois).

proceedings to overturn the adoption. After 4 years, the child's adoption was revoked on the basis that the father had not consented to the process, and that his Constitutional rights had therefore been unjustifiably violated. Additionally, it was held that the best interests principle would only be applied in such cases if a clear finding of 'unfitness' (i.e. to parent the child) was also established.[44] The presumption that fit parents will generally aim to act in the best interests of their children was subsequently endorsed by the Supreme Court in *Troxel v Granville* (2000).[45] As such, the State would only be obliged to

> inject itself into the private realm of the family to further question the ability of that parent to make the best decisions concerning the rearing of that parent's children.[46]

Such a doctrine of 'superior parental rights' whereby genetic parents are essentially presumed to have the best claim to the custody, care and control of their child, may serve to relegate the best interests of the child to a position of secondary concern,[47] with the courts accepting as an initial premise the presumption that it will be generally in the best interests of the child to be brought up by his genetic parents.[48] In *Re RLS a Minor* (2006) however[49] the Supreme Court of Illinois described several aspects of the *Kirchner* decision as 'questionable'. This case suggested that fit parents can and ought to be deprived of their custody of the child, albeit in extreme circumstances. Considerable emphasis was placed however upon the issue of the child's residence, and the question of where the child had actually been living at the time the dispute over parenthood actually arose. Although it does not refer explicitly to the best interests of the child principle, this formula does at least provide some increased scope for the doctrine of child welfare paramountcy to be applied, even if only on a case by case basis.

Similarly, in *Thurnwald v A E* (2007)[50] Utah's Supreme Court held that putative fathers do have a Constitutional-enshrined right to claim custody of their newborn infants.[51] As Clayton has further argued, such a stance may be likely however to conflict with the state's obligations to achieve efficient resolution where disputed

---

[44] Ibid per Heiple J.

[45] *Troxel v Granville* (2000) 530 U S 57, 147 L Ed 2d 49, 120 S Ct 2054.

[46] Paras 68–69. This presumption was used to invalidate the court's statutory power to order contact (visitation) between kin, on the basis that the legislation permitted the court to second-guess the decisions of a fit parent. See further *Wickham v Byrne* (2002) 199 Ill 2d 309.

[47] In *re Marriage of Carey* (1989) 188 Ill App 3d 1040, 1046 544 NE 2d 1293.

[48] see also *In Re Marriage of Sechrest* (1990) 202 Ill App 3d 865, 875 560 NE 2d 1212; where a non-parent lacks standing to petition, they must rely upon proving unfitness of the 'natural parent'.

[49] *In Re RLS* 218 2d 428 per Thomas CJ.

[50] *Thurnwald v A E* (2007) UT 38, 163 P 3d 623.

[51] *Ibid* at p. 635. This case turned on the question of the legality of the statutory time limits which unmarried fathers must satisfy in registering their objection to the adoption (i.e. before the mother has formally agreed to relinquishment). Here, the father was prevented from complying because the relevant offices were closed over a holiday weekend.

relinquishment has occurred.[52] The concept of a near-inviolate right to parent has the potential to seriously compromise a domestic embedding of the best interests principle as the paramount consideration. By subsuming the welfare of the child into the rights of the birth parents (founded on the presumption that quasi-contractual consent has been clearly given or on the basis that law or policy is 'unconstitutional' in nature) the best interests principle may be entirely disregarded or effectively eroded through long delays in respect of court proceedings or via a late removal of the child from his adoptive parents.

Such outcomes do little to promote the concept of genetic relatedness as a recognition-worthy, positive force in law, culture or society. Equally, they do little to frame the need for biological kinship as an important element of children's psychological welfare. More recently, the nature and significance of the blood-tie (as a means to overturning de facto consents to relinquishment) has also been examined in *Re Custody of MCC v Aisha Umer* (2008).[53] Here, as in the earlier case of *Re RLS* [2006][54] the question of disputed custody arose following the death of the birth mother, with the unmarried father clashing with the child's maternal grandparents over the question of who should have custody of the child. The case appeared to turn upon the adult-centric issue of whether custody had been 'voluntarily and indefinitely relinquished.'[55] In what is perhaps an attempt to move away from the inequity of the decision in *Re Kirchner*, it was suggested that

> when determining whether a voluntary relinquishment of physical custody has taken place, the court considers who cared for the child before the custody petition was filed, how the non-parent gained physical possession and the "nature and duration" of the possession.[56]

Here, the father had formally acknowledged his paternity and thus shared the deceased mother's 'superior right' to custodial parenthood, apparently based upon their ties of biological kinship.[57] The child's maternal grandmother had argued however that a de facto relinquishment had occurred, given her agreed role as the child's primary caregiver and the subsequent lack of visitation contact by the child's father. Ultimately the Court found that a relinquishment had not occurred and that a right to custody had thus remained with the father. Detailed discussions of the child's welfare needs and of the best interests principle were absent here, as seems to often occur in cases which are grounded almost entirely upon the issue of parental consent. It is perhaps unsurprising that a quasi-contractual basis for determining the validity or otherwise of relinquishment consents often tends to

---

[52] Clayton (2007–2008), p. 264. See also the Dissenting opinion of Justice Wilkins in *Thurnwald op cit* at n 50, p. 636.

[53] In *Re Custody of MCC (A Minor) v Aisha Umer* (June 27 2008) 06 D 79411 (No 1-08-0203) per Garcia CJ.

[54] *re RLS op cit* n 46.

[55] see *In re Custody of Ayala* (2003) 344 Ill App 3d 574 588 800 NE 2d 524.

[56] *Ibid* p. 7. See also *In Re AWJ* (2000) 316 Ill App 3d 91 96,736 NE2d 716.

[57] See also *In re Custody of Peterson* (1986) 112, Ill 2d 48, 53-54 491 NE2d 1150; *In re Custody of O'Rourke* (1987) 160 Ill App 3d 584 587-88 514 NE 2d 6 (1987).

result in children being described almost as proprietary interests, which require some exclusivity of possession. 'Custody' is a concept that is perhaps much less easily shared between rival parents than the notions of contact, care or parental responsibility might be. Even where there is more substantial discourse on the issue of the child's circumstances, a sort of property law reasoning seems to be apparent. As Cahn has further observed in respect of legal vetoes over disclosure, our society does at times appear to perhaps favour the idea of 'contract over connection.'[58]

Barton too has highlighted the inchoate nature of the 'rights' attaching to unmarried forms of fatherhood, stressing that a 'mere biological link between father and child does not automatically compel constitutional protection.'[59] To avail of such legal safeguards, putative fathers must actively work towards the development of a relationship with their child and be able to demonstrate clear acceptance of some degree of parental responsibility.[60] The 'right' of an unmarried father to veto an impending adoption has on occasion been deemed subordinate to that of the best interests of the child, to allow an adoption to proceed free from paternal objection.[61] 'Putative father registries' in a number of States are also aimed at acknowledging the desire of some birth fathers to retain some form of kinship link with their child whilst still addressing the desire of birth mothers for anonymity.[62] Child welfare interests may be well served by this expedient process of dispensing with paternal consent, but greater mention could perhaps be made of the child's need for knowable ancestry and identity in legal discourses involving parentage and parenthood.

Similarly, in a number of cases little or no reference is made to the issue of potentially allowing for some form of post-adoption contact between separated relatives, despite the obligation on the part of the State to achieve a degree of expedient 'constancy' for the affected child. Presumably, much discretion might be afforded to the eventual custodians of the child, as to whether they will permit, tolerate or actively facilitate some level of contact between the child and their former relatives.[63] Traditional judicial deference towards parental rights, once established, still appears to carry much weight. As the court declared in *Lehr v Robertson* (1983),

> The intangible fibers that connect parent and child have an infinite variety. They are woven throughout the fabric of our society, providing it with strength, beauty and flexibility. It is self-evident that they are sufficiently vital to merit constitutional protection in appropriate cases.[64]

[58] Cahn (2000–2001), pp. 1–27 at p. 26.

[59] Barton (2003–2004), pp. 113–148 at p. 113.

[60] See *Caban v Mohammed* 441 US 380 (1979); *Quilloin v Walcott* 434 US 246 (1978).

[61] *In re Doe* 627 N E 2d 648 [1993] (Ill. App Ct 1993) 656 (as cited by Barton 2003–2004).

[62] Barton (2003–2004), p. 128. Responsibility for registration (and for subsequent consent or objection to the adoption process) is placed upon the birth father; this aims to protect the privacy rights of the birth mother, the father's interest in receiving notice of the proceedings and the child's right 'to a secure adoption'.

[63] *Troxel v Granville* 530 US 57 (2000).

[64] *Lehr v. Robertson* (1983) 463 U.S. 248, 256 [77 L.Ed.2d 614, 103 S.Ct. 985].

Such a 'doctrine' of near-inviolate parental autonomy again received strong endorsement in the case of *Troxel v Granville* (2000). Here, where a mother was opposed to the amount of contact granted to paternal grandparents, the Court looked to the 'constitutionality' of the relevant statute. The legislation in question permitted 'any person to petition for visitation rights at any time'[65] with the court able to ultimately decide if such contact would serve the child's best interests. This provision was found to have the potential to violate the mother's 'due process right to make decisions concerning the care, custody and control of her daughters'[66] and therefore also breached her fundamental right to carry out the role of parent.[67] The court did concede that such intervention would not in itself necessarily lead to a violation of the mother's Constitutional rights; rather it was the judicial failure to grant any 'special weight'[68] to the opinions of the mother which had led to the breach of her parental rights.

Barton has further highlighted the need to 'identify and attempt to protect all the parties' interests before problems arise' in respect of adoption.[69] Generally this might be achieved by attempting to balance the various competing rights and interests of the triad members and by also stressing the child's need for an expeditious placement where possible. It is perhaps unfortunate that this suggested formula does not include a call more significant consideration of child welfare as the paramount concern or for fuller discussion of the need for knowable, accessible genetic ancestry. The emphasis again appears to be on the 'heavier' rights of triad adults and on the duty to avoid an infringement of them. Although a safeguarding of the child's original identity may occur as a side-effect of parental rights protection, it is unclear as to whether the best interests principle is being afforded a meaningful position of primacy. Similarly, where the need for 'stability' of care is cited as being 'essential to a child's physical, mental and emotional development'[70] this might be achieved at the expense of preserving or releasing birth record information. That said, such a system would at least serve to protect against the belated claims of birth parents for the return of the child, post-placement. Where birth relatives are seeking the revocation or overturning of an adoption order or an adoptive placement, an 'all or nothing' approach' to legal parenthood and excusive parenting may produce a highly unsatisfactory outcome with children either being belatedly returned after forming strong psychological attachments to their substitute social parents or perhaps barred absolutely from having any future contact with, or information on, their birth relatives. Many cases will see the child's need

[65] Washington Rev. Code § 26.10.160 (3).

[66] Ibid Per O'Connor J p. 5.

[67] See also *Stanley v Illinois* (1972) 405 US 645.

[68] *Toxel v Granville op cit* at n 63 pp. 8–14.

[69] Barton (2003–2004), p. 140. The four competing party interests are listed here as birth parents, adoptive parents, the State and the relinquished child.

[70] *Ibid* p. 143. See also Hilderbran (1988), p. 1217.

for identity simply outbalanced by adult-centric interests such as new family autonomy or privacy; the best interests principle may be referred to during the discussions, but it often seems to appear more as an afterthought than as the chief priority.

8.4 An 'Enclaved'[71] Approach to Indigenous Child Adoption

A customary, de facto adoption template for substitute social parenting (where birth records are not sealed and the child's original blood-ties are not legally severed) could minimize the harsher effects of origin deprivation, not least by highlighting the psychological welfare needs of adopted children.[72] In contrast to this, the North American 'trans-racial adoption movement' had focused on placing children who had been deemed 'difficult to adopt,' often on the basis that they were of 'mixed race' descent. Arguably, adoptive parents' motives perhaps varied, for example from settling for 'second choice' in terms of adoptee availability (when faced with a shortage of freely relinquished white, newborn infants) to a desire to altruistically embrace a sort of 'rescue fantasy' parenthood, arising from the highly visible social value of adopting an obviously vulnerable child from a socially disadvantaged, ethnic background.

Such adoptions may have served however to effect the assimilation of the child's original heritage, where 'each child "lost" to adoption outside the ethnic group represents an incalculable drain of cultural resources.'[73] The cultural heritage needs of the tribe, with adult members refusing to let their young kinfolk be 'lost' to non-indigenous placements, are generally addressed through the repatriative, family reunification measures of legislation such as the ICWA.[74] This redressive measure does not necessarily in itself always lead to outcomes based upon paramountcy of the best interests of the child principle however,[75] and could potentially

[71] See *Cardinal v AG of Alberta* (1974) 40 DLR (3d) 553 per Laskin CJ at p. 569, for discussion of how Native status often serves to exclude indigenous persons from the remit of provincial legislative provisions, creating a rights 'enclave' in respect of fundamental, aboriginal freedoms (land rights, child welfare) as cited by McGillivray (1983–1985).

[72] *Ibid* p. 83. See also *SKK v JS* [1999] NWTJ No 94 (SC) (QL) [SKK] on the issue of the termination of birth parents' rights and obligations post-customary adoption.

[73] McGillivray (1983–1985), p. 450.

[74] Section 3 of which states that, 'it is the policy of this Nation to protect the best interests of Indian children and to promote the stability and security of Indian tribes and families and the establishment of minimum Federal standards for the removal of Indian children from their families and the placement of such children in foster or adoptive homes which will reflect the unique values of Indian culture..' Pub L No 96-608, 92 Stat 3069 (1980).

[75] See *Mississippi Band of Choctaw Indians v Holyfield* [1989] 490 U S 30 which ostensibly placed tribal interests on a par with the best interests of the child.

expose vulnerable children to abusive or highly inadequate parenting.[76] The significance of customary adoption lies therefore more in its ability to enable the maintenance of original ancestral heritage and cultural integrity, with birth names generally being preserved and kin and tribal contact usually being framed as an essential element of the entire process. The concept of the tribal customary adoption ('TCA') further underscores the fundamental importance of having knowable heritage; the avoidance of origin deprivation is not simply an aspirational desire on the part of decision-makers but can be embedded within domestic laws and policy as a positive, juridical obligation.

Washington has further outlined how the traditional U.S. custom of informally placing relinquished children with 'racially matched' adoptive parents was eventually replaced by a statute-based policy (i.e. under the legislative remit of ICWA) which sought to completely prohibit the consideration of race as a factor in effecting adoptive placements.[77] Neither practice, she argues, serves to truly promote as paramount the best interests of the children involved. The best means of achieving a more balanced policy position might lie somewhere between these two opposing models. It would consist of increased social tolerance for the making of 'trans-racial placements' whilst also requiring that prospective adopters perhaps be required to undertake some 'racial competency training' prior to the child's placement occurring.[78]

Such a regime would be necessary to counteract the 'a-racial' (denial of difference) approach to adoptive parenting that can exist and which may lead to difficulties for trans-racial adoptees in terms of coping with unforeseen racism or prejudices which they may well encounter as they grow older.[79] A lingering 'emphasis on homogeneity'[80] also has the potential to restrict placement opportunities in terms of the adoptive parents' religious beliefs, cultural values or ethnicity. Equally, the best interests principle could be defined simply in terms creating an obligation to achieve a 'speedy placement' of the child with little or no reference to the need to afford recognition or respect to the socio-cultural heritage or ancestry.[81]

[76] See for example the evidence given in *Re Bridget R et al (Minors)* [1995] BO93520 B093520 (Super. Ct. No. BN1980 consol. w/BC114849): 'During these weeks and months, the relationship between Richard and Cindy was deteriorating...Cindy obtained a restraining order, which required Richard to remain at least 100 yards from Cindy and their two sons, Anthony and Richard Andrew. In a declaration in support of her application for the restraining order, Cindy related that on numerous occasions during March, Richard hit and kicked Cindy and pushed her down, broke furniture, and abused the one-and two-year-old children by picking them up by the neck and shaking or dropping them, poking them in the face, or hitting them in the head. On at least one of these occasions, Richard was intoxicated.' Para 42.

[77] Washington (2005–2006), pp. 1–56 at p. 3.

[78] *Ibid* p. 41.

[79] *Ibid* p. 4.

[80] *Ibid* p. 26 Arguably, norms and policies in relation to anonymous gamete donation have also followed this approach, in terms of matching with recipients racially or ethnically.

[81] *Ibid* p. 33. See further Woodhouse (1995), p. 107 (as cited by Washington 2005–2006).

In respect of adoptions involving children who have indigenous ancestry, the issue of framing their best interests as paramount takes on added element of complexity. The question of whether or not a best interests 'language' even ought to be included within relevant legislation or policy documents arose in the United States for example in relation to the drafting and application of the Multi Ethnic Placement Act (1994).[82] It was suggested here that the exercise of social worker discretion could potentially serve to prevent any trans-racial placements, if for example it could not be clearly demonstrated that the child's best interests would be furthered by such an adoptive placement.[83] The enactment of further legislation eventually removed references to the principle of best interest, so that issues of 'race, color or national origin' could simply no longer be cited as factors which might prevent placement.[84] The extent to which this was a response to wider calls for the 'Indian' legislation to be applied to all 'adoptions of colour' is not entirely clear.[85]

As McGillivray has argued, 'adoption, unlike other laws which touch Indians as ordinary persons, intrudes on Indian character, identity and relationship.'[86] In terms of accepting that there is now a positive, juridical obligation upon the state to protect ancestry, it is significant that a number of jurisdictions have enacted domestic legislation aimed primarily at safeguarding heritage, birthright and genetic connections. The United States' Indian Child Welfare Act 1978 ('ICWA') for example, requires that written consents to the adoption process be obtained from birth parents, or from the child's custodian. Hearings may be at tribal courts, applying customary tribal law, with provision made for interventions or objections by the child's tribal kinfolk, in State-led adoption proceedings. A subsequent withdrawal of parental consent might enable the overturning of an adoptive placement especially where some measure of fraud or duress may have featured in the process of obtaining parental consent. Then, unless the welfare interests of the child strongly demand otherwise, he or she will generally be returned to the original parent.[87]

There may be scope too for the tribe to make arrangements for a de facto, tribal customary adoption to occur, so that the child might be placed with substitute, caretaker parents of Indian heritage who will be better placed to keep the child 'in touch with his heritage'[88] than any non-indigenous adoptive parents. This 'Indian

[82] *Ibid* p. 35. The Multi-Ethnic Placement Act 1994 Pub L No 103-382, 551-55 108 Stat 3518, 4056-58 (October 20, 1994).

[83] See further Howe (1999), p. 677 (as cited by Washington 2005–2006, p. 36).

[84] See also 'The Small Business Job Protection Act 1996 Section 1808 ('Removal Of Barriers To Interethnic Adoptions') Pub L No 104-188, 110 Stat 1755 (1997) and 42 USC 671 (a) (18) (A) (2001).

[85] See also Hawkins-Leon and Bradley (2002), pp. 1227–1233; Kennedy (2003).

[86] McGillivray (1983–1985), p. 440.

[87] Hollinger (1989), pp. 451–501 at p. 453.

[88] See also *In Re A M (A person coming under the Juvenile Court Law) Yolo County Dept of Employment and Social Services v M W and Pit River Tribe* (April 11, 2013) C070782) Sup Ct No JV08500 in respect of 'Indian family', available at http://www.courts.ca.gov/opinions/documents/C070782.PDF (accessed 15.04.13).

child exception' (together with the 'benefit exception') creates a 'heightened standards' hurdle for decision-makers under ICWA, with the best interests of the child tying in closely to the need to avoid bringing about the loss of Native heritage, which might occur in the wake of a court-ordered termination of parental rights. That said, the issue of the child's best interests essentially takes second stage to the question of whether parental rights are in existence.

In *In Re Bridget R., et al., Minors* (1995)[89] for example the Californian Court of Appeal relied upon the 1978 Act to gauge whether the indigenous-descent infants ought to remain with their non-Native adoptive parents or be returned to the care of their natal kinfolk.[90] The legislation stated that any adoption applications involving indigenous children must have been preceded by a voluntarily written parental relinquishment. This must in turn have been witnessed by a member of the judiciary and executed not less than 10 days after the birth. Consents to child relinquishment that fail to comply with the statutory rules would be deemed illegal, and the Court could declare this to be so at any stage of the proceedings, with a right to bring an appellant action held by the child, the birth parents (or the child's custodian) and, perhaps most significantly, members of the tribe.[91] Under the 1978 Act, judiciaries must have clear regard to the 'existing Indian family'[92] doctrine to try and establish whether or not the child's birth parents might be protected by its provisions.

The statute also requires Native status parents to show that they have maintained 'a significant social, cultural or political relationship with their tribe'.[93] Here, such a finding would have rendered the relinquishment consents void. The birth father initiated proceedings to have the infants returned from their prospective adoptive parents who had been caring for the children almost from birth, having received them when they were a few weeks old. The birth father had lived several hundred miles away from the reservation and had never sought to participate in tribal life. He had also previously chosen to keep his Pomo tribal ancestry a secret from the adoption social workers so that it would not slow down the adoption of the infants in any way.[94] The Court had to decide whether his retrospective enrolment as a member of the tribe would allow him to invoke the protections of the 1978 Act. If a finding of 'tribal relationship' resulted, a full Guardianship hearing would be needed to discover whether the children should be removed from their prospective adoptive parents and placed instead with their Native American grandmother for tribal customary adoption.

---

[89] *Re Bridget op cit* n 4.

[90] The Indian Child Welfare Act 1978 A 25 U.S.C.A. 1901. The Bureau of Indian Affairs must also be informed.

[91] ICWA (1978) 25 U.S.C.A., Section(s) 1913, subd. (a).

[92] (See, e.g., *In re Adoption of Crews* (1992) 118 Wash.2d 561 [825 P.2d 305]; *Matter of Adoption of Baby Boy L.* (1982) 231 Kan. 199 [643 P.2d 168].

[93] B093520 (Super. Ct. No. BN1980 consol. w/BC114849).

[94] *Re Bridget op cit* n 14 p. 4.

It is note-worthy that the Court (citing the earlier case of *Re Jasmon O* [1994][95]) endorsed the principle that children are not merely the chattels of their parents, but do possess fundamental interests of their own. The children's best interests were not treated as stand-alone welfare-based rights but seem to have been linked to and defined by the notion of family life rights, given that 'an individual's many related interests in matters of family life are compelling and are ranked among the most basic of civil rights.'[96] These would in turn be strengthened by the degree of 'historic respect – indeed, sanctity would not be too strong a term – traditionally accorded to the relationships that develop within the unitary family.'[97] The Court did however also highlight the Supreme Court's attitude towards parental rights (citing *Stanley v. Illinois* [1972][98]) to support the argument that that the interests 'of a man in the children he has sired and raised . . .undeniably warrants deference.'[99] The decision in *Lehr v Robertson* [1983][100] was also alluded to here however, where it was held that the interests of children are paramount and that 'the rights of the parents are a counterpart of the responsibilities they have assumed.'[101] Ultimately, after 5 years of litigation, the adoptive parents won 'custody' of the children on the basis that the ICWA did not apply, by virtue of the birth father's 'lack of connection' with the tribe, rather than on the basis that the child's welfare was the paramount consideration.

Bakeis has suggested that legislation such as ICWA may actually serve to promote unfair or unequal treatment of Native children and of their original parents, given that urgent issues of child protection may undermine or negate the ostensibly welfare-based aim of placing indigenous children with suitable indigenous adopters. Considerable delays or frequent changes of placement may further jeopardize the Native child's best interests, and prevent useful forms of permanency planning.[102] Although the statutory provisions do promote and protect ancestry and cultural heritage rights and recognize the value of maintaining original genetic ties, they may fail to frame the best interests of the child as the paramount consideration, especially where child welfare might be a matter of acute concern.[103]

The aims of the legislation seem to have arisen from a number of assumptions: that child relinquishment will have been consensual and essentially irrevocable post-placement; that the child's original biological heritage (and, to an extent, his wider tribal bonds of social kinship) will be actively preserved or restored; and that

---

[95] In *Re Jasmon O* (1994) 8 Cal.4th 398, 419.

[96] Citing *Quilloin v. Walcott* (1978) 434 U.S. 246, 255 [54 L.Ed.2d 511, 98 S.Ct. 549]; see also *In re Marilyn H.* (1993) 5 Cal.4th 295, 306.

[97] Citing *Michael H. v. Gerald D.* (1989) 491 U.S. 110, 123 [105 L.Ed.2d 91, 109 S.Ct. 2333.

[98] *Stanley v. Illinois* (1972) 405 U.S.

[99] *Ibid.* at p. 651.

[100] *Lehr v. Robertson* (1983) 463 U.S.

[101] *Ibid* at p. 257.

[102] Bakeis (1996). See also Hollinger (1989), pp. 450–501.

[103] See further Hazeltine (2002), pp. 58–84.

the child's welfare needs will be met by the placement. Contact and information availability are regarded as generally beneficial to child welfare, so that a bar upon these would occur only in exceptional circumstances, with the chief focus being on the need to maintain the child's original identity heritage and acnestry. Where however birth parents seek belated revocation of the adoption, perhaps arguing that an ongoing placement is illegal under the terms of the legislation, the notion of framing preservation of the blood-tie as the paramount concern becomes much less straightforward as a model for good practice. Where a child has formed an emotional attachment to the adoptive parents, or faces a clear risk of abuse or neglect, then an argument based upon the principle that the blood-tie must predominate at all costs, clearly fails to address the needs or protect the rights of vulnerable children.

Hollinger has argued that such a legislative pre-occupation with the notion of 'tribal survival' may quite easily serve to completely undermine the best interests principle in many such cases. Attention appears to be focused more upon the issue of remedying past social injustices than on ensuring future child welfare.[104] As McGillivray has further observed in respect of much of the Canadian jurisprudence on customary de facto adoptions, a number of domestic jurists have repeatedly stressed that the 'placement of the child...must not be used as a means of righting past wrongs'.[105] In some areas, the legislative reforms aimed at redressing or preventing further rights violations has sometimes instead led to socio-cultural disenfranchisement of the people most affected by the changes to the legal frameworks. In Saskatchewan for example, persons of indigenous descent could derive considerable material benefit from being able to register as 'status Indians' under domestic law.[106] Individuals adopted by non-status parents lost the right to such benefits, although they could later apply for recognition of their Native ancestry upon gaining the age of majority. The fact that they might not have been informed of their rights to hold Native status by their adopters could however serve to entirely exclude them from their birthright heritage entitlements.[107]

This contrasts with the customary norms of de facto native adoption processes, whereby 'children so adopted grew up with full knowledge of their origins, and of the fact and fate of their biological parents. Disclosure was viewed as their birthright.'[108] Within such an informal model for achieving the ties of social kinship, the child's original birth name was often retained, with some degree of contact between natal relatives also tending to be enabled for the benefit of all

---

[104] Hollinger (1989), p. 453.

[105] *Ibid* p. 459.

[106] *Ibid* p. 438. See also The Indian Act RSC 1970 c 1-6, s 11(1) (e) and 12 (2); The British North America Act 1867 30 & 31 Victoria c.3 (UK). The benefits which attach to being a person '*of Indian blood, reputed to belong to a particular tribe, band or body of Indians..and their descendants*' include rights to own tribal land, to hunt, fish and shoot, to access improved medical benefits and health care, tax exemptions, and greater educational support (to gain voting rights however, 'Indian status' would have had to be renounced).

[107] McGillivray (1983–1985), p. 447.

[108] *Ibid* p. 448. See also Ladner (1977) and House (1997), pp. 333–363.

concerned.[109] This differs from the more formalized models of state adoption, in the sense that absolute secrecy might have traditionally held sway with a significant level of judicial deference afforded to the legally heavier interests, rights and needs of adoptive parents. As Baldassi has further argued in respect of informal indigenous adoptions,

> some customs include additional rights and obligations for all parties and most flow from a different philosophy of family, kinship and child rearing than do standard statutory adoptions.[110]

In respect of Canadian customary adoptions, increasing recognition has been afforded since early cases such as *Re Tagornak (1984)*,[111] with the practice given clear endorsement a decade later in *Casimel v Insurance Corp. of BC* (1993).[112] Here, the plaintiffs were members of the Stellaquo Band of the Carrier People in British Columbia. They had raised their grandson after their daughter abandoned him in infancy. When he died aged 28, having acted as their carer for many years, the Court declared that they were eligible to claim the death benefits due to them as his de facto adoptive parents. The practice of customary adoption is not however a uniform one; as House has reported (in respect of the Mi'kmaq people of Nova Scotia) a variety of processes exists, creating significant problems for non-indigenous lawyers tasked with determining whether de facto adoption has occurred.[113]

In *ANR and SCR v LJW* (1984)[114] a customary indigenous adoption appeared to have been agreed between Métis (i.e. 'non-status' Native American) foster parents and the child's birth mother. The birth mother had rehabilitated herself after having had a long history of problems with alcohol and thus sought the return of the child. The Manitoba Supreme Court considered whether the child's psychological need for knowable ancestry might outweigh the risk of her suffering psychological harm, which was felt to be a likely outcome if she was removed from the care of her social parents, the fostering couple.[115] The Court noted that domestic legislation had dispensed with the need for birth parental consent in cases of de facto adoption, but stressed that

---

[109] See for example *Re Adoption of Katie* E7-1807 (1961) 32 DLR (2d) 686 38 WWR 100 (NWT Terr Ct).

[110] Baldassi (2006), pp. 63–100 at p. 64.

[111] As Aboriginal status rights under s.35 of the Constitution Act 1982; *Re Tagornak* [1984] 1 CNLR 185; (1983) 50 AR 237 (NWTSC); see also *Re Katie* (1961) 32 DLR (2d) 686; 38 WWR 100 (NWT Terr CT) per Sissons J on Inuit customary adoption; Re *Deborah* (1972) 6 RFL 299 [1972] 5 WWR 203 (NWTCA) where Inuit custom adoption was challenged by birth parents, per Morrow J and on appeal at (1972) 28 DLR (3d) 483 (NWTCA) per Johnson J p. 488.

[112] *Casimel v Insurance Corp of BC* (1993) 82 BCLR (2d) 387; 106 DLR (4th) 720 (CA). See also The Aboriginal Custom Adoption Recognition Act SNWT 1994 of the North West Territories.

[113] House (1997), p. 333. See also Lomax 1997, 14.

[114] *ANR and SCR v LJW* (1984) 37 RII 2d 181 Man Civ Supreme Court.

[115] *Ibid.* See also Opinion of Matas J from the lower Court at p. 179.

> this does not mean, of course, that the child's tie with its natural parents is irrelevant...it is obviously very relevant in a determination as to what is in the child's best interests.[116]

The lower court had initially sought a compromise whereby the foster parents were granted custody of the child, with a view to maintaining some form of contact with the birth mother. Wilson J, in affirming the adoption, stressed that the question of whether legal 'abandonment' by the mother had occurred or not was not the central issue. The 'best interests of the child' were key to the eventual outcome: they took precedence over issues of race and identity. Arguably, this decision was more easily arrived at given that the foster parents were themselves of indigenous ancestry, albeit of 'non-status' heritage.

In the case of *H v M* [1999][117] indigenous genetic grandparents (post-reunion with their relinquished adult daughter who had been raised by white adoptive parents in the United States) also sought custody of their grandson. Opposition to their application was made by the child's non-indigenous, adoptive grandparents. By the time the case came before the Canadian Supreme Court, the 4-year-old child had been taken to live with his grandfather on his reservation in northern Manitoba. The Court ultimately awarded custody to the adoptive grandparents (but allowed for contact with the birth kin) essentially ignoring the 'legislative trend to keep aboriginal children in aboriginal families'.[118]

This decision was in keeping however with the decision in *C (JMN) v Winnipeg Child and Family Services* (1998)[119] from a year earlier, which saw the dismissal of cultural heritage arguments by the Awasis Agency (who had sought the overturning of a placement with non-indigenous carers) in favour of 'constancy' as a means of protecting the indigenous child's best interests.[120] Such decisions do not detract from the fact that the long-term effects of origin deprivation, especially in respect of vulnerable members of society, such as indigenous or minority groups, are gradually being afforded a greater degree of judicial recognition however. The legacy of strict, closed records policies impacted profoundly upon adoptees of indigenous and 'mixed race' descent. Their experiences highlight how discriminatory treatment, stigma and basic rights infringements might easily accompany the loss of ancestral heritage.[121]

The adoption of indigenous children by non-indigenous families clearly has the potential to challenge the policy objectives such as secrecy and

---

[116] *Ibid* per Wilson J.

[117] *H v M* [1999] 1 SCR 328; see also [1999] 1 SCR 761 where the Court dismissed an application to permit representations from the birth grandfather's First Nation. See further Lazarus (1997), p. 255 (as cited in Mossman 2004) p. 178.

[118] Mossman 2004, p. 178.

[119] *C (JMN) v Winnipeg Child and Family Services* (1998) 33 RFL (4th) 175; Man. R (2d) 81 (QB).

[120] See also *Natural Parents v Superintendent of Child Welfare* (*Sub Nom Re Adoption Act*) [1976] 1 WWR 699 where the best interests principle was applied by viewing the child as an individual rather than as belonging to a specific racial or cultural grouping, with a 'good home' taking precedence over the need to preserve heritage.

[121] See Baldassi (2006), p. 214. See also Mossman (2004), p. 169.

characteristics-matching that often enabled or assisted the quiet creation of 'don't tell' social kinship bonds. The overturning of such adoptions has often seen judges refer directly to the harms and losses of original heritage and non-ancestry that can easily result from sealed record placements.[122] Complete inversion of this secrecy-led template would see a right to access ancestry become a normative rather than exceptional feature of social kinship. To an extent this has occurred already within the 'jurisdiction' of indigenous and 'international' adoptions, at least insofar it is more difficult for genetic realities to hide behind the 'legal truth' of adoption.

This body of case law is perhaps particularly significant given that it does seem aimed at an attempt to find a resolution in respect of such fundamental issues as the balancing of conflicting rights between the two sets of triad parents and the relinquished or removed child. That said, customary adoption models do not necessarily turn entirely upon a strict adherence to the principle that the child's best interests are the paramount consideration. Wider considerations of adult-centric family and societal issues may be taken into account and are perhaps prioritized over the child's interests when decisions on placement are being made. The social prestige associated with parenthood or with having a large family unit, for example, might be accepted as quite valid reasons for wishing to enact or preserve a customary adoption, as might the parental desire to strengthen social alliances.[123] Even if the child's best interests are not the key concern in such contexts, however, the views and opinions of the child may still factor heavily in the making of important decisions on their future care and welfare. Additionally, the practice of de facto customary adoption often tends to permit adoptees to retain some level of knowledge over the actual identity of their genetic parents, within a more 'fluid' atmosphere of preserved kinship bonds.[124]

In the case of *Re Deborah* [1972][125] for example a birth mother sought the return of her daughter after the child had spent 11 years in the care of her customary adoptive parents. The North West Territories Court endorsed the informal arrangements, on the basis of witness testimonies which had confirmed that customary norms would never have permitted a revocation of adoptive kinship to occur at such a late stage.[126] In *CAD v VG* [1992],[127] a not dissimilar situation arose with the mother claiming that no customary adoption had occurred, and seeking to effect

[122] See for example *Mooswa v Minister of Social Services for the Province of Saskatchewan* (1977) 30 RFL 101 wherein the Court ordered the return of a Native child to her birth mother on the basis that she should be '*raised if possible by (her) natural mother…amongst people of her own race and culture*' See also for example The Ontario Child and Family Services Act 1985 s 53(5).

[123] See for example *CAD v VG* [1992] NWTR 236, 33 ACWS (3d) 90 (NWTSC).

[124] Baldassi (2006), p. 71.

[125] *Re Tucktoo et al and Kitchooalik et al (Sub Nom Re Deborah)* (1972) 27 DLR (3d) 225 [1972] 3 WWR 194 (N W Terr Ct). This was the first case of customary adoption where the court was required to consider the content of the process rather than merely the issue of whether or not the parties' qualified for 'Native status'.

[126] See further Baldassi (2006), p. 82.

[127] *CAD v VG op cit* n 19.

some form of ongoing contact with her child on that basis. The adoptive parents did not raise an objection here, emphasizing that the child clearly had a need to be connected with her ancestral heritage. The Court adopted a passive, declaratory approach to the issue rather than taking an interventionist stance, on the basis that the practice of customary adoption fell within the remit of indigenous matters. The hypothetical need to apply the best interests principle was however, still highlighted, given that the case had effectively centred upon the issues of child care and custody.

Similarly, in *Re Tagornak Adoption Petition* (1983) it was held that customary adoptions were likely to at least be based upon consideration of the following elements: the consent of both sets of parents to the process, a voluntary placement of the adoptee with a new substitute family, and the native ancestry of the adoptive parents.[128] The emphasis here was again mainly upon preserving the child's indigenous cultural heritage and maintaining an original genetic identity, whilst also aiming to prevent situations of conflict arising between the two sets of triad parents. In fairness, although the principle that the best interests of the child are the paramount consideration was not explicitly referred to here, an over-arching need to protect the welfare of the adopted child may be implied. Placement must again be consensual and it was also expected that the terms of the arrangement would be honored by both sets of parents. The child's right to access an ascertainable, cultural heritage would also be promoted here through her placement with adopters from a similar, Native background.

The case of *Adoptive Couple v Baby Girl* (2012)[129] (the 'Baby Veronica' case) ties together a number of key elements of this area of jurisprudence but also highlights the difficulties that can arise in respect of seeking the revocation of an adoptive placement and trying to preserve the child's original indigenous identity while still aiming to protect her best interests. Here, the South Carolina Supreme Court found that the child (who had an unmarried Indian father, non-Indian relinquishing mother) could not be adopted by non-Indian parents unless the parental notification provisions of the ICWA had been fully complied with. She was returned to her birth father after spending the first 2 years of her life with her adoptive parents. On appeal to the United States Supreme Court there seem now to be two key issues to be determined: whether a non-custodial parent can rely upon the rights enshrined in the ICWA to block an adoption which had been underpinned by a voluntary relinquishment on the part of the non-Native birth mother and whether the Act's definition of 'parent' might apply to an unmarried genetic father who had essentially lacked any legal status as a parent under the relevant state legislation.[130] Zug has observed that these are quite strange questions, given that

---

[128] Baldassi (2006), p. 82.

[129] *Op cit* n 4.

[130] Transcript of opening oral arguments available at http://www.supremecourt.gov/oral_arguments/argument_transcripts/12-399.pdf (accessed 17.04.13).

they 'have already been answered and the answer to both is clearly yes.'[131] The Supreme Court has established the remit of tribal jurisdiction in *Holyfield*,[132] confirming that voluntary relinquishments for adoption under state law can be stopped by ICWA, while section 1903 (9) of the Act itself defines a parent as 'any biological parent' of the Indian child. She suggests that the Supreme Court might perhaps use this appeal as an opportunity to examine further the 'existing Indian family exception' which some lower courts have used to deflect ICWA's provisions.[133]

Alternatively, the issue of the Act's constitutionality might come under Supreme Court scrutiny. Because the *Baby Girl* appeal concerns a different statutory provision to the one discussed in the *Holyfield* decision any overturning of the earlier case would possibly have to also involve a finding that all of ICWA was essentially unconstitutional. Zug makes the further point that ICWA is based upon two premises: firstly, that the special laws and enhanced standards for effecting child relinquishment which relate to Indian adoption practices are not simply based upon race and secondly that the legislature has the authority to make such special laws in respect of Native Americans. Were the Supreme Court to reject either of these two propositions, it is possible that 'most of Indian law would fall as well.'[134]

Such an outcome is fairly unlikely however given the political aspects and consequences of such a decision and bearing in mind the underpinning purpose of the legislation, which was enacted 'to prevent the breakup of Indian families, ensure the transmission of Indian culture and heritage, and prevent the extinction of Indian tribes.'[135] There is perhaps much more of a possibility that the Court might instead seek to endorse a sort of exception to the federal remit of ICWA, which could allow states an increased authority in such cases, for example where the relinquished Indian children have been born 'off-reservation.'[136] As Howe observes, opening arguments seem to be focusing on quite a wide range of issues, not least those of potential discrimination. The decision reached by the lower court, if affirmed, might well result in a banning of any interracial adoptions that involve Native children, and could potentially give rise to a sort of second-class citizenship for any non-Native mothers whose children have been fathered by Native men. Adoptive parents, as the appellants seem to be arguing, would possibly also be discriminated against in such scenarios, in favour of the child's original parents on the basis of their racial background. In terms of any free-standing rights of the child, the Court

---

[131] Zug (2013), pp. 46–54 at p. 47.

[132] *Holyfield Op cit* n 75.

[133] See for example *In Re The Adoption of Baby Boy L* 643 P.2d 168 (Kan. 1982); If the child had never been part of an Indian family, the ICWA would not apply. Zug further notes the divisiveness of this test and argues that there is now a near-universal acceptance that it offends against the spirit and legislative purpose of the 1978 Act.

[134] Zug (2013), p. 49.

[135] Ibid p. 51.

[136] Ibid p. 52.

noted that it was basically being tasked with having to effect an interpretation of a federal statute that will essentially serve to remove from such cases 'the ordinary best interests determination' that would usually be made in similar family law hearings, in the lower courts.[137]

The issue of whether the best interests of the child had been fully considered here by the lower court (which had found in favour of the birth father gaining custody of the child under the remit of ICWA) was also raised. It had been held there that this transfer of custody would not cause the child serious emotional or physical damage; whether this had amounted to a decision based squarely upon the best interests of the child was however unclear and seems unlikely. For best interests questions to even arise, parental rights would presumably have first had to be legally terminated; arguably the Native child might therefore have been treated differently from other non-indigenous, vulnerable children (in terms of having her best interests protected or promoted) solely on the basis of her racial heritage.

In terms of the child's genetic father, who is currently parenting the child, being legally recognised as a parent under the provisions of the ICWA, the legislation provided for procedural protections requiring that any applicant who wishes to have the parental rights of another terminated must demonstrate that 'active efforts' were made to preserve the Indian family, for example that remedial services and rehabilitative programs aimed at preventing dissolution were made available but had failed to keep the family intact. Parental rights could be terminated only where there was evidence 'beyond a reasonable doubt' that the continuance of custody (by the parent or guardian) would be likely to cause the child serious emotional or physical damage.'[138] It was argued that because the father had not had any social relationship with the child at the time of the initial proceedings however (beyond that of their genetic connection) there were as such no meaningful ties of relatedness between them to be broken. The fact that he had offered no support, financial or otherwise, to the child or her mother prior to his legal challenge of the adoption was also discussed. The child's genetic heritage (her father was descended from the Cherokee Nation) was also referred to, with the Chief Justice expressing some concern that a 'single drop' of blood could give serve to rise to the various rights enshrined in the ICWA.[139]

The case may yet turn ultimately on the interpretation of the language in the statute and in respect of its underpinning purpose, rather than on the wider question of whether the Act is essentially an unconstitutional instrument. As Justice Kagan observed, careful interpretation would be needed to avoid the creation of two classes of parenthood, one of which would be absolutely protected by the statute and one which would not; it is unlikely that this was the original legislative intention.[140] The issue of whether the genetic father had, or rather could have,

[137] See further Howe (2013), available at http://www.scotusblog.com/2013/04/argument-recap-no-easy-answers-in-indian-adoption-case (accessed 17.04.13).

[138] Section 1912(d) and (f).

[139] *Op cit* n 130 p. 42.

[140] *Ibid* p. 54.

legally 'renounced' his parentage (via text message to the child's mother during her pregnancy apparently[141]) perhaps highlights the degree of confusion that still attaches to the concepts of unmarried genetic fatherhood and legal paternity.[142]

One of the most significant issues of this appeal is perhaps unlikely to be afforded much time or meaningful discussion by the Court; the question of how decision-makers might more fully define and better promote the best interests of the child in such difficult cases. Some discussion of why or whether legal determinations involving the issue of the child's best interests must essentially be either avoided, ignored or subject to the question of whether parental rights were legally existent (or being meaningfully exercised) at the time of the application for custody, would be useful here. It might also be relevant to other areas of law which have seen parental and child rights come into direct conflict with one another. If for example the Court were to make some comment on the general nature of the child's right to genetic ancestry, and whether this should exist irrespective of whether the child is of indigenous or non-indigenous descent, it would possibly lend weight to the argument that harsh severance of the blood-tie can amount to a fundamental rights-violation. If vetoes on genetic, ancestral heritages were found to amount to an unjust and discriminatory occurrence, then it is difficult to see how they can then simply be justified as normative to social kinship creation, and in respect of other rights-vulnerable members of society, such as adoptees and gamete donor children. In other words, anyone who has been so deprived of the fundamental right to access their authentic onomastic identity and genetic birthright should be able to argue that their rights have been outweighed by those of others. This would place an onus upon decision-makers to show that such events occur exceptionally rather than normatively, and to demonstrate that the central issue in such origin deprivation was one of protecting the child's best interests. In terms of harms being suffered by the 'veto-victims' then this too is surely a generic rights-led issue that is not unique to one particular culture or society. To be origin-deprived to the extent that one is permanently orphanised genetically is to be prone to vulnerability; race and ethnicity may add to this inherent susceptibility to harm but they should not in themselves serve to create very narrow precedents that serve to perpetuate rights-violations against other sections of society. Put quite bluntly, it seems now that where it is impossible to hide or assimilate the child's racial or ethno-cultural background, law and policy-makers have had to accept that biological truths do matter. This should be the case for every relinquished child however, whether their cultural heritage matches that of their social parents or not.

Similarly, the question of whether the concept of the best interests of the child should simply be deemed irrelevant on the basis that a parent has not been found to be 'unfit' to parent, perhaps merits fuller discussion. An opportunity to meaningfully

---

[141] See comment by the New York Times http://www.nytimes.com/2013/04/17/us/justices-hear-case-of-adopted-indian-child.html?ref=us&_r=0 (accessed 16.0.13).

[142] In relation to the controversies surrounding the ICWA generally see Atwood (2002), pp. 587–676.

embed child welfare as the paramount consideration in cases that obviously have as their main concern their future care and well-being of the child has perhaps arisen with this appeal. If it is found that the best interests of the child standard may be used against a fit genetic parent, in cases where the child is actually resident with that parent, then this might give rise to a more far-reaching interpretation of the notion of children's rights than has previously been declared by the Court however.[143] Such a decision might open the door to the possibility of the child's best interests becoming the central concern (or at least regarded as a much weightier, rights-bearing factor) in cases that clearly involve their future welfare.

Whether such an outcome might create a situation where, in theory, any prospective adoptive parent could basically just argue that the child's best interests obviously demand a transfer of custody to them, seems unlikely, given that there would surely also need to be strong evidence (the presence of strong emotional attachments, a risk of significant harm if such bonds were broken) that such a decision was definitely the best option for the fought-over child. As the jurisprudence on best-interests from the United Kingdom indicates, framing the child's best interests as the paramount concern still provides scope for the courts to consider other rights-relevant issues, such as the obligation to protect original ancestry, ties of social kinship and the Article 8 ECHR family life rights of parents (for example, via some form of contact). The danger of not having child welfare as the underpinning central concern is that other matters seem to too easily fill any legislative gaps left by its absence. A link to cultural heritage and identifiable ancestry is often a key element of child welfare, but where this must be firstly defined by and then completely dependent upon extrinsic, adult-centric concerns (e.g. protection of the parental right to an exclusive parenthood, tribal preservation of socio-cultural heritages and population numbers, attempting visible redress for past social injustices) it is difficult to see how the best interests of the child can feature as little more than an afterthought, if at all. In other words, the issue of the child's welfare might not really feature in the court's rights analysis, if the question of parental rights (their termination or initial existence) is seen as the key concern.

Settling cases such as this one purely on the basis of which side can demonstrate a 'better' legal claim to parentage (rather than establishing that they would be the best person to serve as parent, and from the perspective of the child) leaves no room for substantial evaluation of the child's welfare-based right to be parented in the best manner possible. From the relinquished child's point of view, her psychological attachments matter; irrespective of whether or not her genetic father had any parental rights, had legally renounced them or deserved to be awarded them belatedly on the basis of a statute aimed at preserving and repairing communal cultural heritages, it is worth remembering that this child's mother elected to firmly reject her motherhood. As such, the child is vulnerable. Her welfare ought to be the paramount consideration with the issue of who should 'win' legal parentage moving away from centre stage.

[143] *Op cit* n 130 pp. 36–37.

Protection of indigenous cultural heritage does not require and should not demand that children be placed at risk of psychological harm. She had, from birth, formed bonds of attachment with her social parents; on the basis that her welfare would have been the paramount concern, she should not have been removed from their care. Equally, some form of direct regular contact should have been agreed to and enabled so that her genetic ties to her father and to her ancestral heritage might have been maintained, or, more accurately in this case, repaired. By the same token, she has now formed psychological attachments to her father; to move her again would be likely to compound the risks of harm to which she has already been unfairly exposed. Again, in terms of finding the best solution for genetically relinquished children, who may have also suffered the added emotional trauma of being separated from those who they have bonded with as parents, the best interests of the child cannot simply be seen as an irrelevant concern.

Arguably, if the Supreme Court were to perhaps glance towards the provisions of the Children's Convention (which the United States has yet to fully embrace) in a bid to find some degree of guidance on where or how the child's best interests might at least feature in such cases, their decision could mark a significant move towards promoting the rights of the child as being at least equivalent to those of parents. Similarly, had the case occurred in a European Convention signatory state, the child would probably have remained with her adoptive parents, with her father possibly being granted a right to maintain significant contact with his daughter, so that his, and indeed the child's, right to family life under Article 8 of the European Convention would have been protected, and so that the child's right (and need) for accurate heritage identity might have been addressed. In terms of process and procedure, Article 6 ECHR rights might well also have arisen in respect of the state having to ensure that the birth father had perhaps been kept informed of the impending adoption proceedings.

It is note-worthy that the Court referred to the possibility of custody occurring 'via estoppel' on the basis of the passage of time.[144] The consideration of a best interests 'calculus' at the time of the initial assertion of parental rights, rather than at the time of the court hearings, might have brought about a different result given how the lapse of time can affect the child in developing strong attachments to her adoptive parents, or, more recently, to her genetic and now custodial father. The term estoppel which the court used in respect of any custody decision which would be based upon time passing, brings to mind two legal concepts from the field of property law, namely the protection of a proprietary, promissory or equitable right and the notion that one side is more deserving than the other of either winning or retaining ownership over the subject of the estoppels-based claim.

If the use of equitable devices and doctrines is appropriate to such cases, why not extend the reasoning to argue that the doctrine of laches applies (so that delay would defeat the equity of the case) or even incorporate other principles from the

[144] *Op cit* n 130 p. 24.

wider field of business law, so that the behaviours of the neglectful or absent parent might be open to analysis as an act of negligence or a breach of the duty of care. Treating the child as an item of property would tie in with the notion that parenthood must be based upon an exclusive right of possession rather than linked to the child's welfare. Principles of good conscience, fairness of treatment, and detrimental reliance upon a promised benefit that has failed to materialize, often contrive to transfer ownership in cases involving equitable interests on the basis of owner 'worthiness'. That the court attached such an adult-centric, intention-heavy property law term to the issue of the vulnerable child's future care is quite telling. If such cases are to be guided mainly by reference to legal principles of property ownership, or of equitable, meritorious modes of possession then it would seem that the best interests of the child will indeed be very far removed from the position of paramountcy (or even relevance) that other jurisdictions have afforded to them. For the child's best interests to only become relevant once or if parental rights have been terminated or established, or for them to be defined essentially by reference to parental or tribal Constitutional rights or cultural needs, is to diminish them. In sum, the child's best interests are either the paramount consideration, the primary concern, or they are prone to being outweighed and dismissed.

The presence of emotional attachments (as opposed to the risk of the child suffering harm) should be central to any analysis of the child's welfare in parentage or parenting disputes such as this one. A child raised from birth with substitute carers is likely to have formed strong psychological attachments to them; from the perspective of the child these are her parents. It is difficult to see how the decision of the lower court in this case paid much if any heed to the child's best interests. As in the Irish 'Baby Anne' case, the over-riding rights issue was adult-centric and aimed at protecting rights to parent and to exist as part of a family. The child's best interests were simply not weighty enough to confer any form of stand-alone rights upon the child.

Had the court been free to suggest or enable some form of compromise, such as shared parenting or frequent contact with the genetic parent, such a solution might well have reflected the child's need for good parenting, maintained attachments and preserved genetic identity. This outcome would of course have required both sets of parents to be willing to afford the child's welfare a position of paramountcy over and above that of their own rights to an autonomous, exclusive parentage or parenthood. It seems ironic that so much value has been placed on the 'at all costs' maintenance of genetic ancestry, in this particular context, in a region that otherwise has an entrenched tradition of severing children's original blood-ties via gamete donor anonymity and closed adoptions. As long as childhood and child welfare remain so closely associated with such contract-led legal notions as intentional parenthood, child 'custody' and consent to relinquishment, and with socio-legal definitions of the child's best interests so utterly reliant upon adult-centric parental rights and interests, it seems likely that cases such as these will continue to happen.

8.5 Conclusion

Although 'the denial of difference is a prevailing parental strategy'[145] in respect of creating and supporting some forms of social kinship, it cannot be denied that genetically relinquished persons continue to form 'one of the most vulnerable demographics'[146] in terms of being afforded adequate levels of child protection, restoring lost or removed attachments, and enabling ancestral heritage and identity. As Washington has argued, some form of compulsory training for prospective adoptive parents in issues of racial competency might help to alleviate the 'color-blindness' that can attach to some placements. Equally, other forms of competency training for other decision-makers could help to ensure that such instances of 'identity-blindness' within social parenting models might be similarly addressed in respect of child welfare. Woodhouse has further suggested that adoption ought to be viewed not as a modern creature of statute but as the formalization of a long-standing, universal customary practice.[147] As such, the process perhaps ought to merit the same level of recognition given to other modes of social kinship creation, such as marriage. In other words,

> adoption is not a benefit created for and conferred by the state, subject to whatever terms and conditions the state may choose to impose; it is the statutory recognition of a fundamental family relationship.[148]

Children would thus have a potentially juridical right to be adopted, independent of the essentially inchoate 'right' to achieve parenthood of their potential adopters. Such a template would differ significantly from the position often assumed by domestic judiciaries[149] where the child's 'right' to family life may be entirely dependent upon the rights of the warring sets of adults to either adopt him or be reunited with him.[150] On this basis, the principles underpinning certain precedents on family life[151] ought to be equally relevant to those situations where 'children who have already formed de facto parent–child relationships with would-be adoptive parents'.[152] To do otherwise might be to expose already vulnerable children to the risk of discriminatory or harmful treatment.[153]

---

[145] Washington (2005–2006), pp. 1–56 at p. 26.

[146] *Ibid* p. 41.

[147] Woodhouse (2005–2006), pp. 297–329 at p. 316.

[148] *Ibid.*

[149] See for example *Lofton v Secretary of the Dept of Children and Family Services* 358 F 3d, 804, 811-12 (11th Cir 2004).

[150] Woodhouse (2005–2006), p. 319.

[151] *Loving v Virginia* (1967) 388 U S (1) on the illegality of miscegenation laws.

[152] Woodhouse (2005–2006), p. 324.

[153] See further *In the Matter of Adoption of J.J.B.* (1995) 119 N.M. 638 [894P.2d 994, 1008-1009]; *Bookert v. Roth* (1995) U.S. [133 L.Ed.2d 110, 116 S.Ct. 168]; *In the Matter of Custody of C.C.R.S.* (Colo. 1995) 892 P.2d 246, 257–258.

By the same token, if it is accepted that other forms of substitute care, such as permanent guardianship, carry the risk of reduced stability or the possibility of multiple placements, then barring an adoption may deprive parentless children 'of something of great value'.[154] Viewing adoption as an essential component of a child's right to family life in certain circumstances, might allow the best interests principle to be defined in a more fully child-centric manner, free from balancing or weighting exercises.[155] As Nagarsheth has argued however, 'the best interests standard was created to be a multi-factor balancing test.'[156] This has perhaps led to much discretion-based judicial ambiguity in respect of its interpretation, especially where parental interests such as privacy have been traditionally regarded as weightier, in a bid to counter competing child interests, such as a right to identity or welfare paramoutncy. Given generally that 'children are the only involuntary parties to an adoption'[157] it is perhaps not unreasonable to expect that states should be compelled to ensure that their best interests are as fully addressed as may be possible in any given case, whether through access to accurate birth information and ancestry, via facilitation of meaningful contact post-adoption, or through specific training for key decision-makers on the specific needs and vulnerabilities of genetically relinquished persons.

As Cabellero has suggested, the state can and should, in its role as *parens patriae*, be obliged to actively promote the child's best interests as the paramount consideration in both the short and long-term.[158] Children should not have to rely upon the goodwill of parents, social workers and judges to be afforded the permission to achieve psychological well-being or the right to accurate identity. Similarly, if 'the court is the nexus in adoption'[159] then a change to some of the current norms surrounding social kinship could perhaps serve to tilt the balance of rights more in favour of the child. Genetic relatedness alone should not be enough to simply negate a previous relinquishment, de facto or otherwise, or to overturn a long-settled adoptive placement; children who are 'repatriated' in such a manner could be at risk of suffering such trauma and loss of attachment that their restored origins might have been achieved at too high a cost. Such a formula for preserving heritage is likely to be insufficient to address the complex issues and difficulties that arise in relation to post-adoption contact, especially where an abusive or neglectful form of relatedness may have been a feature of the child's original familial history. Equally, laws and policies aimed at providing redress for past cruelties and injustices ought not to seek to do so by ignoring the best interests of the child, or by relegating them to the level of irrelevant or secondary concern.

---

[154] *Ibid.*

[155] See further Bojorge (2002), pp. 266–291; see also *In Re RMG* 454 A2d 776, 791 (DC 1982) where a three-part analysis was suggested for determining disputed adoptions.

[156] Nagarsheth (2005–2006), pp. 45–68 at p. 67.

[157] Caballero (2005–2006), pp. 291–313 at p. 312.

[158] *Ibid.*

[159] McGillivray (1983–1985), p. 454.

Many of the cases discussed in this chapter support this argument, highlighting the problems associated with protecting adult-focused family life rights while still ensuring child protection. Judiciaries generally seem to opt for adherence to traditional, adult-centric principles, such as freedom from interference in family life, or find quasi-contractual solutions to either negate or declare parental consent to relinquishment (e.g. through a lack of notice or lapse of time). Where policy-based racial assimilation is a possible factor, domestic legislation may seek to highlight the over-arching need for cultural heritage protections within adoption policy and practice. This too can ultimately hinder the effects of the best interests principle, when a workable child protective placement must be found. In an ideal situation, the burdens and responsibilities that attach to social kinship would be shared between the relevant adults. Unlike the vulnerable child, they have achieved adulthood, will usually have consented to their being involved in the social kinship process, and are not often at risk of total origin deprivation even where their 'right' to parenthood might be curtailed to some extent by their having to share some element of parental responsibility. This would however require a radical change of mindset in terms of understanding what parenthood represents, with the *act* of parenting rather than the *status* of parenthood taking centre stage, and the child's best interests rights being raised to a more meaningful position of paramountcy.

References

Atwood, B. A. (2002). Flashpoints under the Indian Child Welfare Act: Toward a new understanding of state court resistance. *Emory Law Journal, 51*, 587–676.

Bakeis, C. D. (1996). The Indian Child Welfare Act of 1978: violating personal rights for the sake of the tribe. *Notre Dame Journal of Law, Ethics, and Public Policy 10*, 543, 569.

Baldassi, C. L. (2004–2005). The quest to access closed adoption files in Canada: Understanding social context and legal resistance to change. *Canadian Journal of Family Law, 21*, 212–265.

Baldassi, C. (2006). The legal status of aboriginal customary adoption across Canada: Comparisons, contrasts and convergences. *University of British Columbia Law Review, 39*, 63–100.

Barton, K. (2003–2004). Who's your daddy? State adoption statutes and the unknown biological father. *Capital University Law Review, 32*, 113–148.

Bojorge, C. (2002). Inter-country adoptions: in the best interests of the child? *Law and Justice Journal*, *2*(2), 266–291.

Caballero, R. (2005–2006). Open records adoption: Finding the missing piece. *Southern Illinois University Law Journal, 30*, 291–313.

Cahn, N. (2000–2001). Children's interests and information disclosure: Who provided the egg and sperm? Or, mommy, where (and whom) do I come from? *The Georgetown Journal of Gender and Law, 2*, 1–27.

Cahn, N. (2002). Birthing relationships. *Wisconsin Women's Law Journal* 163.

Clayton, B. J. (2007–2008). A day late and a dollar short: Should Utah's unmarried dads get one more chance to claim their newborns? *Journal of Law and Family Studies, 10*, 249–266.

Hawkins-Leon, C., & Bradley, C. (2002). Race and transracial adoption: The answer is neither simply black or white nor right or wrong. *Catholic University Law Review, 51*, 1227–1233.

Hazeltine, S. L. (2002). Speedy termination of Alaska Native Parental Rights: The 1998 changes to Alaska's child in need of aid statutes and their inherent conflict with the mandates of the Federal Indian Child Welfare Act. *Alaska Law Review, 19*, 58–84.

Hilderbran, G. S. (1988). *In re Baby Girl Eason*: Balancing three competing interests in third party adoptions. *Georgia Law Review, 22*, 1217.

Hollinger, J. H. (1989). Beyond the best interests of the tribe: The Indian Child Welfare Act and the adoption of Indian children. *University of Detroit Law Review, 66*, 450–501.

House, J. (1997). The changing face of adoption: The challenge of open and custom adoption. *Canadian Family Law Quarterly, 13*, 333.

Howe, R. A. W. (1999). Adoption law and practices in 2000: Serving whose interest? *Family Law Quarterly, 33*, 677.

Howe, A. (2013). *Argument Recap: No Easy Answers In Indian Adoption Case*. SCOTUSBLOG (Apr. 16, 2013, 1:21 PM).

Kennedy, R. (2003). *Interracial intimacies: Sex, marriage, identity and adoption*. New York: Pantheon.

Ladner, J. A. (1977). *Mixed families adopting across racial boundaries*. New York: Anchor Press.

Lazarus, K. B. (1997). Adoption of native American and first nations children: Are the United States and Canada recognizing the best interests of children? *Arizona Journal of International and Comparative Law, 14*(1), 255.

Lomax, B. (1997). Hlugwit'y, Hluuxw'y-My Family, My Child: The survival of customary adoption in British Columbia. *Canadian Journal of Family Law, 14*, 197.

McGillivray, A. (1983–1985). Transracial adoption and the status Indian child. *Canadian Journal of Family Law, 4*, 437–467.

Mossman, M. J. (2004). *Families and the Law in Canada; Cases and commentary*. Toronto: Harcourt Canada.

Nagarsheth, T. (2005–2006). Crossing the line of color: Revisiting the best interests standard in transracial adoptions. *The Scholar, 8*, 45–68.

Washington, T. (2005–2006). Loving Grutter: Recognizing race in transracial adoptions. *George Mason University Civil Rights Law Journal, 16*(1), 1–56.

Woodhouse, B. B. (1995). Are you my mother? Conceptualizing children's identity rights in transracial adoptions. *Duke Journal of Gender, Law and Policy, 2*, 107.

Woodhouse, B. B. (2005–2006). Waiting for *Loving*: The child's fundamental right to adoption. *Capital University Law Review, 34*, 297–329.

Zug, M. (2013). Two and a half ways to destroy an Indian. *Michigan Law Review First Impressions, 3*, 46–54.

Case Law

United States

Adoptive Couple v Baby Girl (2013) No. 12–399 (The 'Baby Veronica' case) Oral Transcript http://www.supremecourt.gov/oral_arguments/argument_transcripts/12-399.pdf (accessed 16.04.13)

Adoptive Couple v. *Baby Girl*, 398 S.C. 625; 731 S.E.2d 550 (S.C.) (2012)

Caban v Mohammed 441 US 380 (1979);

Ellis v Social Services Department of the Church of Jesus Christ of Latter Day Saints (1980) 615 P 2d (Utah)

Haley A Mark et al v Elizabeth L. (1996) Contra Costa County Superior Court No. 9400055

In Re A M (*A person coming under the Juvenile Court Law*); *Yolo County Dept of Employment and Social Services v M W and Pit River Tribe* (April 11, 2013) C070782)

In re Adoption of Baby Boy L. (1982) 231 Kan. 199 [643 P.2d 168]

In re Adoption of Crews (1992) 118 Wash.2d 561 [825 P.2d 305]

In Re AWJ (2000) 316 Ill App 3d 91 96,736 NE2d 716
In *Re BGC* (1992) 496 NW 2d 239 (Iowa)
In *Re Bridget R et al (Minors)* [1995] BO93520 B093520 (Super. Ct. No. BN1980 consol. w/BC114849):
In re Custody of Ayala (2003) 344 Ill App 3d 574 588 800 NE 2d 524
In *Re Custody of MCC (A Minor) v Aisha Umer* (June 27 2008) 06 D 79411
In re Custody of O'Rourke (1987) 160 Ill App 3d 584 587–88 514 NE 2d 6 (1987)
In re Custody of Peterson (1986) 112, Ill 2d 48, 53–54 491 NE2d 1150
In re Doe 627 N E 2d 648 [1993] (Ill. App Ct 1993) 656
In re Jasmon O. (1994) 8 Cal.4th 398, 419
In re Kirchner (1995) 649 N E 2d 324 (Illinois)
In re Marilyn H. (1993) 5 Cal.4th 295, 306
In *re Marriage of Carey* (1989) 188 Ill App 3d 1040, 1046 544 NE 2d 1293
In Re Marriage of Sechrest (1990) 202 Ill App 3d 865, 875 560 NE 2d 1212;
In Re RLS (2006) 218 2d 428
In Re RMG 454 A2d 776, 791 (DC 1982)
In Re The Adoption of Baby Boy L 643 P.2d 168 (Kan. 1982);
In the Matter of Adoption of J.J.B. (1995) 119 N.M. 638 [894P.2d 994, 1008–1009] *Bookert v. Roth* (1995) U.S. [133 L.Ed.2d 110, 116 S.Ct. 168];
In the Matter of Custody of C.C.R.S. (Colo. 1995) 892 P.2d 246, 257–258
Lehr v. Robertson (1983) 463 U.S. 248, 256 [77 L.Ed.2d 614, 103 S.Ct. 985]
Lofton v Secretary of the Dept of Children and Family Services 358 F 3d, 804, 811–12 (11th Cir 2004)
Loving v Virginia (1967) 388 U S (1)
Michael H. v. Gerald D. (1989) 491 U.S. 110, 123 [105 L.Ed.2d 91, 109 S.Ct. 2333
Mississippi Band of Choctaw Indians v Holyfield [1989] 490 U S 30
Quilloin v. Walcott (1978) 434 U.S. 246, 255 [54 L.Ed.2d 511, 98 S.Ct. 549];
Re Bridget R et al (Minors) (1995) 25 USCA 1901 BO93520
Sanchez v LDS Social Services (1984) 680 P 2d
Stanley v Illinois (1972) 405 US 645
Thomas v Children's Aid Society of Utah (1961) 364 P (2d) 1029
Thurnwald v A E (2007) UT 38, 163 P 3d 623
Troxel v Granville (2000) 530 U S 57, 147 L Ed 2d 49, 120 S Ct 2054
Wells v Children's Aid Society of Utah (1984) 681 P 2d (Utah)
Wickham v Byrne (2002) 199 Ill 2d 309

Canada

ANR and SCR v LJW (1984) 37 RII 2d 181 Man Civ Supreme Court
C (JMN) v Winnipeg Child and Family Services (1998) 33 RFL (4th) 175; Man. R (2d) 81 (QB)
CAD v VG [1992] NWTR 236, 33 ACWS (3d) 90 (NWTSC)
Cardinal v AG of Alberta (1974) 40 DLR (3d) 553
Casimel v Insurance Corp of BC (1993) 82 BCLR (2d) 387; 106 DLR (4th) 720 (CA).
H v M [1999] 1 SCR 328
Mooswa v Minister of Social Services for the Province of Saskatchewan (1977) 30 RFL
Natural Parents v Superintendent of Child Welfare (Sub Nom Re Adoption Act) [1976] 1 WWR 699
Nelson v Children's Aid Society of Eastern Manitoba [1975] 5 WWR 45, 56 DIR (3d) 567, 21RFL 222 (Man CA) [1974] 5 WWR 449, 46 DIR (3d) 633 (Man QB)
Re Adoption of Katie E7-1807 (1961) 32 DLR (2d) 686 38 WWR 100 (NWT Terr Ct)
Re *Deborah* (1972) 6 RFL 299 [1972] 5 WWR 203 (NWTCA)
Re Tagornak [1984] 1 CNLR 185; (1983) 50 AR 237 (NWTSC);
Re Tucktoo et al and Kitchooalik et al (Sub Nom Re Deborah) (1972) 27 DLR (3d) 225 [1972] 3 WWR 194 (N W Terr Ct).
SKK v JS [1999] NWTJ No 94 (SC) (QL) [SKK]

Ireland
In Re JH [1985] IR 375
N v Health Service Executive & Ors [2006] IESC 60
Northern Area Health Board v An Bord Uchtála [2002] 4 IR 252

England
Re F (Infants) (Adoption Order: Validity) [1977] Fam 165
Re F (R) (An Infant) [1970] 1 QB 385; *Re RA (Minors)* (1974) 4 Fam Law 182;
Re G (Children) (residence: same sex partner) [2006] 4 All ER 241
Webster (The Parents) v Norfolk County Council & Ors (Rev 1) [2009] EWCA Civ 59

Chapter 9
Guiding Principles, Hard Cases

9.1 Introduction

> The guiding principles should be a commitment to truth, individual autonomy and priority for the rights and interests of those primarily affected.[1]

The notion that a type of 'difference perception'[2] can adversely impact upon adoptee self-concept or self-esteem[3] seems to be a generally accepted one; as Powell and Afifi have argued for example, a harmful uncertainty over one's heritage may, in terms of risk and harm, be highly 'recursive in nature'.[4] Grotevant's study of 'harmful behaviours'[5] has also highlighted how a long-term state of 'unresolved grief' may lead in turn to chronic feelings of 'uncertainty and loss'.[6] Differing judicial interpretations of the best interests principle in cases involving 'fragmented biological processes'[7] also seem to point to an 'ever-increasing acknowledgement of (the) complexity and multifaceted nature'[8] of legal relatedness and social kinship.

This chapter will argue that a number of guiding principles do now exist in respect of the blood-tie; these strands of 'blood-tie jurisprudence' cut across a fairly

[1] Bainham (2008), p. 324.

[2] Carsten (2000), pp. 687–703 at p. 689.

[3] Ryburn (1995), pp. 41–64 at p. 41. See also Verrier (1993). See also, on 'self worth' Verrier (2003).

[4] Powell and Afifi (2005), pp. 129–151 at p. 148.

[5] Grotevant (2003), p. 753 at p. 755.

[6] Powell and Afifi (2005), n 4, p. 129.

[7] Mykitiuk (2002), pp. 771–781 at p. 772.

[8] Horgan and Martin (2008), pp. 155–162 at p. 156 available at http://www.coe.int/t/e/legal_affairs/legal_co-operation/family_law_and_children%27s_rights/Documents/Rosemary%20Horgan%20Article.pdf (accessed 01.05.11).

A. Diver, *A Law of Blood-ties - The 'Right' to Access Genetic Ancestry*,

DOI 10.1007/978-3-319-01071-7_9, © Springer International Publishing Switzerland 2014

wide variety of legal contexts.[9] Although these laws and policies on the 'right' to not be deprived of ancestry and genetic connection appear at times to be so ad hoc as to amount to a sort of 'birthplace lottery', it may still be argued that a number of key, underpinning principles are becoming common to most of these bodies of law. As these fundamental elements emerge and, to an extent, crystallize, they seem to be either endorsed or denounced across the various jurisdictions. Taken together, they at least seem to confirm that the biological birth right heritage increasingly matters, even if only in terms of placing legal barriers between genetic relatives. It may be argued that as a whole they represent a sort of template for inverting the 'severance as best option' reasoning employed by some jurisdictions (i.e. in respect of sealed records, donor anonymity or the absolute parental veto on information disclosure) but may also be used in preventing or tempering the 'preserve at all costs' approach to family reunification that exists in other regions.

Neither of these two models for kinship (genetic severance or reverence) can be said to be truly child-centric, either in terms of outcome or outlook; the principles that are suggested in this chapter, are aimed instead towards finding where possible a child-protective, compromise-grounded alternative, where the loss of genetic connection is a highly exceptional rather than normative event, yet still compatible with the key principle that the child's welfare must be the paramount consideration. As a sort of template for decision-makers, it is suggested that these guiding principles may provide useful direction, not least in those situations where 'hard cases' have arisen, and especially where those solutions which appear to be the most workable might actually be quite questionable in terms of protecting or promoting the child's best interests as the paramount consideration. Similarly, where a closed-records or donor-anonymous jurisdiction might be seeking to effect meaningful, non-discriminatory legislative or policy changes, these guiding principles might be useful in terms of preventing the worst excesses of traditional, adult-centric rights-veto systems. They could perhaps allow for a wider range of outcomes, not least by endorsing the concept of a best-interests-led need for 'fluent narrative'[10] as essential to child welfare.

9.2 Child Welfare and Best Interests as Paramount: Domestic Embedding via Checklists?

In terms of such fundamental human rights 'filtering down' from the persuasive, paper-based principles and norms of international law, a number of jurisdictions have seen their domestic courts increasingly accept that the best interests of the

[9] International law e.g. UNCRC Country Reports, Concluding Observations etc.; (see Chap. 4); European Court of Human Rights case law involving Articles 6 and 8 of the European Convention (see Chap. 5); domestic law (e.g. Canada, Ireland and the United Kingdom) on genetic parentage and parenthood; see Chap. 6); domestic law on contact post-adoption and disputed relinquishments (e.g. the United Kingdom; see Chap. 7); domestic law where blood-ties appear to be the paramount consideration (e.g. Ireland, the United States, Canada, see Chap. 8).

[10] Kirkman (2003), pp. 2229–2242 at p. 2229.

child ought to be framed as the paramount concern in child welfare cases, and as a free-standing, basic right, not constrained or defined according to the 'weightier' interests or desires of parents or other adults. Where child welfare paramountcy has been endorsed as the over-riding, central issue, aspects of child protection and family life rights seem more likely to encompass and accept the notion of a need for kin contact and accurate information, as essential elements of protecting or recovering knowable ancestry. The 'right' to genetic identity in domestic contexts, if it exists at all, does so as a highly qualified, quasi-justiciable entitlement; complete severance of the biological connection does seem however to be gradually becoming more exceptional than normative where cases involve parentage disputes, revoked adoption consent, kin contact or information disclosure.[11]

That said, in some jurisdictions, the right to genetic connection has been taken as an adult-centric, absolute norm for settling disputes over parenting or parenthood. This can work directly against the best interests of the child. Where the blood-tie has been framed as the paramount consideration (for example through legislation that seeks to protect or restore tribal cultural heritages or promote rights to parenthood or family autonomy) domestic jurists may have little or no option but to return a child to the parents who had perhaps originally relinquished or abused him, or to remove a child from her psychological parents in direct opposition to the recommendations of expert witnesses, and possibly in contravention of the best interests principle itself. If such cases constitute a 'victory' for those arguing that a juridical right to genetic connection exists, it is a pyrrhic one. The child's best interests must be central to any solution or settlement of a dispute involving heritages, whether these have been lost, preserved or restored post-severance. Judicial decisions based mainly or entirely upon adult-centric concerns such as privacy rights, the maintenance of tribal cultural heritage or a 'contractual' lack of informed consent to relinquishment, may leave little or no room for proper consideration of the child's welfare needs; such cases seldom give rise to comfortable precedent.

In terms of defining, embedding and promoting the best interests of the child at the level of domestic proceedings, a number of jurisdictions have introduced statutory child welfare checklists. Whether these represent a best-interests generic model or one aimed more specifically at the adoption process, such checklists generally seem to emphasize that child welfare must be the primary or paramount

[11] See for example the UK cases *Re G* [2013] EWHC 134 (Fam) (sperm donor fathers may bring proceedings in terms of seeking contact); *K (Children) v Sheffield City Council* [2011] EWCA Civ 635 (25 May 2011) ('exceptional circumstance: parental contact curtailed); *H (Children), Re* [2012] EWCA Civ 743 (16 May 2012) (on non-separation of siblings and the importance of the maternal bond); *S v L* [2012] UKSC 30 (11 July 2012) (on the best interests of the child as paramount concern); see also the Irish authority *M.R & Anor-v-An tArd Chlaraitheoir & Ors* [2013] IEHC 91(genetic connection between child and egg donor mother recognised, legal presumption *mater camper certa est* abolished).

consideration[12] in all proceedings that might impact upon the child's 'upbringing'.[13] Within the United Kingdom for example, Section 1 (1) of The Children Act 1989 cited seven key factors;

> (a) the ascertainable wishes and feelings of the child concerned (considered in the light of his age and understanding)
> (b) his physical, emotional and educational needs;
> (c) the likely effect on him of any change in his circumstances;
> (d) his age, sex, background and any characteristics of his which the court considers relevant;
> (e) any harm which he has suffered or is at risk of suffering;
> (f) how capable of meeting his needs is each of his parents and any other person in relation to whom the court considers the question to be relevant;
> (g) the range of powers available to the court under this Order in the proceedings in question.

Section 1 of The Adoption and Children Act 2002 went further in terms of enshrining a similarly drafted, but more specific 'adoption welfare' checklist which decision-makers, agencies and courts should refer to when engaged in any decision-making that might affect the welfare of the child in question.[14] It stresses the needs of the child 'throughout his life' and highlights the duty on all concerned to avoid unnecessary procedural delays where possible. It refers specifically to the adopted child's 'particular' needs and to the 'likely effect' of his 'having ceased to be a member of the original family and becoming an adopted person.'[15] In keeping with the Strasbourg jurisprudence on Article 8 of the European Convention on Human Rights' the checklist places much emphasis upon the value of the child's current relationships, allowing decision-makers to gauge whether these are 'relevant.'[16] The significance or otherwise of relatedness may be determined by reference to a variety of factors, such as 'the likelihood of any such relationship continuing and the value to the child of its doing so.'[17]

The key question may however be whether the child's relatives 'or any such person' are willing or able to offer appropriate substitutive kinship care within a 'secure environment.'[18] The checklist stresses that the concept of 'relationships'

---

[12] Article 3 (1) of The Children's Convention refers to the best interests of the child as '*a primary consideration*' rather than the paramount element.

[13] See also the welfare checklist in Article 3 (3) of the Children (NI) Order 1995. 'Relevant matters' are those which fall within the remit of private law 'Section 8' Orders (Article 8 Orders in Northern Ireland) i.e. Residence, Contact, Specific Issues and Prohibited Steps. Other proceedings on child upbringing include Care and Supervision Orders, Guardianship, Adoption, and Section 30 of the Human Fertilisation and Embryology Act 1990. The Adoption and Children Act 2002 reiterates the need for child welfare paramountcy: available http://www.opsi.gov.uk/acts/acts2002/en/02en38-c.htm (accessed 24.08.11).

[14] See Section 1 (4) available at http://www.opsi.gov.uk/acts2002/ukpagea_20020038 (accessed 29.07.10).

[15] *Ibid* S 1 (4) (b) and (c).

[16] *Ibid* S 1 (4) (f).

[17] *Ibid* S 1 (4) (f) (i).

[18] *Ibid* S 1 (4) (f) (ii) The third and final element is that of '*any of the child's relatives wishes and feelings.*'

should not simply be defined solely by reference to the issue of 'legal relatedness'.[19] Such a provision may place a considerable burden upon domestic decision-makers in respect of determining how the child's future bonds of kinship might have to be arranged, maintained, severed or replaced. In any event, a not inconsiderable degree of discretion exists here which could at least encourage jurists, social workers and parents to examine the extent to which non-ancestry might impact adversely upon the longer term welfare of relinquished or removed children.

Given that genetic kinlessness can arise in circumstances other than the traditional adoption triad (for example via gamete donation) decision-makers might also seek guidance from the key principles that have emerged in respect of the best interests-led 'right' to genetic identity.[20] A useful template for domestic checklists could, for example, require decision-makers to particularly address;

(a) the *relinquished or removed* child's particular needs making special reference to *genetic and cultural heritage and to the child's long term psychological welfare needs*; *particular reference should be made to the child's right to maintain or achieve kinship ties and access original identity information as enshrined within the provisions and principles of international and human rights law*
(b) the likely effects of any changes to circumstances, with *particular reference to the effects of severing original biogenetic links or those social bonds of kinship and attachment which may have arisen post-placement with substitute carers. Ties of kinship between other persons (such as siblings) should similarly be taken into account with a view to ensuring that they are protected*
(c) age, sex, *ethno-cultural and genetic heritage* and any *other* characteristics considered relevant, *with particular reference to the unique socio-cultural position and psycho-social needs of the child who might be at risk of being rendered genetically 'kinless'*
(d) how capable of meeting his needs *all* of his parents or other substitute carers, *natal or social* and any other relevant persons *might be but without prejudicing the child's right to be protected from harm*

The aims of such a welfare checklist would be three-fold: to acknowledge the unique socio-cultural position of the genetically kinless child, to prevent where possible the adverse effects of non-ancestry and to promote best interests-led welfare paramountcy in both the short and longer term. Additional terms that might be included in an origin deprivation prevention checklist might be framed as follows:

(a) Decision-makers should where possible refrain from automatically balancing, weighing or weighting the rights of the child against those of their parents or care-givers (natal or social) and other relevant adults (such as tribal elders or other third parties)
(b) Where a pre-existing adult veto attaches to issues of information disclosure or contact (via domestic legislation or previous Court proceedings) the child should not be deprived of or denied access to a domestic right of appeal against such vetoes

---

[19] S 1 (8)(a).

[20] The most relevant provisions would perhaps include: Articles 8, 9, 20, and 21 (a) of The Children's Convention 1989; Articles 16 (1) (a) and 30 (1) of The Hague Convention 1993; Article 27 of The ICCPR 1966; Article 3 of The ICESCR 1966; Articles 2 and 25 (2) of the UDHR 1948 and Article 6 of the Universal Declaration on Human Genomes and Human Rights (1997).

(c) Non-ancestry or genetic, origin deprivation ought only to occur in highly exceptional circumstances, such as where the best interests of the child clearly demand such an outcome.

Within the United Kingdom, recent domestic case law on the making of parental orders (for example, in surrogacy cases) have confirmed that child welfare is now the paramount consideration in cases involving the child's care and upbringing.[21] The European Court of Human Rights also appears to have endorsed the principle that domestic decision-making processes must actively and fully consider the best interests of the child. Failure to do so may result in a violation of Article 8 ECHR, as has occurred in cases such as *Anayo*.[22] In respect of this case, it is worth noting that, under the German civil code, 'the legal parents' motives for refusing contact did not necessarily have to be based on considerations relating to the children's best interest' in such circumstances.[23] 'Sufficient' reasons for refusing contact should however be given in such a scenario, and adequate consideration should also be afforded to the fundamental issue of whether contact would be in the best interests of the child. That child-centric reasoning should be applied to such rights-balancing analyses, reflects the fact that child welfare paramountcy ought now to be at the centre of such disputes. Children should not be vulnerable to simply having their best interests overlooked, outweighed or sidelined at the level of domestic proceedings, for example to instead promote or protect rights to parental autonomy or family privacy.

Perhaps the most significant aspect of the *Anayo* decision was that the European Court of Human Rights appeared to afford a greater level of recognition to the 'potentiality' of relatedness that existed between the genetic father and his children, albeit on the basis of his intention to commit to family life. Arguably this represented a departure from earlier jurisprudence which often required the presence of a more established social relationship before Article 8 rights might be considered to have arisen.[24] Whether a new rights-bearing, juridical category of 'potential family life' now exists, remains to be seen. Such a development could, in theory at least, lend support to the contention that absolute bars on genetic kinship (via loss of kin contact or through complete vetoes on identifying information) may represent unjustified, perhaps unjustifiable, rights violations under Article 8 of the European Convention. To an extent, at least from the perspective of the disappointed, contact-denied genitor, a justiciable right to genetic kinship is at least quite strongly suggested here.

[21] See *Re L (Commercial Surrogacy)* [2011] 1 FLR 1423 per Hedley J at para 9; see also *A v P* [2011] EWHC 1738 (Fam).

[22] *Anayo v Germany* [2010] ECHR 2083 (21 December 2010).

[23] *Ibid* para 69.

[24] See for example *Marckx v Belgium* (1979) (application no 6833/74) (1979); *Keegan v Ireland* (1994) (App no 16969/90) 18 EHRR 342; *Kroon v The Netherlands* (1995) 19 EHRR 263; *X v Croatia* (app no 12233/04) (17.07.2008); *Johansen v Norway* [1996] EHRR 31 (app no 17383/90).

Similarly, decision-making processes apart from the court proceedings in such scenarios (such as pre-hearing, informal negotiations on post-adoption contact for example) ought now to include, if not actively address, those issues that might touch and concern the best interests of the children affected. Whether some 'potentiality' for relatedness or an 'intention' to forge kinship bonds can be pleaded by other members of the triad (or by grandparents, estranged siblings or disappointed adoptive parents post-reunification) or by veto-affected individuals who are no longer in childhood, is more difficult to predict. As Sloan has observed, a close relationship falling somewhere short of familial life might still fit within the remit of Article 8 private life rights and thus merit domestic legal protection against unjustified State interferences.[25] Recent decisions of the European Court, though not entirely unequivocal, at least indicate 'an admirably flexible approach'[26] in respect of the question of whether potentiality of relatedness merits greater recognition within domestic legal frameworks. Traditional rights-balancing exercises perhaps become much more difficult when the more common dyads (parent, child) or social kinship triads (birth parents, adoptive parents, child) are absent: placing the child's best interests at the centre of such equations however, and thus affording them more meaningful levels of welfare-led paramountcy, might offer the best means of finding workable, Convention-compliant solutions to disputes over parentage and parenthood.

Arguably, a checklist whose chief purpose was to more fully embed child-welfare paramountcy principles within domestic legal and policy systems (by reminding decision-makers of the potentially harmful consequences of sidelining or ignoring the child's best interests) might go some way towards redressing any domestic legislative gaps. As Bainham has also recently argued, positive state obligations to avoid harm and rights infringements do increasingly seem to surround such difficult private family law issues as contact-maintenance and familial reunification (albeit more in the context of emergency care or removal) with the domestic margin of appreciation appearing slightly narrower here than it might be in relation to the initial stages of emergency, child protective intervention.[27] In respect of the United Kingdom's domestic welfare checklists, judges already must have explicit 'regard to every factor' contained therein and are required to 'give full expression to the terms of Article 8 of the ECHR' when granting court orders with potentially far-reaching effects.[28] Although there is no absolute obligation to effect family reunion, simply leaving matters to be settled by the passage of time clearly risks violating the rights of the child and possibly also infringing the rights of estranged relatives. It may be further argued that such developments signify a gradually growing awareness of the value of genetic connections as an element of the right to identity or at least an increased judicial

[25] Sloan (2011), pp. 314–317 at p. 316.

[26] *Ibid.*

[27] Bainham (2011), pp. 312–314 at p. 314.

[28] *Ibid* pp. 313–314.

regard for the need to afford some meaningful measure of legal protection to the welfare-paramount best interests of vulnerable children being raised within such scenarios.

9.3 Blood-Tie Jurisprudence: Emergent Principles

In terms of linking the need to avoid psychological harm to a positive, justiciable right to not be deprived of origin, it is necessary to tie together the various strands of blood-tie jurisprudence. International law provisions do generally suggest that there is at least a fundamental right to identity: in terms of possessing name, nationality and the notion of authentic heritage, such concepts are valuable items, worthy of juridical protections and clearly tied to other meaningful rights (access to genitor-identifying information, contact with relatives) that ought to be fully justiciable at the level of domestic courts. Interpretations of instruments such as the Hague Convention on International Adoption (2008) seem for example to have been influenced by the emergent norm that the child's best interests ought to include aspects of genetic ancestry, free from the limitations of adult-centric, parental interests.[29]

The notion of a free-standing, dignity-based right to not be deprived of authentic heritage can be found within the jurisprudence. The context generally tends to highlight how profoundly the consequences of lost identity might impact upon the genetically kinless person.[30] The Children's Convention has also served to gradually inform understanding of the concept of a welfare needs-based right to genetic ancestry (via the principle of the child's best interests) within a number of domestic legal frameworks, aided by increasing reference to such provisions as Articles 6 and 8 of the European Convention on Human Rights.[31] A number of cases from the European Court of Human Rights on family and private life rights, and on 'fair hearing' processes, clearly tie the concept of the child's right to authentic identity to such issues as best interests-led welfare paramountcy. That a child may derive benefit from (and more significantly, be seen to need) some level of genetic information, and possibly some form of kin contact, seems now to be the normative approach to cases where the central question is one of defining their best interests in opposition to parental desires.[32] Contact especially seems to be an important component, as a means to ensuring that identity heritage is not simply removed or lost through the passage of time and that the right to family or private life is not violated.[33] Involvement of persons such as grandparents or siblings (who would

[29] *Neulinger v Sweden* [2010] 54 EHRR 1087.

[30] See further Oren (2001), pp. 123–195.

[31] See for example *R and H v United Kingdom* [2011] 35348/06 ECHR 844; *Uyanik v Turkey* (3 May 2012) (app 60328/09).

[32] See for example *Godelli v Italy* ECHR 347 (2012).

[33] See for example *Anayo v Germany* [2010] ECHR 2083 (21 December 2010).

traditionally have fallen outside of the traditional, triadic remit for inclusion) in, for example, pre-hearing processes, on the basis that they too may have rights worthy of Article 6 protection, is perhaps one of the most significant aspects of this move towards greater recognition of the importance of genetic connectedness. Arguably, the following key principles have emerged from the various strands of jurisprudence on blood-ties:

1. The fundamental right to an authentic, ancestral identity is often a key component of child-welfare led best interests and the right to family and private life[34]
2. In terms of realising the right to identity, the best interests of the child may be closely tied to such issues as contact, 'potentiality' of relatedness or relationship, or to procedural matters such as passage of time[35]
3. The welfare of the child is the paramount consideration[36]
4. The best interests of the child may require informational disclosure given the significance of the genetic connection and that a positive, juridical obligation to preserve or repair this connection exists[37]
5. Loss of genetic connection ought to occur only in exceptional circumstances[38]
6. Preserving genetic connections should not compromise the principle of child welfare paramountcy[39]

Taken together, this list of guiding principles may provide persuasive direction at the level of domestic decision-making in cases involving genetic connection and child welfare. Its child-centric focus could allow for greater adherence to the recommendations of expert witnesses, for example in relation to the child's rights, psychological needs and in terms of protecting their short and longer term welfare. Similarly, jurists and social workers might be less inclined to delegate to parents those tasks that are often most closely associated with the child's welfare, such as making determinations on whether or not information disclosure or kin contact

[34] See in particular articles 3, 7, 8, 9,20 and 21 of the Children's Convention (1989); articles 16 and 31 of the Hague Convention (1993); articles 9 and 4 of European Convention on Nationality (1997) and articles 10, 14 and 22 of the European Convention on Adoption (2008).

[35] See for example *Anayo v Germany* [2010] ECHR 2083 (contact); *Neulinger v Sweden* [2010] 54 EHRR 1087.

[36] See for example *Y C v United Kingdom* [2010] (App 4547/10) ECHRR; *R and H v United Kingdom* [2011] 35348/06 ECHR 844.

[37] See for example *A v P* [2011] EWHC 1738 (Fam); *M.R & Anor-v-An tArd Chlaraitheoir & Ors* [2013] IEHC 91; *N v Health Service Executive* [2006] 4 IR 372; *Re S (Parental Order)* [2009] EWHC 2977 (Fam).

[38] See for example *Re B (A child)* [2009] UKSC 5; *Neulinger v Sweden* [2010] 54 EHRR 1087; *S v L* [2012] UKSC 30; *Re L2 and O* [2007] NI Fam 12; *SEHSST v LS* [2008] NI Fam 12; *Re K* [2007] EWHC 393 (Fam); *K v Sheffield City Council* [2011] EWCA Civ 635; *Re H* [2012] EWCA Civ 743; *Re K* [2011] EWCA Civ 1064; *Re A* [2012] EWHC 1689 (Fam).

[39] *N v HSE* [2006] IESC 60; *Thurnwald v A E* (2007) UT 38, 163 P 3d 623; *Re Custody of MCC v Aisha Umer* (2008) 06 D 79411; *Re RLS (a minor)* (2006) 218 2 d 428; *In re H.R.* (2012) 208 Cal. App.4th 751; *In Re A M (Yolo County DESS v Pitt River Tribe*, (2013) CO70782 (SC JV08500) available at http://www.courts.ca.gov/opinions/documents/C070782.PDF (accessed 14.04.13).

might be appropriate, what form contact should take, and how little or much of it might be needed to protect the child's best interests. In terms of maintaining, repairing or perhaps suspending the genetic connection, this too could be approached from the perspective of the child rather than from that of the parent who might be claiming an over-arching right to privacy, parentage or parenthood.

This could perhaps see other aspects of origin deprivation processes inverted: birth records might for example remain open until sealed (under 'good cause') and unaltered birth certificates could be available to all adoptees and donor children unless highly exceptional circumstances demanded otherwise. The harsher forms of origin deprivation (parental vetoes on disclosure, gamete donor anonymity, sealed birth records and complete bans on contact) might be seen to occur only in exceptional circumstances rather than normatively, and then solely or primarily on the basis that they must occur to protect the child's welfare or that it is impossible to have an alternative outcome. For example, it may be the case that the child's genetic parent simply cannot be found or that they were abusive, their parenting so dangerous that contact must be prevented or identifying information withheld until the child reaches the age of majority. Equally, there is little to be done where a genetic parent has chosen to deliberately effect a complete relinquishment, for example via a refusal to name a co-genitor or by the use of some 'baby-nest' or safe-haven baby-drop system.[40]

It would be perhaps be equally unlikely that adoptive placements would be overturned, given that the current reasons for doing so tend to involve such purely adult-centric, paper-based rights issues as flawed or incomplete consents to adoption, infringements of parental constitutional freedoms or entitlements (such as privacy or family sanctity) or political, socio-cultural factors such as the need for tribal heritage preservation. As such, it seems highly unlikely that children would be removed from their social, psychological parents once they had formed emotional attachments to them, on the sole basis that an adult-centric issue required the child's repatriative return. Possibly, the principles allow for a measure of 'shared' care, or parentage (if not communal parenting[41]) on the basis that the child's welfare requires some level of connection to be maintained. This would be likely to perhaps involve an arrangement enabling some level of contact rather than expecting parents in such situations to simply and amicably divide their responsibilities equally in relation to the child and their upbringing; case law already exists however in respect of surrogacy and gamete donation which might essentially provide a useful comparator for such a model.

By applying these principles on preventing origin deprivation to a small selection of the 'hard cases' examined in earlier chapters (i.e. on sealed records, donor

[40] See further The Adoption Council of Canada's position statement on safe haven law available at http://parentfindersottawa.ca/theadoptioncouncilofcanadajan2013.pdf (accessed 31.03.13).

[41] See further the Norgrove Report (Family Justice Review November 2011) available https://www.education.gov.uk/publications/eOrderingDownload/FJR-2011.pdf (accessed 12.12.13); Munro Review of Child Protection (May 2011) available at http://www.official-documents.gov.uk/document/cm80/8062/8062.pdf (accessed 22.02.13).

anonymity and overturned adoptions) it should become possible to gauge more accurately whether changes to domestic legislation and policy based upon these principles might provide a more coherent approach to the issue of genetic connection. Although hard cases generally tend to not make 'good law' they do often provide useful insights into the less pleasant realities of family life, especially where child welfare is the key issue. Equally, it may be argued that much child protection law is driven by reactive, legislative responses to scandals, where for example media-led calls for law reform are underscored by dire warnings to parents that their rights are being eroded, or that child protection services and systems are in danger of meltdown or no longer fit for purpose. Hard cases do represent tragic realities but the question of whether reform of the current systems must come from 'above' (for example via human rights law or international law principles), 'within' (reform of domestic laws and policies) or somewhere else entirely (changes to parental attitudes) is most relevant to these difficult scenarios involving severed or restored blood-ties. They arguably share one key, common feature: in not one of these hard cases can it be said that child welfare was the paramount consideration for the domestic court. Had it been so, the outcomes might perhaps have differed.

9.4 Sealed Records, 'Clean-Break' Severance

In the case of *Kearns v France* (2008)[42] the focus of the French courts was almost entirely on the adult-centric issue of the potentially flawed, subsequently revoked maternal consent to the infant's relinquishment. In respect of the subsequent appeal to the European Court of Human Rights, the case appeared to largely turn on the fact that the child's birthmother had actively chosen to render her genetically kinless by opting for the French legal device of anonymous birth registration. That she had prepared a small parcel of non-identifying, mnemonic keepsakes (such as letter and photographs) was also deemed to be evidence of her willingness to have their genetic connection completely and permanently severed via a clean-break relinquishment, rather than as a means of maintaining some small measure of connectedness or possibility of a reunion at some stage in the future.

[42] *Kearns v France* ECHR 10 January 2008 (Application no 35991/04). The Irish birth mother sought to revoke her consent on several grounds: that she had not fully understood the consequences of her actions; the strictness of the time limits had breached her Article 6 and Article 8 European Convention rights; eight weeks was too brief a 'cooling-off' period, and she had signed consents the day after the birth with no official interpreter available, in the post-natal ward. (She was however, accompanied by her mother and a French lawyer who she had instructed earlier as to relinquishment.) The child's genetic father had not consented to the adoption, and had instituted paternity proceedings in Ireland; her husband (in Irish law, presumed to be the child's legal father) subsequently 'recognised' the child as being legally his, to no avail.

Had the case arisen and been heard in the Irish mother's country of residence (leaving aside the fact that the time limits for the revocation of her consent would also not have applied here) the eventual outcome would potentially have been entirely different.[43] As demonstrated by the Irish 'Baby Anne'[44] case, the mother's marriage (even though her husband was not the genetic father of the child) would have almost certainly provided her with sufficient Constitutional protections to prevent the State from interfering with her rights to family life. Similarly, had such a case occurred in the United States, the lack of fully informed consent on the part of the mother, genetic father (and indeed the legal father, the mother's husband) might well have served to prevent or annul the adoption.[45] Had the child been of 'mixed-race' or 'trans-racial' origin, the French or European Courts might have shown a little more reluctance to endorse the time limits for revoking consent and the system of absolute severance from ancestry. Some consideration was given to the issue of her psychological welfare needs, and her attachment to her new caregivers, but neither the French nor the European Court explicitly stated that the best interests of the child principle should be paramount in such a case. Article 8 was not infringed, given the wide margin of appreciation that attached to such issues (there is no consensus on time limits amongst member states) and Article 6 rights were not engaged.

Arguably, had this case been heard after the *Anayo* decision, the French courts might have demonstrated a little more reluctance to bring about such a harsh, clean-break split between the genetic mother and her child, on the basis that a child's right to identity may require some degree of informational disclosure. The lack of a domestic remedy, although grounded in the French Civil Code, is not entirely dissimilar to the legislative gap which was so relevant to the later German case. Potentiality for relationship and an intention to commit to relatedness were clearly shown by the birth mother here, albeit belatedly. The failure to consider more fully and centralise the child's best interests at the level of domestic proceedings perhaps also mirrors the statutory omissions in *Anayo*. That said, it seems unlikely that the European Court would have ordered an overturned adoption outcome or declared the French system of anonymous birth to be incompatible with the provisions of the European Convention, given the wide margin of appreciation that seems to attach generally to matters of private family law. Potentially, it might have simply found that the child's or the mother's Article 8 rights had been unjustifiably interfered with, but would most likely have stopped well short of suggesting that some identifying information should be made available to the child on reaching the age of majority. Some comment by the Court upon the issue of the child's best interests as they relate to the right to identity would have perhaps been useful, given the trends in a number of other member states towards increased openness in adoption, fostering and gamete donation generally.

[43] The Irish authorities made no representations to the European Court, despite the fact that the birth father had begun proceedings to establish parentage in Ireland.

[44] *N & Anor v Health Service Executive & Ors* [2006] IESC 60 (The '*Baby Anne*' case)

[45] See for example *Lehr v Robertson* (1983) 463 us 248, 256.

Applying the principles on preventing origin deprivation to the *Kearns* case (at domestic level) it might be argued that the eventual outcome could have differed, although not necessarily in respect of overturning the adoption itself, given the child's attachments to her social parents and the passage of time since proceedings first began. As *Neulinger v Switzerland*[46] established, timing may be a key factor in determining and defining a child's best interests; a well-settled child ought not to be put through harmful emotional upheavals in terms of breaking his or her psychological attachments. In this 2011 case, the child's Article 8 ECHR rights had effectively served to defeat the Hague Convention, so that the return of the child in question was not allowed. Best interests, it was held, ought to be gauged by reference to the present situation, and not through a rights-balancing analysis which referred largely to past events or parental rights. Such a principle would potentially see children such as Baby Anne perhaps remaining with their social parents (arguably, even where the psychological parents were foster carers, or prospective adopters, rather than actual adoptive parents) but equally might allow for some form of contact between the estranged relatives. It would also almost certainly permit, if not actually require, the release of identifying information in later life on the basis that potentiality of legal relatedness (and actuality of genetic connection) should at least allow for some possibility of eventual reunion.

Taking the best interests of the child principle as the paramount factor, greater judicial consideration of the child's welfare, and of her right to identity, in cases such as *Kearns* might now have to occur, not least in respect of such issues as short-term child protection (e.g. where for example birth mothers plead fears over birth father behaviour or a privacy-based right to relinquish) but also the potentially permanent effects of rendering someone genetically kinless. Enforced repatriation or reunion with birth family would clearly not have been in the best interests of the child in *Kearns* but it is difficult to see how some measure of kin contact or informational release would have been so detrimental to her welfare as to require the complete legal ban on it.

Maternal vetoes on disclosure or contact, and the parental right to completely relinquish one's parenthood, cannot be seen as anything other than firmly adult-centric in nature. The adoptive parents in *Kearns* may well have felt somewhat threatened by the lingering presence of the child's genetic relatives abroad, but their unease at this should not be enough to justify placing a permanent, absolute legal bar on their child's right to achieve or access a knowable ancestral identity. The state's duty to protect the child's longer-term psycho-social welfare ought to rest upon the principle that the child's best interests are paramount and therefore not subordinate or subject to parental or government vetoes, as occurred here, successively.

---

[46] *Neulinger & Shuruk v. Switzerland* (Application no. 41615/07) (2011) 1 FLR 122 available http://www.incadat.com/index.cfm?act=search.detail&cid=1001&lng=1&sl=2 accessed 10.02.12.

Genetic kinship should be preserved, but not via a harsh, enforced 'repatriation' with stranger birth-kin, which can plainly work against the best interests of the vulnerable child. Such an outcome would also accord with at least some of the recent Strasbourg jurisprudence on family life and the best interests of the child. *Clemeno v Italy* (2008)[47] for example looked at the issue of absolute bars on kin contact, the effect of such vetoes on genetic ties and the issue of whether the State had failed to take appropriate steps to enable an active 'reconstitution' of the separated family.[48] It was noted that 'future relations between parent and child should not be determined by the mere effluxion of time'[49] and that there was at least an aspirational obligation on signatory states to create the proper conditions' to enable estranged family members to effectively 'communicate with each other'.[50]

Similarly, the increasing importance of 'biological realities' as discussed in *Kroon v The Netherlands* (1994)[51] perhaps means that Article 8 rights to respect for family life need not necessarily always be limited to 'marriage-based relationships'. Ties 'amounting to family life' may exist between child and unmarried genetic parent by virtue of their biological connection, despite the 'exceptional' nature of the circumstances in many cases. *X v Croatia* (2008)[52] also acknowledged to an extent that significant harm can attach to the loss of genetic relatedness. The issue of maternal consent was also highlighted here, with the Court finding that a failure to inform the mother of the adoption proceedings constituted a violation of her Convention 8 rights.[53] Although the mother in *Kearns* did sign consent forms, the question of whether she was fully informed as to the absolute nature of her consent to relinquishment was clearly very relevant, given the language barrier, her lack of legal representation and the fairly short period of time which had elapsed since the birth of the baby.

---

[47] *Clemeno v Italy* (2008) Judgment 21.10.2008 [Section II] No 19537/03. See also *Jucius and Juciuvière v Lithuania*—(2008) App no 14414/03 wherein in was held that the domestic legislation failed to place sufficient emphasis upon the wishes of the adopted child in such proceedings. The principle of 'non-separation of siblings' should also have been adhered to.

[48] Here, a prohibition on contact arose as a result of the domestic authorities failing to consider sufficiently the child's best interests, including her need to maintain relations with her birth family.

[49] *Clemeno op cit* at n 47 para 32.

[50] *Ibid.*

[51] *Kroon v The Netherlands* (1994) 19 EHRR 263 para 11.

[52] *X v Croatia* [2008] (app no 12233/04) (17.07.2008).

[53] See also the Motion of The Council of Europe Parliamentary Assembly Doc 11742 (9th October 2008) para 4 which noted that the use of mental incapacity to exclude parental responsibility in *X v Croatia* also potentially violated Article 6 ('*fair trial*') and Article 13 ('*remedy*'), that judgments in family proceedings in England are often not made publicly, in conflict with Article 6 (at para 9) and requesting an investigation into 'forcible' adoptions in respect of Portugal, Croatia and the United Kingdom (at para 12). Available http://assembly.coe.int/Main.asp?link=/Documents/WorkingDocs/Doc08/EDOC11742.htm accessed 10.01.11.

The case of *Mikulić v Croatia* (2002)[54] further examined the child's need to access the level of information 'necessary to uncover the truth about an important aspect of ... personal identity'.[55] Here the Court discussed Articles 6 (1) and 8 of the European Convention noting the positive obligations that require signatory States to actively protect private and family life rights.[56] Here, 'uncertainty as to ... personal identity' had to be addressed.[57] *Odièvre v France* (2003) similarly concerned the inability to 'find out ... origins'. Here again, the issue of whether the French anonymous birth registration system had breached the provisions of Article 8 in respect of private life was discussed.[58] The issue of lost 'emotional ties' included those that might arise with one's genetic relatives, including siblings and half-siblings.[59] The applicant argued that French law created an unfair 'legal fiction' which served to hide the fact that she had been born; maternal anonymity, she argued, amounted therefore to 'an act of violence' against her.[60]

Given the margin of appreciation attaching to family law matters, the perceived social need to protect third parties (such as birth fathers, siblings and other genetic relatives[61]) and the lack of European consensus on questions such as the opening of sealed birth records, it seems unlikely that the anonymous birth system would have been found to violate Convention rights, even if the Court in *Kearns* had been focusing solely on the concept of child welfare paramountcy.[62] It is beyond the Court's remit to order that some form of contact between the child and her genetic relatives must occur; they may however have felt more inclined to scrutinize the legal processes underpinning such a decision (as occurred in *Anayo*) to highlight the potential harms of such a system, or to expand upon exactly how the best interests of children might be compromised by the loss of original, genetic identity, especially given the 'privileged position' of children in international law.[63] Arguably, on becoming an adult, the child's Article 8 rights might carry equal weight to those of other triad members. Whether these would be sufficiently robust to permit her to

---

[54] *Mikulic v Croatia* (2002) (no 53176/99) 54 ECHR; See also *Human Rights Case Digest* (2002) (13) 1–2, pp. 55–95.

[55] *Ibid* para 64.

[56] *Ibid* para 57. Lack of domestic remedy (in respect of delays) also amounted to a violation of Article 13.

[57] *Ibid* para 56.

[58] *Odièvre v France* 1 FLR 621 (Application no 42326/98) (2003) para 3.

[59] *Ibid* para 28.

[60] *Ibid* para 31.

[61] *Ibid* para 44.

[62] *Ibid* para 43. The Court distinguished the applicant's situation from those of the claimants in *Gaskin* and *Mikulić*, stating that 'access to information about one's origins and the identity of one's natural parents is not of the same nature as that of access to a case record concerning a child in care or to evidence of alleged paternity.'

[63] *Ibid*. Concurring Opinion of Rozakis J at para 3.

access her identifying information seems unlikely. In *P C & S v United Kingdom* (2002)[64] the mother's lack of representation (during legal proceedings) breached Article 6(1) while the newborn infant's removal into care a few hours after delivery was deemed to be a non-substantive violation of both child's and parents' Article 8 rights, and not dissimilar to cutting a baby off 'from its roots'. Such intervention would only be justified in exceptional circumstances or where the best interests principle demanded such an outcome.[65]

In *Gorgülu v Germany* (2004) the Court looked to the 'long term effects' of separation, asking whether domestic courts ought to consider such matters[66] and noting that a parent should not be allowed to argue Article 8 rights to 'have such measures taken as would harm the child's health and development.'[67] The Court referred to the possibility of reunion and to the need for an outcome which reflected a 'fair balance between the interests of the child and those of the parents.'[68] If courts and decision-makers in such cases were to include the prevention of origin deprivation within their best interests analyses, rights-balancing exercises might tend less towards adult-centric outcomes. As the Recommendations of the Parliamentary Assembly of The Council of Europe (2000) noted, albeit in respect of international adoptions, a right to learn the truth about original ancestry should be freely available.[69] Domestic laws and policies should not obstruct this aim via limited, 'rights-based' approaches, aimed more at respecting adult wishes than protecting the welfare of the origin-deprived child.[70]

Decisions such as the one reached by the French Court in *Kearns* do not fit particularly well into the template for preventing origin deprivation proposed by the six principles. A domestic system of anonymous birth registration:

a) Denies that there is any fundamental right to an authentic, ancestral identity
b) Does not interpret the best interests of the child as comprising of a need for contact; 'potentiality' of relatedness or relationship is similarly irrelevant, although the passage of time in this case was potentially a relevant factor in terms of best interests
c) Does not elevate the welfare of the child to a position of paramountcy

---

[64] *P C and S v United Kingdom* (2002) 35 EHRR 1075.

[65] *Ibid* at para 118. See also *Gnahoré v. France*, (2000) no. 40031/98, § 59, ECHR 2000-IX.

[66] *Gorgulu v Germany* (Application No 74969/01) [2004] 1 FLR 894 para 46.

[67] *Ibid* para 43. See also *Johansen v Norway* [1996] EHRR 31 (app no 17383/90); *Scozzari and Giunta v. Italy* (2000) [GC], nos. 39221/98 and 41963/98, § 169, ECHR 2000-VIII.

[68] *Ibid* para 43.

[69] Parliamentary Assembly of the Council of Europe (2000) para vii—available at http://assembly.coe.int/Main.asp?link=/Documents/AdoptedText/ta00/EREC1443.htm accessed 02.01.09. It was further proposed that all member states should move to ratify and observe the principles of The Hague Convention on Adoption (1993) as a matter of urgency, to monitor the 'psychological well-being' of children in international adoptions.

[70] See further Herring (2009) available http://www.newlawjournal.co.uk/nlj/content/family-revoking-adoptions accessed 10.08.11.

d) Does not tie the best interests of the child to the need for informational disclosure nor does it afford any legal or social significance to genetic connections; no obligation to preserve or repair blood-ties exists
e) Renders the loss of genetic connection a normative rather than exceptional occurrence
f) Preserving genetic connections in this case (via informational disclosure at a later stage) need not have compromised the principle of child welfare paramountcy

It is ironic that the child's right to genetic identity was sidelined not by the birth mother's right to anonymity (which she then sought to forego) but by social concerns i.e. to preserve the anonymous birth system via permanently sealed records. With the child's fundamental right to identity not being afforded any degree of legal recognition here, the remaining principles have little relevance. A right to contact is irrelevant where there is no information legally available on the genetic parent; there is little point in asking whether such contact might represent a rights-bearing element of the child's best interests. The child's welfare was clearly not the paramount concern here, although the court's decision to not overturn the adoption could have been justified on the basis of the child's best interests requiring such an outcome, so as to prevent psychological harm. Depriving her completely of her genetic ancestry was not a best interests-based act, but one aimed at preserving a long-standing social custom. It is difficult to see how the child would have been left open to the risk of harm had she been able to one day receive an accurate, unaltered birth certificate or have the occasional visit or letter from her natal relatives.

Cases such as this highlight how easily adult-centric concerns such as privacy may translate into such weighty, rights-bearing elements which leave little room for the child's best interests to arise or be addressed. At the level of the Strasbourg Court's involvement, the wide margin of appreciation afforded to such adult-centric private, family life matters meant that the French system of vetoing birth records was found to not offend against the provisions or spirit of the European Convention. Although Article 8 was not violated here, the decision seems a little out of step with some of the recent cases on kin contact, child welfare and matters of procedure.[71]

9.5 A Juridical 'Right' to Knowable Parentage

Domestic courts have, on a number of occasions, been asked to decide whether paternity testing would be in the child's best interests. This issue was examined, for example, in the English Court of Appeal case *Re P (A Child)* [2008].[72] Here, the

[71] See in particular *Godelli v Italy* (2012) ECHR 347 (app no 33783/09) where the Court that found a violation of Article 8 ECHR had occurred in respect of completely closed birth records; Italian law did not allow for a fair balance to be struck, where a child not been formally recognized at birth and then adopted.

[72] *Re P (A Child)* [2008] EWCA Civ 499.

putative father sought contact with the child, although his desire for kinship was completely conditional upon whether or not they were actually genetically related. Wilson J decided that genetic testing could harm the child's psychological welfare, by perhaps undermining her current sense of identity and threatening her social kinship attachments.[73] Failure to test for paternity would prevent the Court from deciding on the issue of paternal contact however.[74] Child welfare was discussed, with the issue of parental interests perhaps being evaluated here to a greater extent than the psychological needs or welfare rights of the child. The question of whether a sufficiently close relationship had existed between the parties to enable 'family life' rights to arise was also examined.[75] By ordering the DNA test to proceed, the Court arguably afforded a significant degree of priority to the 'potentially present' blood-tie, perhaps including it amongst other less inchoate elements of the right to family life. If it existed here, it would perhaps over-ride the wishes, and possibly the welfare-based rights, of both the child and her resident parent, in relation to enabling a decision on paternal contact.

This case seems to imply that a need for genetic accuracy could form part of a domestic checklist for promoting and protecting children's long-term psychological welfare needs. Had a more specific welfare checklist possibly been referred to, the eventual outcome may have differed however. The child's age and understanding, and her 'right to be heard' (not to mention her need for stability and permanence) could perhaps have carried a little more weight here. The applicant father's somewhat disingenuous wish to engage in kin contact only if his parentage was first proven might have merited being outweighed by the paramount best interest rights of the child. Had a child-led veto on contact or paternity testing been enabled here, the applicant might have argued however that he was being subjected to an unjustifiable, indefinite wait for testing, which in turn might have been conditional upon the child ever deciding to lift her ban on releasing information. Such a situation mirrors the one faced by many adoptees and gamete donor children when they are attempting to find information on estranged genetic relatives in sealed records jurisdictions.

The issue of vetoes on ancestry was also central to the English case of *Re C*: (*a child*) *v XYZ County Council & Anor* (2007).[76] Here, maternal refusal to identify the child's genetic father, or to inform him of her pregnancy, was examined in the light of recent legislative reforms.[77] The Court stressed that the collecting of identifying information was not in itself a priority. Determination of the child's parentage

---

[73] *Ibid* para 12.

[74] *Ibid* para 14.

[75] *Ibid* para 13. The decision turned upon the fact that the granting of a Contact Order would depend upon whether the child was the genetic descendant of the applicant. Issues of age and understanding were not expanded upon, nor were the child's wishes discussed in connection with the tests, or ongoing contact. See also *Re D* [2006] EWHC 3545 (Fam) which was distinguished here from *Re P* [2008] *ibid.*

[76] *Re C (A Child) v XYZ CC* [2007] EWCA Civ 1206.

[77] The Adoption and Children Act 2002.

seemed to relate mainly to the need to find suitable long-term substitute carers for the child.[78] Where genetic parents have not married, the nature of their relationship may be a key concern for the court, especially where decision-makers are tasked with having to gauge the level of commitment that has been or might yet be shown by absent fathers. This issue carries much weight both before and after birth, to demonstrate that a relationship, or perhaps some form of rights-bearing legal relatedness might come to exist.[79] Paternal 'consent' to adoption proceedings may be sought, but this is not deemed to be a necessary step where meaningful relationships are not found to have existed in the first place.[80] The longer term consequences of such an 'information-free' model of adoption practice ought to be included in any child welfare paramountcy rights-analysis however.[81]

The Court exercised their statutory power to dispense with birth parental consent here, despite the genetic relatives' claims that their Article 8 rights were at risk of being infringed.[82] 'Natural family preference'[83] was not held to be part of the current judicial policy on child adoptive placement however.[84] The Court's primary role in such contexts was to enable maternal vetoes over information release, given that a number of 'valid social policy considerations' arose.[85] Setting aside one half of the child's genetic ancestry has serious rights-implications however insofar as it surely diminishes whatever significance the blood-tie might have potentially had.[86]In both of these cases, the interests and wishes of a birth parent (i.e. to either have the child's DNA tested against her wishes or to completely veto the discovery of paternal parentage) seem to have prevailed over the longer-term psychological

[78] See also *Re H*: *Re G* [2001] 1 FLR 646 and *Z County Council v R* [2001] 1FLR 365 where the Court found that because of the mother's attitude, it would not have been in the child's best interests to inform the father of the adoption.

[79] *Re C (A Child) op cit* at n 76 para 51. Arden LJ noted the birth mother's young age, her unequivocal desire for relinquishment and the fact that she had 'never cared for the child' (at para 9).

[80] *Ibid* para 51 per Collins LJ, distinguishing between genetic kinfolk and those 'relatives with whom the child has a relationship'.

[81] See s.52 Adoption and Children Act (2002).

[82] See also The Rules of Procedure (2005) whereby a birth father who has subsequently gained Parental Responsibility may be deemed to have consented to relinquishment, if the child has since been placed with adopters. Notice need not necessarily be given to objecting birth parents (Regulation 13) or to absent ones (Regulation 27).

[83] *Re C op cit* n 76 at para 43.

[84] See Thorpe LJ's comments on France's anonymous birth registration system and his argument that abortion and infanticide rates might be lower as a result of having such a process. *Ibid* (para 59).

[85] *Ibid* para 61. See however the more child-focussed approach of *Frette v France* [2004] 38 EHRR 21 (42) wherein it was established that the process of adoption now serves to give a family to a child rather than the other way around. See also *Marckx v Belgium* [1979] ECHR 2, wherein it was held that natal grandparents may be able to establish Article 8 rights.

[86] This is perhaps at odds with the decision in *Re P*[2008] EWCA Civ 535 heard a year later, which seemed to emphasize the need for 'knowable ancestry'.

welfare needs or best interests of the child. The Court's inherent and statutory discretion (for example to 'make no order') was not exercised here in favour of promoting as the paramount concern the welfare of the child. The need to prevent placement delays in the latter case is an entirely valid one but does not fully address the issue of the child's loss of one half of her genetic heritage. Had a rule of 'blood-tie paramountcy' been adhered to here however, the outcome for the child might have been equally unsatisfactory. Had the genetic link been deemed inviolate, on the basis of the rights of the adults involved, social workers might have been obliged to place the child with birth family members, rather than with possibly more suitable prospective adopters; the question of the child's best interests might have been completely irrelevant or sidelined, never mind being the paramount consideration. Similarly, despite the mother's straightforward, quick relinquishment, other natal relatives might have, on being informed of the child's birth, sought to legally challenge or overturn a settled placement, leading to protracted legal proceedings and a possible removal of the child long after she had formed emotional ties of attachment to her adoptive parents.

The issue of enforcing such an outcome is also a key one. Where relinquishing birth mothers have simply refused to identify the child's genetic father, there is little that can realistically be done to force them to do otherwise. Proceedings for contempt of court for example would not be likely to win widespread public support, nor would some form of injunction or financial sanction for non-compliance with a request for disclosure be appropriate here.[87] Arguably, some form of compromise might be possible in such cases, whereby mothers could perhaps be encouraged to name the birth father unless some 'good cause' exception is found, with identifying information not released until the child has attained the age of majority. Such an approach could potentially also clarify other issues, such as whether newborn infants have any right to family life, or to a greater recognition of their genetic kinship bonds.

Ensuring that the rights of the child are afforded a more meaningful level of domestic protection is especially important in situations where the balance of power has traditionally been adult-centric. In other words, if child welfare paramountcy was confirmed as the core value in such cases, decisions to completely deprive children of their genetic identity would perhaps be unlikely to occur other than in exceptional circumstances.[88] Additionally, cases on disputed parentage may

---

[87] See however The Child Support Act 1991 s.46 under which benefits paid to a child's mother could be frozen if she refused to name the father of the child.

[88] See also *Re L* [2007] EWHC 1771 (Fam) (20 July 2007) where the birth mother refused to identify the child's biological father. The court declined to cross-examine her, or make a finding of contempt of court, stating that such an action would not be 'justified merely because we believe, however strongly, that what we are doing is being done in the best interests of a child.' Per Munby J at para 38. See however para 30 on the likely consequences of the birth mother's actions. See also *Z County Council v R* [2001] 1 FLR 365, on the 'historic' purpose of adoption, in enabling mothers to conceal both pregnancy and birth; per Holman LJ at page 367; see also *Re R* [1993] 2 FLR 762 per Butler-Sloss on the duty of the mother to inform her child.

perhaps serve as useful precedent for cases involving assisted reproduction. Questions over the issue of 'whether to tell' remain significant here, especially where a 'medical' device of characteristic-matching might have been employed. In *Re R* (*a Child*) [2003][89] for example the Court of Appeal noted that the statutory provisions on gamete donation were unusual in so far as they conferred

> the relationship of parent and child on people who are related neither by blood nor by marriage. Conferring such relationships is a serious matter involving as it does not only the relationship between father and child but also between the whole of the father's family and the child.[90]

Here, the mother's former partner, who was not genetically related to the child, had claimed legal paternity (via a Parentage Declaration in the High Court[91]) on the basis of his having previously consented to her receiving infertility treatment via donor insemination. The child's mother had not informed the clinic that she was no longer involved with the respondent. The court acknowledged the difficulties connected with establishing legal relatedness, noting that the law could render children legally fatherless where, for example, a lesbian couple or single woman had availed of anonymous sperm donation or a child had been conceived using the sperm of a deceased donor.[92]The consequences of granting legal paternity to the respondent were limited; even the presence of 'sensible' contact arrangements would not necessarily have the effect of making the child a legally recognised member of his family.[93] The court explicitly stated that no Article 8 family life rights arose between the putative father and child in this case, observing that, 'family life does not inevitably flow from genetic fatherhood although it often does.'[94] Because there was neither a genetic tie nor a previous relationship history of having cared for the child, no family life rights arose. It was further suggested that had the mother been seeking child support from the respondent, 'the court would have been slow to adopt a construction which would allow her to do so.'[95]In respect of determining parentage it was noted that such a finding could also be made through substantive proceedings to perhaps determine issues such as maintenance or upbringing.[96] Difficult cases involving law or public policy, such as this one would perhaps be better dealt with by a formal parentage declaration however.

Noting that the High Court did not have jurisdiction to make the Contact Order, Hale LJ suggested that this issue might be reconsidered by the lower court, given

---

[89] *Re R (A Child)* [2003] EWCA Civ 182.

[90] *Ibid* at para 20.

[91] Under S 55A of the Family Law Act 1986.

[92] *Re R op cit* n 89 para 27.

[93] *Ibid* para 27.

[94] *Ibid* para 28.

[95] *Ibid* para 29.

[96] *Ibid* para 33 (such as The Children Act 1989 s.4).

the Official Solicitor's opinion that 'the risk to the stability of the family unit would outweigh the benefits to the child.'[97] That said, given that indirect contact had already been working for 18 months, 'the respondent might have an important role in helping the child to understand and come to terms with her origins.'[98] Under the domestic legislation that was then in place,[99] the issue of child welfare had to be afforded due consideration; the Court avoided a detailed discussion of the child's needs, but did however suggest that the parties ought to be able to reach a solution themselves on the question of further contact, by reflecting 'on where this little girl's best interests lie.'[100]

The respondent was no longer a legal parent of the child, nor did he have any right to continue indirect contact with her. He was however essentially set the task of finding a child welfare-led solution to the problems facing the family and, to an extent the genetically fatherless child. Arguably, the court seemed to be saying that decisions involving 'contact, if any' were at times best left entirely to parents or those who best fill the role of parent; they reserved the right to intervene in the event of subsequent disputes, but remitted the matter to the lower court for resolution should the child's mother and the respondent fail to reach a workable agreement. A blood-tie paramount model would have been unlikely to provide a much altered solution however, given that the child's genetic father was an anonymous donor. A revised checklist would similarly present very few other options here in the face of donor anonymity (which has since been abolished in the United Kingdom). The case is significant however in respect of the court's clear acknowledgement of the long-term problems that the child could potentially face as she grows older. The decision does also clearly stress that the adults concerned should place the child's welfare above their own difficulties and interests. If such reasoning were applied to other cases concerning parentage, where, for example, a mother has refused to identify the genetic father, then perhaps origin deprivation might eventually come to be more generally regarded as a socially unacceptable outcome for children. In any event, such cases serve to highlight not only the problems associated with applying legal fictions to biological realities, but the difficulties involved in legislating for and around such elaborate devices aimed at erasing genetic ancestry.

In relation to cases such as *Pratten* (where a policy of donor anonymity was not found to give rise to questions of unjustifiable discrimination or inequality of treatment for gamete donor children) application of the six principles could, in theory at least, allow for a different outcome.[101] Again, the key hurdle was to perhaps initially convince the court that donor gamete children have a juridical right

---

[97] *Ibid* para 34.

[98] *Re R op cit* n 89 para 34.

[99] See The Human Fertilisation and Embryology Act 1990. See also however The Human Fertilisation and Embryology Act 2008 s.24 and s 54 and Schedule 6; see further http://www.hfea.gov.uk/112.html accessed 31.05.010.

[100] *Re R op cit n 89* para 35.

[101] *Pratten v British Columbia (Attorney General)* [2012] BCCA 480.

to genetic identity; the issue of whether or not some form of best interests-led kin contact was an essential element of this fundamental right to identity would however have depended upon whether the court accepted that the welfare of the child was the paramount concern. As the plaintiffs in this class action were adults, it could be argued that there were no child rights as such to be argued over, balanced our or even considered. There are perhaps two responses to this: sealed records, parental vetoes and gamete donor anonymity generally serve to permanently infantilise those who are by affected them, for example by making their right to identity completely subordinate and subservient to the permanent privacy rights of their sperm donor fathers. If this type of parental right does not end when the child reaches adulthood, then a sort of enforced childhood arises, in the sense that such veto-victims may never be afforded the same level of rights-protection as other adults who can easily access an authentic birth certificate or plead the absolute (in this context) right to privacy.

Alternatively, the court might have looked to the issue in terms of whether a duty arose to protect the future best interests of those who were still in childhood, and whether their welfare would require access to identifying information at a future point in time. Such a question would perhaps have been quite relevant to the case's core issue of whether or not donor children were being discriminated against unjustifiably in terms of not being able to access identifying information on genetic ancestry. Again, the genetic connection itself, let alone the right to have it maintained or restored, has here been afforded no degree of meaningful recognition, meriting no legal protections whatsoever. In relation to the court's rather convoluted reasoning on the issue of discrimination,[102] it seems difficult to see how this rights-analysis afforded any degree of priority to the question of the plaintiffs' welfare or best interests, or to that of other donor children who might have similarly wished to access their genetic heritage, had some measure of law reform been brought about in the wake of the Court's decision. That said, potentially this outcome might have been justified here on the grounds that an exceptional circumstance did apply, for example that the original birth records no longer existed in respect of the gamete donation, but with some judicial guidance being set down in terms of the inequality of treatment afforded to donor children on the basis of their birth status, in terms of their triadic position, rather than by a comparison to adoptees.

Had the Court hearing the *Pratten* appeal been willing to accept that donor children and adoptees face analogous, if not identical, difficulties in terms of origin deprivation and rights infringements, it might perhaps have been willing to find that donor children, had, like generations of sealed-records or veto-bound adoptees, been:

---

[102] *Ibid* para 38 They found that adoptees who could access support services and birth records, subject to the absence of a parental veto, were in need of such special treatment on the basis of their vulnerabilities and their long history of having faced socio-cultural stigma, disadvantages and prejudice.

a) denied a right to an accurate, authentic genetic identity
b) deprived of any chance to engage in future kin contact or realise a potentiality for family life or genetic relatedness
c) treated very differently to their biological parent(s), entirely on the basis of their birth status (i.e. lower position in the triad rights hierarchy) insofar as their right to identity is permanently outweighed by their genitor's absolute right to privacy and to relinquish their genetic parentage
d) placed within a normative rather than exceptional framework for loss of ancestry
e) afforded no recognition whatsoever in respect of their lost genetic connections and unknown ancestral heritage
f) seen their best interests completely outweighed, sidelined or ignored by state-sanctioned vetoes on genetic connection

Had the Court found that the welfare of donor children was the paramount consideration in cases involving access to genetic identity, this case might have been very significant in terms of impact upon other areas of adoption law. It might for example have proven quite difficult for the provincial government to justify their continued use of the disclosure and contact vetoes that prevent some adoptees from accessing their birth records, especially given that the case seemed to turn on the issue of legislative constitutionality and Charter-protected rights such as equality. Arguably the case also touched upon the notion of human dignity, given the Court's preferred focus on the more civil, political Canadian Charter section 7 rights to 'life, liberty and security'[103] rather than framing the right to genetic identity as more socio-economic in nature, as the Strasbourg Court has largely interpreted it (via Article 8 ECHR). It may be further argued that, in doing so, the Court here has at least highlighted that the concept of genetic identity may more appropriately amount to a fundamental, 'first generation' right, rather than existing amongst the highly qualified, potentially non-justiciable (and often very resource dependant) socio-economic rights associated with private and family life. As more regions gradually open sealed birth records and work towards making gamete donor anonymity illegal, it will perhaps become much more difficult for courts to claim that there is no social consensus on the issue of the need for genetic ancestry, and that origin–deprived persons have not been subject to unequal treatment.

9.6 Adoption as Child Protection

The *Webster* case (where acute concerns over child protection resulted in removal of the children from their birth parents) differs from the other disputed relinquishment cases in this chapter because here the original parents had never provided consent nor accepted that their children's health problems were due to deliberate

[103] *Ibid* para 54.

neglect on their part. Domestic legislation within the United Kingdom regards adopted children as 'having been born' into their replacement families.[104] Only in highly exceptional circumstances are original parents likely to be able to overturn an adoption order.[105] The expert evidence here was fairly equivocal, and it was argued here that cases such as this ought to be open to at least some possibility of legal challenge.[106] As with the decision in *Kearns* however, the loss of their parenthood did not amount to an unlawful interference with the birth parents' Article 8 ECHR rights.[107] The fact that 4 years had elapsed since the proceedings had first begun, with no kin contact having been permitted during that time, meant that the passage of time had effectively rendered the Webster family's genetic ties almost redundant. The court seemed to evade the issue of post-adoption contact, even insofar as its absence might have adversely impacted upon the family life rights of the separated Webster siblings. No balancing of the conflicting familial rights occurred, with the question apparently being pretty much settled on the basis that all Article 8 ECHR rights (parental and possibly those of the children) had simply passed from the natal parents to the adoptive families when the Adoption Orders were granted by the court.[108]

This case was controversial given the presence of some procedural irregularities and possibly mistaken professional opinions on whether the removed children had been deliberately abused by their original parents.[109] These were not found to be sufficiently serious however to prevent or annul the making of the Adoption Orders.[110] Domestic legislation to 'free' the children for adoption looked only to the actions of the parents, with the key question being whether they had 'behaved unreasonably' in withholding their consent to the infants' placements.[111] As with *Kearns*, a fairly pragmatic approach was taken, focussing on the ultimate outcome

---

[104] See section 67(1) of the Adoption and Children Act 2002.

[105] *Webster (The Parents) v Norfolk County Council & Ors* (Rev 1) [2009] EWCA Civ 59 (11 February 2009).

[106] *Ibid* para 65. See however Wall LJ at para 141 on the interests of birth parents. see also *Pini v Bertani v Romania* (2005) 2 FLR 596; See also *Re K (Adoption and Wardship)* [1997] 2 FLR 221 where the process was described as 'fatally flawed' (per Butler-Sloss LJ at p 228H).

[107] *Ibid* para 175.

[108] *Ibid.*

[109] In previous, highly exceptional UK cases, this had served to annul an Adoption Order. See for example *Re F (R) (An Infant)* 1970] 1 QB 385. See also however *Re RA (Minors)* (1974) 4 Fam Law 182; *Re F (Infants) (Adoption Order: Validity)* [1977] Fam 165 where Adoption Orders were held to be valid (if still theoretically voidable) as a result of the adopters not being legally married at the time of the hearing.

[110] *Webster op cit* n 105, para 162 per Wall LJ. The Court distinguished the case of *Re K (Non-accidental injuries: Perpetrator: New Evidence)* [2004] EWCA Civ 1181, [2005] 1 FLR 285 on the basis of timing; i.e. the adoption hearing had not yet been held in *Re K.*

[111] Placement Orders had not yet been introduced by the Adoption and Children Act (2002); Freeing Orders fell within the remit of The Adoption Act 1976, mirroring Northern Ireland's current legislative position under the Adoption (NI) Order (19897).

rather than revisiting the very start of the adoption process to re-examine the facts surrounding the relinquishment or the grounds for intervention.[112]

In fairness, the best interests of the child principle seemed to be at least an implicit consideration, with the relevant European Convention rights discussed to some extent. The English Court of Appeal seemed more willing than the French court to extend the notion of family life rights to all members of this adoption triad, rather than simply defining them as adult interests; they linked the children's Article 8 ECHR rights very clearly to their need for new family permanence but noted that the reassurance of the prospective adoptive parents was also an entirely necessary, wider policy aim.[113] Had such a case perhaps occurred in a jurisdiction where the duty to preserve genetic ties was enshrined in statute, either on the basis of parental rights-paramountcy or Constitutional protection of the family unit, the Webster children might well have been returned to their birth parents, or at least permitted to engage in some meaningful level of contact with them, perhaps on the basis of fresh, exonerating evidence subsequently arising over the issue of the alleged child abuse.

Neither of the two 'kinship rights templates' (to sever or preserve genetic ties) would however have produced an entirely satisfactory result in terms of making child welfare the paramount consideration. Within the clean-break, blood-tie severance model, adult vetoes on kin contact are capable of completely negating genetic connections; until the Freeing Order was granted, the children were not deemed to be genetically kinless in law. From that point onwards however, they lost their natal links, including the ties between siblings, some of whom had been placed in separate homes. Had biological relatedness been framed instead as an essential aspect of their heritage, the children might have perhaps been 'repatriated' with their original parents, but would then have possibly suffered the loss of the emotional attachments that they had formed with their substitute parents, and maybe been exposed to further instances of (albeit unintentional) neglect.

Under a revised child welfare checklist, aimed specifically at preventing origin deprivation, a number of different outcomes might have been possible here. The Court could perhaps have examined in more detail the rights and vulnerabilities of the children, in isolation from the needs and wishes of the two sets of parents, or maybe felt inclined to expand upon the current remit of the best interests principle, to include the long term psychological welfare considerations. Genetic ties might have been found to warrant an increased level of preservation or repair, unless exceptional circumstances existed to sanction their complete curtailment. The best interests of the affected children would most likely have warranted more substantial discussion, especially in respect of how an absolute bar on kin contact might fit within the child welfare paramountcy principle. The lack of a domestic mechanism

[112] This was so, despite the French courts' initial focus on the question of whether or not informed consent had been provided by the birth mother and whether or not it was needed, given the child's condition of legal 'kinlessness'.

[113] Webster *op cit* n 105 para 89.

for challenging the loss of Article 8 family life rights could, on balance, maybe be justified on the basis that the children might have still been vulnerable to parental abuse or neglect if the family was reunified. It is much more difficult to justify a complete bar on contact however, given that indirect modes would seem unlikely to carry grave risks; the loss of sibling kinship is especially difficult to defend, when a child-centric, welfare paramountcy approach to decision-making is employed.

If the concept of 'potentiality' of family life (as evidenced by an intention to commit to a relationship with one's relatives) is added to such a consideration of the children's best interests, then the outcome in *Webster* seems unlikely to accord with the reasoning behind *Anayo*. In fairness, the *Anayo* case concerned a private family law dispute, at least initially, whilst *Webster* arose from state intervention and child protection concerns. From the perspective of the origin deprived children however, and in terms of promoting as paramount their best interests, the central issues are essentially the same: the harms of genetic kinlessness and the losses of original kinship. Some level of familial contact or information release could serve to protect the children at the centre of such litigation from origin deprivation.

Had a more specific checklist been available in *Webster*, some level of post-adoption contact might have been enabled (e.g. by remitting this matter back to the lower court for adjudication or by encouraging the adoptive parents to allow it via an informal agreement, subsequently endorsed by the court). As is now increasingly the case, decision-makers might have to consider more fully how bars on contact, especially between separated siblings, potentially impact adversely upon adoptees. Courts and social workers in the United Kingdom do seem to accept that access to genetic relatives often helps vulnerable children gain a sense of identity and selfhood, and perhaps minimise feelings of loss or rejection which a 'child who remembers her birth family may feel if she is completely cut off' from them.[114] Psychological evidence in such cases could however be afforded a greater degree of deference, especially where experts have suggested that psychological scars might result from estrangement. Expert opinion on adoptees' unique needs should also not be overlooked in favour of other non-specific reports, as was noted in *Down Lisburn Health and Social Services Trust v H* (2006).[115]

As was also declared in *South Eastern Health and Social Services Trust v LS* [2008][116] focusing too much on past issues and insufficiently considering present ones, is not necessarily the best course of action for decision-makers to follow.[117] In *Re EFB* [2009][118] for example it was argued that sibling contact could help 'consolidate a positive genealogical identity' and that the decisions made at this stage could affect the child for life.[119] The court still warned however against

[114] *Down Lisburn Health and Social Services Trust v H* [2006] UKHL 36 per Baroness Hale (Dissenting Opinion) para 7.

[115] *Ibid* para 13.

[116] *South Eastern Health and Social Services Trust v LS* [2008] NI Fam 12 (27 October 2008).

[117] *Ibid* para 23 per Weir J.

[118] *Re EFB* (Freeing Without Consent) [2009]NIFam 7.

[119] *Ibid* para 13.

burdening adopters 'with too many arrangements' less they would erode 'the control of the adoptive parents.'[120] In *J* (*Care Order*) (2008),[121] the procedures of the Foster Care Panel were discussed, with the court asking whether these had complied fully with Article 8 of the European Convention, in respect of 'ensuring inclusive involvement in the decision-making process' given that the grandparents had not been involved.[122] Also noted here was the onus on adoptive parents to recognize the child's need for 'ongoing contact'.[123]

That said, some recent Northern Ireland case law seems to have avoided simply enshrining child welfare paramountcy as 'the only consideration'.[124] This region of the United Kingdom still awaits legislative reform; thus a Freeing Order was refused in *Re JJ* [2009] on the apparent basis that the Court ought to have 'take (n) into account and weigh proportionately'[125] parental rights to family life, against the prospects of achieving 'secure attachment'[126] for the child, especially where long-term fostering might have represented an appropriate alternative outcome.[127] Although in *SEHSST v LS* [2009][128] the 'likelihood test'[129] was applied, the court noted that a 'compelling reason' ought to exist for a Freeing Order to be granted.[130] In England, the case of *SB v A County Council*; *Re P* [2008][131] saw the statutory welfare test itself scrutinised, with the court asking whether this was a 'simple welfare test' or an 'enhanced' one. If parents' rights were protected by Article 8 ECHR, then domestic legislation perhaps did have to require something more than a basic definition of welfare in decision-making.[132] This may be especially so where the issue of genetic parentage rather than good parenting is in dispute, given that significant challenges tended to arise in such cases, in terms of determining how best to meet child welfare needs. Similarly, applying the six principles to the decision in *Webster*, it could be argued that the children in this case:

a) were not absolutely denied the fundamental right to authentic, ancestral identity, given that they could presumably receive their original birth certificates at age eighteen, but

[120] *Ibid* para 42 per Stephens J.

[121] *Re J* (*Care Order*) [2008] NIFam 11 (1 August 2008).

[122] *Ibid* para 46.

[123] *Ibid.*

[124] *Re JJ* [2009] NIFam 2.

[125] *Ibid* para 15 per Morgan J (now LCJ).

[126] *Ibid.*

[127] *Ibid* para 24.

[128] *SEHSST* v LS [2009] NI Fam 14.

[129] Per Weir J (at para 21, citing Lord Nicholls of Birkenhead *In Re G* (*Children*) (*FC*) [2006] UKHL 43).

[130] *Ibid* at para 2 '*In the ordinary way the rearing of a child by his or her biological parent can be expected to be in the child's best interests.*'

[131] *SB v A County Council*; *Re P* [2008] EWCA Civ 535.

[132] *Ibid* para 154 per Wall LJ.

b) had seen their rights interfered with, in terms of kin contact being relevant to the best interests of the child, and to the promotion of a right to identity and Article 8 ECHR family and private life rights,
c) had not seen their welfare afforded positional paramountcy; given however the length of time that they had spent in the care of their adoptive parents, it is unlikely that the adoptions would have been overturned
d) In terms of preserving or repairing the genetic connection, should have had some form of kin contact enabled, not least between the estranged siblings
e) Ought not to have seen this outcome occur as a normative feature of being placed with a substitute family; the medical evidence, although later seen to be ambiguous, did however demand some form of protective intervention
f) Had the adoption orders been overturned, it would very likely have offended against the children's best interests, given the time elapsed. Some form of contact would not necessarily have been contrary to their interests however

In fairness, the legislative reforms in England, Wales and Scotland (and, it is hoped, those that are pending in Northern Ireland) have gone some way to towards preventing such a harsh model of clean-break origin deprivation. The courts have confirmed that the child's welfare is the paramount consideration and have demonstrated reluctance to order placement with a view to adoption unless there is some scope for post-adoption contact agreed beforehand. The focus now seems to be more on procedural issues, such as for example the need to ensure that birth relatives have been informed of impending adoption proceedings and of their potentially profound nature in terms of losing their parental responsibility or contact with the placed child.

9.7 Blood-Tie Paramountcy

Cases involving annulled adoptions and those which effectively 'transplant' the right of veto from adopter to genetic parent, often turn on the question of whether consent to the relinquishment of parenthood was properly obtained.[133] Preservation of the blood-tie seems paramount,[134] with the Courts generally bound by domestic statute[135] or Constitutional provisions to maintain it at all costs, even where principles of child welfare might be seriously compromised. Where an adoptee has formed attachments to his social parents, or faces the possibility of abuse if reunited with genetic parents, the need to preserve genetic relatedness can be said to

[133] See for example *In re Adoption of Crews* (1992) 118 Wash.2d 561 [825 P.2d 305]; *Matter of Adoption of Baby Boy L.* (1982) 231 Kan. 199 [643 P.2d 168].

[134] See *Michael H. v. Gerald D.* (1989) 491 U.S. 110, 123 [105 L.Ed.2d 91, 109 S.Ct. 2333; *Stanley v. Illinois* (1972) 405 U.S. at p. 651.

[135] See The Indian Child Welfare Act 1978 25 U.S.C.A. 1901; See also Section(s) 1913, subd. (a).

have sidelined the child's welfare-led best interests as the key consideration in such cases.

The case of *N v Health Service Executive* (2006)[136] (the 'Baby Anne' case) illustrates this, with the Irish Supreme Court conceding that it had no other option but to return the placed child to her birth parents. This decision was grounded upon the birth mother's belatedly revoked consent (and hastily arranged wedding) which conferred the Constitutional protections of a legally recognised family life upon her and the child's father.[137] The court was bound to accept the presumption that the child's welfare needs, if not her family life rights, would be best met within the 'traditional', biology-based family unit. It rejected the respondents' arguments (which the High Court had initially endorsed) that such a presumption was rebuttable in exceptional circumstances, for example where parents had failed to care adequately for their children.[138] Following an earlier precedent, *In Re JH* (1985)[139] the Court had stressed that a child might be subsequently 'legitimated' where birth parents had later married. As well as gaining the Constitutional protections due to 'family members' a child in such a situation would also be afforded a wider degree of protection than simple 'balance of welfare'[140] tests might have provided, as courts could then take into account a wider range of factors than the child's basic welfare needs.

Had *N v Health Service Executive* occurred in a jurisdiction where the principles of customary indigenous adoption could have been applied, the prospective adopters would have been quite likely to have been deemed *de facto* parents, given the length of time that had passed and the fact that relinquishment had to a significant extent occurred. The expert witness' evidence outlined the harmful, likely effects of a sudden change of primary care-giver in such circumstances.[141] Such risks to the child's welfare were not capable however of outweighing the original parents' Constitutional right to have their child returned to their

---

[136] *N & Anor v Health Service Executive & Ors* [2006] IESC 60.

[137] Article 40.4.1 and 2 provide that '*No person shall be deprived of his personal liberty save in accordance with the law*'. Articles 41 and 42 provides that '*The State recognizes the Family as the natural, primary and fundamental unit group of Society...possessing inalienable and imprescriptible rights*' Had the couple remained unmarried, the Constitutional protection of their family life would have not been applicable; Article 8 ECHR 'family rights' rights might also, arguably, have been beyond their reach, given their lack of 'relationship' (i.e. contact) with their infant daughter. See also *Northern Area Health Board v An Bord Uchtála* [2002] 4 IR 252. The unmarried mother in this case could claim a right to be reunited with her child under Article 40, but had no entitlement to the '*protections and presumptions which benefit the married family under Articles 41 and 42*' per Murray CJ.

[138] *N v Health Service Executive* [2006] *ibid* page 15. This model for rebutting the presumption mirrors this thesis' suggested 'inverted veto' template, which presumes that blood-ties ought to be preserved but not at all costs, so that exceptional circumstances can activate a child-welfare centred veto on contact, residence or access to information.

[139] *In Re JH* [1985] IR 375.

[140] s.3 Guardianship of Infants Act 1964.

[141] *In Re JH ibid* n 127 p. 23.

'custody'.[142] Such rights of veto are in many ways as adult-centric and as potentially absolute in nature, as the anonymity rights of fully relinquishing genetic parents in other regions. Where adoptions (or adoptive placements) have been so overturned, children may be reunited with their genetic relatives but at the expense of having strong social kinship ties severed, and of potentially suffering psychological trauma at the loss of such attachments.

Had this case occurred in another jurisdiction where, for example, the approach taken in *Webster* was adhered to, the eventual outcome might have again differed significantly. The Irish Supreme Court did not for example cite the 'vast social importance of not undermining the irrevocability of adoption orders'[143] as the judges in *Webster* did. Although the Adoption Order had not yet been made in this case, the child had, (unlike the children in *Webster*) known no other parents since her birth. Psychological attachments with the substitute parents had clearly been formed and, it was argued, harm would very likely result if these were to be severed. Equally, the child in such a situation has genetic connections that cannot be simply ignored, even where the law has served to create elaborate fictions to disguise them or minimise their importance. Post-adoption or post-reunion contact between child and 'lost' parent is often not a feasible option in either situation, especially where the successful parent might be vehemently opposed to any form of communal parenting or shared parentage. It can also be assumed that courts and social services agencies would also be unlikely or unable to actively enforce any direction for contact in such circumstances.

Had a revised statutory welfare checklist been available here, a wider range of outcomes might have perhaps been possible. Arguably, such a device could have enabled the Irish Supreme Court to at least suggest that the child's best interests in such situations can carry equal or greater weight to the 'family life' rights of original parents, given that child welfare paramountcy ought to be afforded the highest priority in situations where vulnerable children are exposed to significant harms. One potential outcome could have seen the child remain with her social parents, on the basis of preventing psychological harm, if necessary in a long-term fostering placement, so as to not offend against the Constitutional rights of family units. Contact with her birthparents could have been ordered by the Court. Such an outcome would require however that the conflicting sets of parents be prepared to focus on the act of parenting rather than demand as of right the enhanced, exclusive status of parenthood. It would also require ongoing involvement by the authorities to ensure that the welfare of the child was not being prejudiced by such an arrangement.

Such a decision would be in keeping with the Strasbourg jurisprudence, given that a juridical form of Article 8 family life possibly existed between the child and

---

[142] *N v Health Service Executive* [2006] *op cit* n 136 p. 22. Nor was it found that the birth parents had failed in their duty of care towards their child, despite having relinquished her shortly after birth.

[143] *Webster op cit* n 105 para 204.

her prospective adopters, based on the care they had given her and the attachments that had been built up between them. Intentionality of relatedness might also have been included in any such rights-analysis. Domestic legislative drafters seeking to reform the law might also plead that the prevention of psychological harm in such cases falls clearly within the persuasive remit of the Children's Convention. Again, under a revised checklist, a right to veto blood-tied kinship could perhaps rest with the child rather than with the warring triad adults. Given the young age of the child in this case, the State or the judiciary might be obliged to act instead as *parens patriae*, to exercise its powers on the basis of child welfare paramountcy. The court would perhaps be equally unlikely to seek any outcome that left the child origin-deprived to the extent that she could have no direct contact whatsoever with her birth kin.

In respect of the six principles, it may be argued that the decision in Baby Anne:

a) Did acknowledge that there is a fundamental right to genetic connection and to family life; this was not however framed in terms of child welfare or best interests
b) Defined the best interests of the child via reference to adult-centric rights issues; contact (either with the birth parents, or with the social parents post-reunion) was not relevant to the question of child welfare, nor was the passage of time
c) The welfare of the child was not the paramount consideration
d) The best interests of the child seem to have been only narrowly addressed i.e. in terms of preserving the genetic link
e) Loss of genetic connection was an exceptional occurrence (provided the parents of the child were married to one another)
f) Preservation of the genetic connection in this case clearly gave rise to serious concerns over child welfare being paramount

It is difficult to see how the best interests of the child, in terms of losing all contact with her social parents, were considered in the *Baby Anne* case to any real extent, let alone afforded any degree of parity with the rights of her genetic parents. It could be argued that a decision allowing some form of shared parenthood or contact might have better promoted the child's best interests here. The subsequently-married birth parents' Constitutional rights essentially outweighed any rights (e.g. family life) of the child's social parents and also prevented even a basic consideration of the child's welfare or best interests. This case raises an interesting point: had the disappointed, prospective parents sought to challenge the Irish Supreme Court's decision to return the child to her genetic parents, they might have been able to rely upon the argument that their Article 8 rights had been unjustifiably interfered with. They had clearly established a potentially juridical level of family life with her, as she had with them, given their strong social attachments built up over several years of actual parenting i.e. daily care and residence.

Such a bond is more obviously akin to Strasbourg-defined family life than the merely 'potential' family life model that has been highlighted as capable of violation and worthy of domestic protections. In terms of protecting the child's best interests, and from the perspective of the child, the breaking of her established attachments does not sit easily alongside the European Court's subsequent support for preserving potentiality of relatedness. Again, cases involving blood-ties seem to easily give rise to bifurcated judicial approaches, not least where kin contact, and the need to enable or veto it, are at issue. Where legal relatedness is absent, potentiality of relationship and a clear commitment to establishing familial ties, might be worthy of protection. Had the case progressed to Strasbourg, the European Court would probably have been unlikely to declare provisions of the Irish Constitution incompatible with the European Convention. It might perhaps have found however that the lack of domestic remedy for the disappointed prospective adoptive parents to an extent mirrored those aspects of the German system that were criticised in *Anayo*.

9.8 Conclusion

The predominant framework for 'genetic identity rights' seems to be founded on the presumption that parental vetoes do not necessarily offend against the principle of child welfare paramountcy. Child welfare may therefore essentially be subsumed into the 'heavier' interests and rights of triad adults, rather than viewed as a stand-alone or paramount concept. The alternative model, which frames preservation of the blood-tie as the paramount concern, might also sideline child welfare in favour of promoting adult interests. Neither template produces an entirely satisfactory outcome in terms of enshrining the best interests of the child principle in domestic law and policy.

In a number of jurisdictions, genetic connections may be weakened, obscured or severed by legal fictions aimed at enabling the various modes of social kinship. Origin deprivation can occur where decision-makers have then vetoed contact or the release of identifying information. The 'right' to genetic identity in such contexts exists (if at all) as a highly qualified, probably exceptional, non-juridical privilege rather than as a meaningful, justiciable aspect of child welfare paramountcy. Where blood-ties are treated as paramount however, an automatic and near-inviolate form of legal relatedness may arise, which can withstand challenges based on child protection concerns. Put simply, vulnerable children may be exposed to dangerous situations involving abusive or neglectful birth relatives, exposed to unwanted contact with genetically related yet estranged or indifferent kin, or suffer the loss of strong emotional attachments that have arisen between them and their substitute parents.

Adult-centric considerations such as lack of consent, flawed notice, time limits, cultural heritage and parental rights, may relegate child welfare to a subordinate position. The six principle 'checklist' proposed here aims to re-frame child welfare paramountcy in blood-tie cases as a 'heavier' stand alone rights concept. This could perhaps have the effect of re-balancing the rights of triad members, in a manner that better favours child welfare, especially in respect of meeting the unique vulnerabilities of relinquished or removed children, who tend to enter the triad as its weakest members. Their welfare-based needs should, where appropriate, include access to an authentic genetic identity, the maintenance of strong attachments to their substitute carers and the achievement of psychological well-being.

As Strathern has recently argued, children should be regarded in law as a 'category with special interests.... a particular nexus of problems (arises) where children are concerned'.[144] Relinquished children should be given sufficient support to cope long-term with their lack of authentic 'narrative'.[145] They should not feel compelled to engage in the construction of an 'artificial self'[146] simply by virtue of their birth status. A domestic checklist aimed at reminding jurists, social workers and parents that genetically kinless children have unique vulnerabilities, and therefore often require extra protection, ought not to be perceived as a radical change to law and policy. It would be unlikely to alter significantly those outcomes that are currently likely under the various legal frameworks for enabling social kinship. It might however enable a wider range of possible outcomes in certain situations, such as allowing for information release, long-term fosterage or indirect forms of kin contact, which might in turn lead to less adult centric solutions to disputes involving two sets of conflicted parents.

Although this might be seen as adding a degree of uncertainty to the process, for example by granting decision-makers more discretion than they might already have, the checklist is nevertheless aimed at specifically focusing more attention on the unique difficulties that tend to attach to genetic relinquishment. In some of the cases referred to here, contact or information release would have been entirely appropriate and in keeping with the best interests of the child principle, had courts not been constrained by domestic statues and precedents that placed a higher priority on adult interests such as privacy. Complete genetic relinquishment should be a rare occurrence, confined to exceptional circumstances; identifying information and original birth certificates ought to be available normatively on gaining the age of majority. Kin contact should be framed as beneficial, subject only to the best interests of the child rather than the need to effect parental reassurance or preserve finite resources. Long-term fosterage should be considered as a workable alternative for resolving those disputes that have become so intractable that the welfare of the child has become jeopardised.

In some cases, judges have already shown a willingness to delegate responsibility for finding workable, compromise-based solutions to parents, in the hope that some agreement might perhaps be reached in respect of sharing some of the parental responsibilities involved in caring for the child. As the court noted in *Re W* a reasonable parent is one who 'places great weight on what will be best for the child'.[147] A specific checklist aimed at protecting the origin deprived child could

[144] Strathern (2005), p. 3.

[145] Dunbar and Grotevant (2004), pp. 135–161 at p. 135.

[146] Lifton (1990), pp. 85–92 at p. 85. See also Patel (2005), p. 332 on how 'data based on assumptions' might be used to fill identity narrative gaps.

[147] *Re W* [1971] AC 682. This was followed in *Re D* [1977] AC 602 and withstood the challenge posed to it by the case of *Re C* [1993] 2 FLR 260.

serve as a reminder that, in many circumstances, the needs of the child demand good parenting rather than simple exclusivity of parenthood. Equally, whilst parentage may be relinquished, this should equally not absolve genetic parents of their obligations to prevent origin deprivation in their descendants.

Arguably, such a checklist may have the effect of 'labelling' some children, or of promoting the assumption that they might be generally 'at risk' and in need of special domestic law or policy protection.[148] Given the nature of modern adoption practice however, many such children might have already spent time on a child protection register before being finally relinquished, removed, freed for adoption, or placed with their substitute carers or prospective parents. On this basis, it does not seem entirely inappropriate that they might be afforded a similar degree of protection to other children potentially facing origin deprivation, such as those within systems of indigenous, mixed-race or 'international' adoption. Placing donor gamete or surrogacy children automatically into such a *de facto* 'at risk' category might be more problematic; recipient parents may regard this as an unnecessary, unjustified intrusion into their private family lives. Difficult issues surrounding enforcement and monitoring would also be likely to arise. Categorising such children as automatically prone to the vulnerabilities related to genetic kinlessness might also be seen as implying that families formed in this manner are perhaps inherently prone to the 'triadic problems' that have arisen in adoption practice. A number of adoption law reform NGO's have effectively assumed this stance in respect of donor anonymity,[149] given the parallels between it and sealed records adoption.[150] From the vetoed perspective of the origin-deprived individual, inability to access biological ancestry or identify genetic relatives is clearly the same, not least in relation to questions of best interests and welfare.[151] Birthplace or birth status should not be the sole or chief determinant of whether authentic genetic ancestry or natal kinship will be completely denied or actively protected in domestic law. As was recently argued in the European Court, 'very weighty reasons need to be put forward before a difference in treatment on the ground of birth-out-of-wedlock could be regarded as compatible with the Convention.'[152]

[148] Such an outcome is not however entirely out of keeping with the wording of Article 20 (1) of the Children's Convention.

[149] See for example the pressure group work of the Canadian NGO '*Parentfinders*' in lobbying for legislative change on closed records and the donor anonymity surrounding reproductive and genetic technologies and in drafting Recommendations on compliance with the various provisions of human rights law. See for example http://www.gov.mb.ca/legislature/hansard/3rd-36th/la_005/la-005.html (accessed 12.08.09).

[150] See further Blyth (2002), p. 797; Brewaeys (2001), pp. 38–46. Grace et al. (2008), pp. 301–314; Jones (2007).

[151] This question may be largely moot given that the majority of donor-recipient families are unlikely to require a Court Order to achieve legal relatedness.

[152] *Pla and Puncernau v Andorra* [2004] 2 FCR 630. See also the Dissenting Opinion of Garlicki J and the partly Dissenting Opinion of Bratza J.

References

Bainham, A. (2008, July). Arguments about parentage. *Cambridge Law Journal, 67*(2), 324.

Bainham, A. (2011). Damned if they do and damned if they do not: holding social workers to account. *Cambridge Law Journal*, 312–314.

Blyth, E. D. (2002). How it feels to be a child of donor insemination. *British Medical Journal, 324*, 797.

Brewaeys, A. (2001). Review: Parent–child relationships and child development in donor insemination families. *Human Reproduction Update, 7*, 38–46.

Carsten, J. (2000). 'Knowing where you've come from': ruptures and continuities of time and kinship in narratives of adoption reunions. *Journal of the Royal Anthropological Institute* (*NS*), *6*, 687–703.

Dunbar, N., & Grotevant, H. D. (2004). Adoption narratives: The construction of adoptive identity during adolescence. In M. W. Pratt, & B. H. Fiese (Eds.), *Family stories and the life course: Across time and generations* (pp. 135–161). Erlbaum, Mahwah.

Grace, V. M., Daniels, K. R., & Gillet, W. (2008). The donor, the father, and the imaginary constitution of the family: Parents' constructions in the case of donor insemination. *Social Science and Medicine, 66*, 301–314.

Grotevant, H. D. (2003). Counselling psychology meets the complex world of adoption. *The Counselling Psychologist, 31*, 753.

Herring, J. (2009, April 6). Revoking adoptions. *New Law Journal.*

Horgan, R., & Martin, F. (2008, September) The European convention on adoption 2008: Progressing the children's rights polemic. *International Family Law*, 155–162.

Jones, C. (2007). *Why donor insemination requires developments in family law—the need for new definitions of parenthood.* Lampeter: Edwin Mellen.

Kirkman, M. (2003). Parents' contributions to the narrative identity of offspring of donor-assisted conception. *Social Science and Medicine, 57*, 2229–2242.

Lifton, B. J. (1990). The formation of the adopted self. *Psychotherapy in Private Practice, 8*, 85–92.

Mykitiuk, R. (2002). Beyond conception: Legal determinations of filiation in the context of assisted reproductive technologies. *Osgood Hall Law Journal, 39*(4), 771–781.

Oren, L. (2001). Righting child custody wrongs: The children of the "disappeared" in Argentina. *Harvard Human Rights Journal, 14*, 123–195.

Parliamentary Assembly of the Council of Europe. (2000) *International adoption: Respecting children's rights.* Doc 9377 Recommendation 1443 (26 January 2000, 5th sitting).

Patel, T. (2005). The usefulness of oral life (hi)story to understand and empower: The case of trans-racial adoption. *Qualitative Social Work, 4*, 327.

Powell, K., & Afifi, T. D. (2005). Uncertainty management and adoptees' ambiguous loss of their birth parents. *Journal of Social and Personal Relationships, 22*, 129–151.

Ryburn, M. (1995). Adopted children's identity and information needs. *Children and Society, 9*(3), 41–64.

Sloan, B. (2011). Unmarried fathers and the frustration of family life. *Cambridge Law Journal*, 314–317.

Strathern, M. (2005). *Kinship, law and the unexpected: Relatives are always a surprise.* New York: Cambridge University Press.

Verrier, N. (1993). *The primal wound: Understanding the adopted child.* Lafayette, CA: Verrier.

Verrier, N. (2003). *Coming home to self: The adopted child grows up.* Baltimore: Gateway Press.

Cases

Ireland

In Re JH [1985] IR 375
M.R & Anor -v- An tArd Chlaraitheoir & Ors [2013] IEHC 91
N & Anor v Health Service Executive & Ors [2006] IESC 60
N v Health Service Executive [2006] 4 IR 372
Northern Area Health Board v An Bord Uchtála [2002] 4 IR 252

United Kingdom

A v P [K v Sheffield City Council [2011] EWCA Civ 635
Down Lisburn Health and Social Services Trust v H [2006] UKHL 36
In Re G (Children) (FC) [2006] UKHL 43
Re A [2012] EWHC 1689 (Fam)
Re B (A child) [2009] UKSC 5
Re C (A Child) v XYZ CC [2007] EWCA Civ 1206
Re C [1993] 2 FLR 260
Re D [1977] AC 602
Re D [2006] EWHC 3545 (Fam)
Re EFB (Freeing Without Consent) [2009] NIFam 7
Re F (Infants) (Adoption Order: Validity) [1977] Fam 165
Re F (R) (An Infant) [1970] 1 QB 385
Re H [2012] EWCA Civ 743
Re H: Re G [2001] 1 FLR 646
Re J (Care Order) [2008] NIFam 11 (1 August 2008)
Re JJ [2009] NIFam 2
Re K (Adoption and Wardship) [1997] 2 FLR 221
Re K (Non-accidental injuries: Perpetrator: New Evidence) [2004] EWCA Civ 1181; [2005] 1 FLR 285
Re K [2007] EWHC 393 (Fam)
Re K [2011] EWCA Civ 1064
Re L (Commercial Surrogacy) [2011] 1 FLR 1423
Re L [2007] EWHC 1771 (Fam) (20 July 2007)
Re L2 and O [2007] NI Fam 12
Re P (A Child) [2008] EWCA Civ 499
Re R (A Child) [2003] EWCA Civ 182
Re R [1993] 2 FLR 762
Re RA (Minors) (1974) 4 Fam Law 182
Re RLS (a minor) (2006) 218 2 d 428
Re S (Parental Order) [2009] EWHC 2977 (Fam)
S v L [2012] UKSC 30
SB v A County Council; Re P [2008] EWCA Civ 535
SEHSST v LS [2008] NI Fam 12
SEHSST v LS [2009] NI Fam 14
Uyanik v Turkey (3 May 2012) (app 60328/09)
Webster (The Parents) v Norfolk County Council & Ors (Rev 1) [2009] EWCA Civ 59
Z County Council v R [2001] 1 FLR 365

European Court of Human Rights

Anayo v Germany [2010] ECHR 2083 (21 December 2010)
Frette v France [2004] 38 EHRR 21 (42)
Gnahoré v. France, (2000) no. 40031/98, § 59, ECHR 2000-IX

Godelli v Italy ECHR 347 (2012)
Gorgulu v Germany (Application No 74969/01) [2004] 1 FLR 894
Johansen v Norway [1996] EHRR 31 (app no 17383/90)
Jucius and Juciuvienè v Lithuania (2008) App no 14414/0
Kearns v France (2008) ECHR (Application no 35991/04)
Keegan v Ireland (1994) (App no 16969/90) 18 EHRR 342
Kroon v The Netherlands (1994) 19 EHRR 263
Marckx v Belgium (1979) 2 EHRR 330 (application no 6833/74)
Mikulic v Croatia (2002) (no 53176/99) 54 ECHR
Clemeno v Italy (2008) Judgment 21.10.2008 [Section II] No 19537/03
Neulinger v Sweden [2010] 54 EHRR 1087
Odièvre v France 1 FLR 621 (Application no 42326/98) (2003)
P C and S v United Kingdom (2002) 35 EHRR 1075
Pini v Bertani v Romania (2005) 2 FLR 596
Pla and Puncernau v Andorra [2004] 2 FCR 630
R and H v United Kingdom [2011] 35348/06 ECHR 844
S H v Austria [2011] 57813/00 ECHR 1878
Scozzari and Giunta v. Italy (2000) ECHR 39221/98 and 41963/98, 169
Uyanik v Turkey (3 May 2012) (app 60328/09) ECHR 196
X v Croatia [2008] ECHR 641(app no 12233/04) (17.07.2008)
Y C v United Kingdom [2010] (App 4547/10) ECHRR

United States
In Re A M (Yolo County DESS v Pitt River Tribe, (2013) CO70782 (SC JV08500)
In re Adoption of Crews (1992) 118 Wash.2d 561 [825 P.2d 305]
In re H.R. (2012) 208 Cal.App.4th 751
Lehr v Robertson (1983) 463 us 248, 256
Matter of Adoption of Baby Boy L. (1982) 231 Kan. 199 [643 P.2d 168]
Michael H. v. Gerald D. (1989) 491 U.S. 110, 123 [105 L.Ed.2d 91, 109 S.Ct. 2333 *Stanley v. Illinois* (1972) 405 U.S.
Re Custody of MCC v Aisha Umer (2008) 06 D 79411
Thurnwald v A E (2007) UT 38, 163 P 3d 623

Chapter 10
Conclusion: Preventing Origin Deprivation

> Definitions of kinship and attitudes toward adoption in society at large have never ceased to emphasize the blood relation.[1]

10.1 Introduction

The blood-tie matters, in terms of law, social policy, identity rights, socio-cultural customs and attitudes, and psychological welfare. The argument that an accurately 'narrativized'[2] background may help to prevent or minimise the worst harms of origin deprivation and promote outcomes based upon child welfare paramountcy, also seems to be gaining ground. Genetic connection increasingly features in judicial discourses on the best interests of the child, across a variety of contexts, not least in respect of 'the right to know' one's original heritage and ancestry. As Strathern has observed, many of the modern dilemmas created by the new reproductive technologies represent a 'recent version of a long-standing preoccupation with knowledge.'[3] The mythologies and mistakes of 'classic' adoption practice (i.e. permanently sealed records, complete relinquishment) could easily be mirrored by some of the modern, scientific methods of creating kinship.

In relation to the origin deprived child's emotional welfare needs for example, some decision-makers seem at times content to largely avoid the issue of how the potentially harmful presence of a 'hereditary ghost'[4] might serve to impact upon families formed via social rather than biological kinship. This is so despite the fact

---

[1] Wegar (1997), p. 43.

[2] See White (1980), pp. 1–23.

[3] Strathern (2005), p. 3.

[4] Levy-Schiff (2001), pp. 97–104 at p. 103; see further Morgan (1877); Lifton (2010), pp. 71–79 on how all members of the triad might be affected by unresolved grief over the losses associated with adoption.

A. Diver, *A Law of Blood-ties - The 'Right' to Access Genetic Ancestry*,

DOI 10.1007/978-3-319-01071-7_10,

that profound 'developmental deficits'[5] may occur in the absence of a strongly developed sense of authentic identity. The ability to successfully 'forget origin'[6] and cope with the losses of genetic heritage and ancestry may well depend on how families have managed to address the 'mental-emotional presence'[7] of absent or unknown genetic relatives. Acknowledgement of the 'genetic gap'[8] has been suggested by a number of theorists as the best way of preventing nameless ancestors or unwanted ancestry from assuming a potentially malign 'phantasmatic role'[9] within the new, socially 'constructed' substitute family unit.[10] Suffering 'stigma potential' throughout one's lifetime might also be harmful to other members of the triad.[11]

As many of the customary, de facto adoption models described in chapter one plainly demonstrate (through the use of healing or cleansing magic, nurturing or protective rituals, or contract-binding fluids) a significant level of loss, pain and fear might attach to the concept of substitute parenthood and to the act of social parenting. The child may for example have been orphaned, abandoned, abused or unwillingly relinquished by the original family; genetic parents may regret their decision to forsake their child (and their biological parenthood) while substitute parents might have had to face their own disappointments, losses and fears, perhaps in terms of facing their biological infertility or in overcoming some sense of 'otherness' that might attach to their non-genetic form of parenthood. Substitute, social parenting may demand that a high level of exclusivity is achieved or maintained in terms of care-giving; this might easily be challenged when children emerge from childhood and perhaps start to seek out information on their genetic origins and ancestry. Curiosity over biological heritage might be taken by some parents (or outside observers) as a sign of ingratitude or as a form of rejection of the social family. As Raynor observed,

> In our society, love is very much bound up with exclusive possession..it hurts most adoptive mothers to acknowledge that another woman has a claim to their child and that the child might one day recognise her claim.[12]

Even where social kinship has not been underpinned by a sense of fear, loss or grief, there is every chance that an absence of 'biological mutuality'[13] may serve to create if not embed a permanent 'marker for difference.'[14] Where, for example,

[5] Johnson (2002), pp. 39–54 at p. 39.

[6] Brandon and Warner (1977), pp. 335–341 at p. 338.

[7] Bonney (2002), pp. 201–208 at p. 207.

[8] Franklin (2003), pp. 65–85 at p. 83.

[9] Grace (2008), pp. 257–262 at p. 261.

[10] Haimes (1988), pp. 46–61 at p. 47; see further Hargreaves (2006), pp. 261–283; McGee et al. (2001), pp. 2033–2038.

[11] Miall (1987), pp. 34–39 at p. 34.

[12] Raynor (1980), p. 37.

[13] Levy-Schiff (2001), n 3, p. 102; see also Weider (1978), pp. 793–811.

[14] Wegar (1997), n 1; see also Carsten (2000), pp. 687–703 on 'difference perception' p. 689.

genetically connected 'intentional families'[15] tend to see many landmark, 'clan-gathering' events such as weddings and birthdays as highly affirmational, happily shared moments of family life, some origin deprived people may regard these as unfamiliar, emotionally unsettling 'non-events'[16] which can bring with them a profound sense of 'alienation'[17] from one's social family or add to the 'threatening landscape'[18] of obvious otherness. This feeling of apartness may be reinforced by a variety of customs, laws and policies aimed at keeping some degree of distance between genetic relatives or at vetoing completely any right to onomastic identity. Verrier puts it succinctly;

> Ours is not a society which understands loss very well. Adoption is seen as a happy event which calls for celebration. . .but it is also true that there is loss involved for everyone, loss which needs to be mourned. Loss which is not mourned can be debilitating. . . . understanding, acceptance, empathy and communication are the keys to the beginning of healing.[19]

In acknowledging that a certain darkness in terms of losses, fears and 'otherness,' can attach to being genetically relinquished, the need for grieving, healing, and extra protections (rituals, laws and psycho-social support) may become clearer to those who perhaps have little or no personal experience of triadic issues or who might see adoption or gamete donation as an entirely positive experience, grounded in a sort of trouble-free 'permanence panacea.' Equally, the notion that resilience and good coping skills[20] on the part of veto-victims can fill in for their lack or loss of authentic, accurate birth records surely echoes the rather narrow argument that origin-deprived persons ought simply to be entirely grateful for their experiences and aim instead towards quickly 'getting over' any issues that they might have with the process, rather than seeking to challenge the laws and policies that have denied them ancestry. Informational disclosures (like kin reunions) do not guarantee an instant, satisfactory resolution to the many issues that can arise from origin deprivation, but they at least provide an alternative to the models for managing harmful losses based upon silent gratitude and a gentle acceptance of this unique, rights- deprived status.

---

[15] Cahn and Singer (1999), pp. 162–165 at p. 162.

[16] Lifton (1994), n 3, p. 37.

[17] Ziller (1969), pp. 287–300 at p. 287.

[18] Yngvesson and Mahoney (2000), pp. 77–110 at p. 77.

[19] Verrier (1993), p. 109; see also Verrier (2003).

[20] See for example Freidlander (2003), pp. 745–752 at p. 749. See also http://www.bastards.org (accessed 21.02.13) for an insightful, useful resource in terms of vetoes, adoptee search information, law reform updates and pithy adoptee humour.

10.2 Preventing Origin Deprivation?

There remains much scope for researchers and reformers to focus less upon the individual psychological shortcomings of the adoptee, and more on the flaws inherent to adoption processes.[21] The long-standing, socio-cultural 'norms of distance'[22] that have traditionally attached to child relinquishment and substitutive parenthood in a number of jurisdictions merit re-examination, given that an increasingly mutable concept of social kinship itself seems to be enabling more child welfare focussed models of parenting. The 'voluntarily surrendered' or orphaned newborn model of adoption, where an anonymous birth mother seeks social rehabilitation for example, bears little relevance to modern scenarios involving older, neglected or abused children.[23]

In terms of preventing origin deprivation in closed records, donor-anonymous regions, a number of frameworks are already in existence that might be potentially useful. Most domestic policies that involve blood-tie issues have traditionally been closely tied to the need to achieve workable models of child protection. The emphasis in adoption and fostering is now ostensibly on finding families for vulnerable children who have been either relinquished or removed.[24] This is inverted in respect of gamete donation, where the focus is on 'curing' adult infertility by creating families.[25] A number of key, common issues are relevant to both adoption and gamete donation however: genetic relinquishment may have long-term consequences for all parties,[26] psychological attachments between family members may be facilitated by having shared physical traits,[27] broken attachments may lead to emotional harm,[28] and identity formation may require that informational disclosures occur, perhaps against the wishes of the social or genetic parents.[29]

To an extent, the darker elements of 'adoptionology' have perhaps served as the main catalyst for informing thinking in this area. Domestic laws or policies which frame post-adoption contact as an essential pre-requisite to ordering that a child should be freed for an adoptive placement or which encourage gamete recipient parents to use non-anonymous donor systems for conception and to tell their

[21] Wegar (1997) n 1, p. xii.

[22] Alber (2004), p. 34.

[23] See further Talbot and Kidd (2004), p. 273.

[24] See further the proposed English legislative reforms http://www.education.gov.uk/a00221161/children-families-bill (accessed 23.04.13).

[25] See further http://www.ngdt.co.uk/ (accessed 12.03.13) on the UK's National Gamete Donation Trust.

[26] Triseliotis (1973); Blyth (2002), p. 797; Bonney (2002), pp. 201–208.

[27] Brown (2001), p. 538; Sandler and Joffe (1969), pp. 585–595.

[28] Baran and Pannor (1989); Powell and Afifi (2005), pp. 129–151.

[29] See further Turner and Coyle (2000), pp. 2041–2051; Turkington and Taylor (2008), pp. 21–38; Neil (2007), pp. 1–19; Carp (1998).

children of their genetic history, are clearly grounded in the belief that genetic non-ancestry can cause the child harm. Seminal studies involving children who were orphaned or unwanted[30] and attachment experiments based on situations that were essentially quite cruel, if fleeting in nature, did much to highlight the nature of adoption losses.[31] Recent scandals[32] and state apologies[33] have further highlighted how adoption 'industries' and the various institutions tasked with protecting parentless children (and 'rehabilitating' unmarried mothers) have failed triad members, not least by creating structures and sanctioning systems that often relied upon falsified records, secrecy, legal fictions, and an outright denial of the best interests of the vulnerable child.[34] In respect of finding child protective solutions, media-led 'panic legislation' has often had limited or unintentional impacts, especially where conflicting rights issues are present.[35]

In some jurisdictions increased public awareness or sympathy for the genetically kinless has translated into policy changes and legislative reforms, with many adoptees now able to access previously-sealed, original and unaltered birth certificates upon reaching adulthood, and gamete donor anonymity either being overturned or banned in respect of future practice.[36] Although the socio-cultural stigma associated with unmarried parenthood or illegitimate birth have, in most jurisdictions, generally diminished,[37] there may still however be a sense of 'otherness' or social disadvantage attached to having partial or absolute gaps in one's genetic background.[38] In closed records jurisdictions for example, origin deprived persons might have to demonstrate some manner of 'good cause' to be able to apply

[30] Bowlby (1980).

[31] Ainsworth et al. (1978).

[32] On Ireland's 'Magdalene Laundries' see further The Department of Justice and Equality Report http://www.justice.ie/en/JELR/Pages/MagdalenRpt2013 (accessed 21.04.13); see further http://www.magdalenelaundries.com/ (accessed 22.04.13).

[33] See for example Australia http://www.guardian.co.uk/world/2013/mar/21/julia-gillard-apologises-forced-adoptions (accessed 31.03.13).

[34] An absolute right to genetic relinquishment is to an extent reminiscent of Roman family law doctrine, with children seen as chattel, subject to the authority or goodwill of their parents. '*Tollere Libere*' for example permitted abandonment via exposure; a father's refusal to 'raise up' such a child visibly endorsed relinquishment. Where another man lifted up the unwanted child, adoption was effected. Genetic fathers retained a power of redemption over children subsequently sold into slavery however. See also '*patria potestas*' (family as discrete legal unit, virtually immune from State intervention) '*ius abutendi*' (the right to destroy one's 'chattels') and '*paterfamilias*' (the family patriarch's absolute power). See further Nicholas (1962).

[35] See further Freeman (1996), pp. 273–297 at p. 297; Bainham (2008), pp. 260–262.

[36] See for the UK's Adoption Act 1976 and the Human Fertilisation and Embryology Act 2008.

[37] See for example *Pratten v British Columbia (Attorney General)* [2012] BCCA 480 on gamete donation; *Kearns v France* ECHR 10 January 2008 (Application no 35991/04 on closed birth records.

[38] See for example Cotter-Busbee and De Meyer (2012).

to receive information,[39] or simply hope that some form of non-disclosure or no-contact veto has not been attached to their 'right' to access genetic ancestry.

In terms of establishing and actually embedding in domestic law a juridical right to genetic connection, a number of possibilities do exist, given the various legal frameworks that have also served to define the remit of blood-tie jurisprudence. International law instruments, such as The Children's Convention, have at least established the right to identity as an ostensibly justiciable concept, despite reluctance on the part of some domestic jurists to be fully persuaded that a right to genetic connection ought to exist.[40] The European Court of Human Rights has essentially tied the right to authentic identity heritage to the best interests of the child concept and confirmed that child welfare must be the paramount consideration in such cases.[41] Contact with natal relatives may also be a significant aspect of the right to identity, and of family and private life.[42] As such, the processes that attach to decisions on ending the ties of familial life ought not to offend against the principles of 'fair hearing' in terms of procedure.[43]

Principles of non-discrimination and equality of treatment are clearly also relevant to the argument that a normative approach to the sealing of birth records and absolute vetoes on genetic ancestry can amount to a human rights violation; such devices have led to the creation of a category of persons who have been discriminated against because of their birth status, in terms of having their rights outweighed by the interests of others on the basis of jurisdiction, birth place, birth date or triadic position. That said, decisions such as the one arrived at in *Pratten*[44] demonstrate the significant limitations of taking such an approach to the issue. Inequalities can be viewed as legally justifiable if they are presented as an essential,

---

[39] See for example *Ross v PEI* (Supreme Court, Family Division, Registrar) (1985) 56 Nfld & PEIR 248 [1985] PEIJ o 1 (PEISC) (QL) on the need to identity to enable inheritance claims, medical information, and avoidance of incest; Re *Baby Boy K* (1999) 701 NY S 2d 600 NY Fam Ct; see also however *Re Adoption of Baby Boy* S (1997) 705 A.2d.822 (NJ) Super.Ct.Ch.Div; *Kelly v Superintendent of Child Welfare and Williams* (1980) 23 BCLR 299 (SC) and *Re Adoption of BA* (1980) 17 RFL (2d) 140 (Man Co Ct) where neither applicant succeeded in showing that 'identity issues' were sufficient to open their sealed birth records; *Ross v PEI* (Supreme Court, Family Division, Registrar) (1985) 56 Nfld & PEIR 248 [1985] PEIJ o 1 (PEISC) (QL) where it was accepted that the adoptee could access her records on the basis of the need to identity to enable inheritance claims, medical information, and avoidance of incest.

[40] On the limitations of International law in this area see further Blyth and Farrand (2004), pp. 89–104.

[41] See for example *Yousef v Netherlands* [2003] 1 FLR 210; *Neulinger and Shuruk v. Switzerland* (2010) [GC], no. 41615/07, ECHR.

[42] *Anayo v Germany* [2010] ECHR 2083 (21 December. See further '*The Best Interests Of The Child In The Recent Case-Law Of The European Court Of Human Rights*' Franco-British-Irish Colloque on Family Law (May 2011) available http://www.echr.coe.int/NR/rdonlyres/E6F5D437-C49A-47C5-9772-578F54FB5C86/0/20110514_COLLOQUE_Dublin_FR.pdf (accessed 02.03.13).

[43] A fairly wide margin of appreciation not withstanding; see for example In *Re L* (*Care: Assessment: Fair Trial*) [2002] 2 FLR 730 on the need to view proceedings holistically rather than singling out particular aspects of process such as parental exclusion from meetings.

[44] *Pratten op cit* n 36.

redressive element of law or policy reform. It is difficult to see however how the distinction between adoptee non-ancestry and that arising via gamete donation is so clear-cut. Missing biological ancestry is clearly relevant to physical and psychological welfare whether it is partial or absolute in nature. If closed-records adoptees are regarded in law and policy as a socially disadvantaged group, in need of 'extra' protection and support,[45] then surely this applies to the entire cohort and not just to those individuals who were lucky enough to not have a non-disclosure veto appended to their file. If it is their lack of access to genetic truth which has made them so vulnerable, then it is difficult to see how gamete donor-children can be categorised so differently in terms of risk and potentiality of harm.

Arguably the central concern in cases such as *Pratten* may have more to do with the need to preserve parental privacy, genetic or social. Whether or not this is an underpinning feature of cases which involve a bid to access genetic ancestry, it seems that the right to privacy in this context has achieved a near-inviolate status.[46] This contrasts quite sharply with how the concept of a right to privacy in other contexts seems to exist as a highly qualified entitlement which can be side-lined by weightier interests (the need to prevent crime, guard public morals or protect the rights of others for example) or set aside where the best interests of the child clearly demand such an outcome.[47] Whilst no-one should be forced to parent against their will, not least for the sake of protecting unwanted (and thus highly vulnerable) children, the act of denying genetic parentage has essentially been enshrined as an absolute right in a number of jurisdictions. This right is founded upon a highly elaborate legal fiction and frequently enabled by a permanent subjugation of one section of the population; such templates for enabling social kinship have also tended to assume that an abdication of the responsibilities of legal parenthood will include an unqualified right to nullify the biological reality of one's parentage.

In the interests of achieving absolute equality of treatment, it could be hypothetically argued that *all* birth records in such contexts ought to perhaps be permanently sealed, with such a decision being justified on the basis of all, or indeed perhaps any, of the following presumptions:

a) adoptees and origin deprived persons may be better off simply not knowing about their genetic ancestry;
b) genetic parents have an unqualified, absolute right to privacy that simply outweighs the various rights of their descendants;
c) substitutive parenthood and legally created forms of parentage carry with them the inviolable right to parent autonomously, free from such extraneous considerations as the child's 'need' to know biological origins

It is very unlikely that such a blunt template for sealing the birth records of all adoptees and gamete donor children across the various open-records jurisdictions

[45] *Pratten op cit* n 36.

[46] *Kearns op cit* n 36.

[47] See for example *Re G* [2013] EWHC 134 (Fam).

would garner much widespread societal support. In terms of promoting child welfare and determining or balancing out more fairly triadic rights however, these seem to be the main arguments that have traditionally served to keep birth records permanently closed or subject to adult-centric vetoes in a number of regions. Arguing a right to genetic identity purely on the basis of the principle of equality, seems to offer an unlikely means of challenging those law and policy norms that prize traditional, triadic rights hierarchies.

Similarly, as individual rights-claimants, many origin-deprived persons cannot seek to argue that they have been dispossessed of shared birthright or common cultural heritage in the way that indigenous groups might be able to do. Profound, permanent loss of genealogical heritage seems to be more easily framed as an unjustifiable rights violation where, for example, non-consensually assimilated indigenous peoples have been permanently cut off from their socio-cultural origins or clan links.[48] Even if adoptees were theoretically able to attain the rights-bearing status of, for example, a sort of politically or territorially displaced *ethnie*[49] they would probably still have to demonstrate some link between the violation of their individual heritage identity rights and the harmful losses of some tangible benefit, such as traditional territorial or nationality claims. A simple history of cultural or social disadvantage seems unlikely to generate the same level of urgency or need for redress.[50] The effects of suffering genetic non-ancestry in private, family law contexts, often tend however to be just as irreversible as the losses brought about by civil or political rights violations.

Embedding a norm of preserving genetic connection (unless exceptional circumstances demand a child-centric veto on the blood-tie) clearly requires changes at the level of domestic law and policy. Interpretation of the best interests principle by national courts, legislators and policy-makers, is often subject to the wide discretion of domestic decision-makers, who may perhaps opt to delegate the more difficult choices on information release or kin contact to the child's parents or to social workers. Child welfare may simply not be seen as the paramount consideration or it may be defined by reference to what the parents feel to be the best outcome in any given case. Judicial reluctance to intervene too much in such private law matters or to be seen to be interfering with the sanctity of family life and the notion of parental responsibility may further compound this.

Any human rights-led argument for removing such absolute bans on genetic ancestry will most likely also be subject to some degree of rights-balancing exercises by domestic decision-makers. As the case law from the European Court of Human Rights further suggests, a considerable margin of appreciation also attaches to such issues as the need for genetic connection and domestic interpretations of the best interests principle. Private, family law matters are arguably perhaps also subject to a further layer of protection against state interference by virtue of their tendency to be

[48] Samson (1997), p. 226.

[49] Eriksen (2004), pp. 49–62.

[50] See *Pratten op cit n 36*.

categorised alongside qualified socio-economic 'second generation' entitlements rather than as the less easily interfered with civil or political rights.[51]

The United Kingdom's system of open records and sperm donor registration could perhaps serve as a model for the unsealing of birth records and genetic information in other jurisdictions. At the level of domestic proceedings and court hearings, cases involving such matters as familial contact and information release have increasingly adhered to the principle that the child's welfare is the paramount consideration. This has meant that birth-parental rights can in exceptional circumstances be removed, to protect the best interests of the vulnerable child and allow for substitute care-giving.[52] Contact post-adoption is also not automatically excluded as a possible outcome in many cases. Although the UK system is far from perfect,[53] it is at least largely underpinned by an essentially child-centric focus rather than simply tying the genetic identity rights of the child to such 'transactionary contract'[54] issues as adult consent, time limits, constitutionally enshrined, inviolate parental rights or a redress-led obligation to achieve preservation of indigenous cultural heritage.

These adult-centric models for achieving some level of 'blood-tie paramountcy' offer one means of enshrining the right to genetic connection in domestic law but they do so in some cases at too high a price. An absolute right to be genetically connected could easily be deemed to operate as a 'two-way' system of entitlement; such a framework would be likely to place domestic judiciaries under a positive obligation to overturn many, if not all, contested (or otherwise 'unconstitutional') adoptive placements. Courts should not be compelled to facilitate potentially dangerous instances of direct or indirect contact between birth relatives simply on the basis that the right to genetic connection demands it, any more than they should assume that genetic relatedness implies an ability to parent adequately. Child welfare must not be subordinated to adult interests; the best interests of the child are either the paramount consideration for decision-makers or they are not. If they are not, they run the risk of being completely side-lined, outweighed or vetoed by adult-centric rights concepts.

The establishment of a domestically justiciable right to genetic connection, as a qualified, yet still fundamental right (which might be outweighed only by an absolute obligation of adherence to the child welfare paramountcy principle) could however serve to prevent the harsher forms of origin deprivation that can flow from absolute vetoes on ancestry. Definitions of child welfare should also aim to encompass more fully the psycho-social aspects and longer term consequences of

[51] On the different categories ('generations') of human rights see Alston and Robinson (2005); Alston (2005), pp. 755–829. See also Alston (1984), pp. 607–615.

[52] See for example *S v L* [2012] UKSC 30 (11 July 2012); *R and H v United Kingdom* (2011) 54 EHRR 28; *K (Children), Re* [2011] EWCA Civ 1064 (20 July 2011).

[53] See for example *A & S (Children) v Lancashire County Council* [2012] EWHC 1689 (Fam) (21 June 2012).

[54] Demian (2004), n 22, p. 100.

origin deprivation, including the unique psychological needs of biologically relinquished children. The duty to prevent where possible the harmful 'loss of genetic continuity'[55] ought to be a priority for all involved in child welfare decision-making processes. The six principles for guidance outlined in the previous chapter might go some way towards enabling a more child-centric approach to the issue of origin deprivation at the level of domestic decision-making.[56] Hard cases do not make for easy authority; in a number of the cases referred to earlier, domestic courts may have been bound by their legislative remit as to what outcomes could be ordered. The presence a Care Order for example may significantly curtail possible solutions. Similarly, kin contact may simply not be a workable option; it might well be unethical to impose certain sanctions, for example in respect of mothers who have refused to name the father of a child or where birth parents do not wish to have any contact whatsoever with their relinquished child.

Equally, children may express a desire to remain or stay in contact with abusive genetic relatives in situations where their best interests clearly demand otherwise. Affording participation rights to vulnerable children in such contexts only becomes a viable option once the child has gained a sufficient level of understanding. That said, in a number of the cases referred to, an order for some level of contact might have served to maintain basic ties of genetic kinship and information. Similarly, in those cases where the child had to be returned to an estranged parent after years spent in an adoptive placement, an order for contact could have been appropriate to avoid the child suffering the loss of her attachments to her psychological parents. Such an arrangement is likely to be viewed as tantamount to shared parenting and therefore unwelcome and unacceptable. From the child's point of view however, had it been in their best interests, such a solution might have represented a workable compromise, preferable to the 'all or nothing' stances taken by the courts in some of the King Solomon-like cases involving disputed parenthood or parentage. A considerable degree of maturity on the part of the two sets of warring parents would be required however to allow such a result.[57]

In respect of defining what is meant by the phrase 'best interests,' Miller has sounded a further note of caution, arguing that the psychological values underpinning the notion of the child's best interests principle often differ from those that attach to its legal definition. Expert witnesses may, for example, feel restricted

[55] Lifton (2010), n 3, p. 37.

[56] 1. The fundamental right to an authentic, ancestral identity is often a key component of child-welfare led best interests and the right to family and private life. 2. In terms of realising the right to identity, the best interests of the child may be closely tied to such issues as contact, 'potentiality' of relatedness or relationship, or to procedural matters such as passage of time. 3. The welfare of the child is the paramount consideration. 4. The best interests of the child may require informational disclosure given the significance of the genetic connection and that a positive, juridical obligation to preserve or repair this connection exists. 5. Loss of genetic connection ought to occur only in exceptional circumstances. 6. Preserving genetic connections should not compromise the principle of child welfare paramountcy.

[57] See however *Re Bridget R et al (Minors)* (1995) 25 USCA 1901 BO93520.

during legal proceedings to examining only what might best protect the child's short-term psychological welfare or 'emotional well-being'. Judiciaries however are permitted to examine a wider range of secondary welfare issues, such as the child's opportunities for education, financial benefit, or the current state of an adoptive parent's health.[58] Psychologists might pragmatically find that 'the effects of poverty on a child are more damaging than the benefit of visitation.'[59]

This is a telling observation; it reveals the more powerful socio-cultural position that an adoptive family might hold in comparison to natal relatives. Through a more secure financial position, they might be able to overcome infertility and achieve parenthood. Genetic relatives might be regarded as able to provide only 'visitations', which carry no discernible material gain (and, in fairness, may actively serve to upset the child or harm the security of the adoptive placement). Equally, 'legal interpretations should (not) cloud psychiatric judgments' on important child welfare matters such as disputes over kin contact.[60] Neither scenario of 'forever family' or birth family reunion necessarily provides a complete solution to all the ills of origin deprivation,[61] nor will all adoptees or donor children choose to seek out information on (or make contact with) their genetic relatives.[62] Even if only a minority of relinquished children are likely to search as a result of curiosity or through experiencing 'fractured',[63] 'discontinued'[64] or 'dual identities,'[65] a largely preventable form of loss is still occurring for a significant section of the population.[66] Irrespective of whether reunions turn out to be positive or negative experiences,[67] in many jurisdictions origin deprived persons are being actively prevented from accessing essential information, often with no legal means of challenging this loss or of seeking redress for the harms that tend to be associated with it. The state's positive obligation to protect child welfare may yet prove to be the main catalyst for reform of this area of law, provided that jurists and policy-makers accept that the psycho-social significance of genetic identity merits greater recognition than it currently tends to receive. If it is accepted that significant harm can flow from genetic kinlessness or a lack of accurate information, then it follows that the loss of genetic identity should be framed in law and policy as an exceptional rather than normative occurrence.[68]

---

[58] Miller (2002), pp. 196–200 at p. 197. See further Eekelaar (1994), pp. 42–61.

[59] *Ibid.*

[60] *Ibid* p. 198.

[61] See for example Rushton (2007), pp. 305–311.

[62] See for example Schooler (2008), pp. 111–123.

[63] Parkes (2004), pp. 587–615 at p. 600.

[64] Ryburn (1995), pp. 41–64 at p. 42.

[65] Levy-Schiff (2001), n 3, p. 102.

[66] See further Feast and Howe (1997), pp. 8–15; Schechter and Bertocci (1990).

[67] See for example Kirton et al. (2000), pp. 6–18 on the motivation behind searches.

[68] See for example Kim et al. (2010), pp. 179–190 on the perception of group meanings as to race and culture; Hill and Edwards (2009), pp. 45–53 on the potentially adverse consequences of a lack of information on the child's health history and the ability of adopters to offer 'therapeutic parenting'.

There is clearly a limit as to what law and policy reforms can achieve in terms of protecting vulnerable children, given that human nature at times seems to tend to prize possessive, exclusivity of parenthood over outcomes that might serve to better further child welfare paramountcy. Where courts have looked primarily to the behaviour of the warring parents to gauge which side might better merit custody of a fought-over child, a sort of adult-centric *terra nullius* approach to parenthood and parenting perhaps seems to apply, with parental rights only really arising where they have been actively 'earned' through the worthy endeavour of caring for the child or demonstrating an ability to commit to relatedness and to providing a 'better' form of parenting. The loss of a child 'to adoption' has many causes; child abuse or abandonment are often key issues but these are not necessarily the only factors that can lead to the removal or curtailment of parental rights, as state apologies on 'forced adoptions' demonstrate.[69]

The achievement of meaningful reforms in respect of embedding child welfare paramountcy and banning vetoes on genetic ancestry is perhaps likely to involve a much wider range of 'subtle and sophisticated factors'[70] than law and policy changes alone might be able to effect.[71] Within a rights-balancing template, the needs of the child might easily become obscured; as King and Trowell observed (albeit in respect of relationship breakdown hearings) child welfare might be better protected via non-legal processes, especially where wide ranging judicial discretion or issues such as finite resource-provision seem likely to feature.[72] Where courts have some degree of discretion over child welfare issues, domestic checklists might at least serve to focus legal analyses on finding solutions based squarely on the best interests of the vulnerable child, rather than on weighing out conflicted sets of rights. Arguably, decision-makers (whether in the courtroom, at pre-hearing negotiations or in the family home) might also be less able or inclined to disregard the psychological evidence of expert witnesses or to find that the notion of new family 'permanence' carries with it an all-encompassing form of perpetual stability that will simply yet adequately address the unique needs of the origin-deprived child throughout their lifetime.

The option of simply letting the passage of time erode meaningful ties between genetic relatives might also become less of a politically correct option, especially in situations where it would take very little to maintain some basic level of connection and potentiality for eventual reunion, for example via indirect contact, or through access to an unaltered birth certificate or accurate record-keeping. A 'regime of secrecy which could only be removed by court order or, occasionally, through the

---

[69] See the Australian Government Apology in respect of forced adoptions available http://parlinfo.aph.gov.au/parlInfo/download/chamber/votes/2013-03-21/toc_pdf/RVPF160.pdf;fileType=application%2Fpdf#search=%22apology%22 (accessed 23.04.13).

[70] Carsten (2004), n 13, p. 183.

[71] Mnookin and Kornhauser (1979), p. 950.

[72] King and Trowell (1992).

discretion of social work agencies and their directors'[73] is likely to continue for as long as birth records are being normatively sealed rather than left open, to be closed only in exceptional circumstances and on the basis of protecting child welfare. Such an inverted template could in theory shift the balance of power in favour of the triad child, with relinquishing parents having to show 'good cause' to maintain their anonymity or veto informational disclosure. Social parents could similarly find themselves legally obliged to inform their children of their genetic origins and possibly facilitate some form of contact between natal relatives unless the best interests of their child vetoed such an outcome.

The key issue is perhaps not one of rights but of parental expectation of entitlement. The concept of having rights 'to' and 'over' our children (access, custody, control) does not leave much room for the idea of somehow sharing them with others, especially those unseen or unwanted others who might only have an 'invisible' genetic connection to the child, which nevertheless represents an unknown or perhaps undesirable past. This original heritage may be heavy with meaning and exert a powerful influence over the child, the created family and their collective future, on both psychological and socio-cultural levels. The concepts of shared parenthood and communal parenting perhaps require thinking that looks beyond that of traditional, Westernised legal and social frameworks. To an extent, notions such as that of the 'allomother'[74] have already been introduced into some domestic law frameworks in terms of defining shared legal maternity post-surrogacy or egg donation.[75]

Similarly, a number of customary, de facto adoption models allow for a degree of communal parenthood, and do not necessarily require total severance of the genetic tie but are aimed at keeping the child's connection to heritage intact; they also generally endeavour to further the welfare of the child, even if they do not explicitly highlight the best interests principle as a central, defining concept.[76] In respect of minimising origin deprivation in other settings, a long-term fosterage or guardianship model, not dissimilar to the *Kafalah* system of substitute kinship, might allow for better protection of the child's best interests, given that these types of substitute care-giving do not demand complete cauterisation of the child's original blood-ties.[77] Special Guardianship, long-term fosterage and kinship care models do already offer workable alternatives to clean-break adoptions and potentially dangerous non-intervention although, as with every system that aims to

---

[73] Baldassi (2004–2005), pp. 212–265 at p. 218.

[74] Blaffer (2003) (as cited by Woodhouse (2005–2006), pp. 298–329 at p. 309.

[75] The Human Fertilisation and Embryology Act 2008, ss 42, 45, 48 allows birth mothers in civil partnerships at the time of conception, to be named alongside their partner as legal parent to the child.

[76] See for example Baldassi (2006), pp. 63–100 at p. 64; Halbmeyer (2004), n 22, p. 147.

[77] On Kafalah and best interests see *Harroudj v. France* (October 2012) App No 43631/09 available at http://hudoc.echr.coe.int/sites/eng/pages/search.aspx?i=001-113819 (accessed 1.04.13).

replicate natal ties of attachment and affection, these arrangements too are not free from difficulties in terms of ensuring that the child's welfare is the paramount consideration.[78]

10.3 Conclusion

Human nature is such that the law cannot force people to parent, nor can it easily forge a meaningful relationship between 'strangers' whose only connection might be one of biology. This does not mean however that it should simply disregard all ties of blood as being unworthy of protection, nor presume that 'placement permanence' will automatically bring about complete child protection, a sense of identity or long-term psychological well-being for all genetically kinless persons. Equally, it should not be assumed that donor children are very far removed from adoptees in terms of the losses and vulnerabilities that they might experience as a result of their unique form of origin deprivation.[79]

Preventing origin deprivation involves maintenance of the genetic connection. This may come about through domestic laws or policies, human rights balancing exercises or via parental commitment to enabling ancestral truths. Vetoes on information should be child-centric, occurring only in exceptional circumstances, and serving to promote the best interests of the child rather than the security of adults. From the perspective of the genetically-vetoed adoptee or gamete donor child, the state of 'childhood' can essentially last into adulthood, in the sense that vulnerability to harm and rights-displacement can be permanent features of their existence. Advice aimed at developing 'coping skills' and 'emotional resilience' or at promoting 'healing' is often very well-intentioned, but would perhaps be less necessary if accurate, authentic birth records were more openly available instead. Arguably, the law is unlikely to provide a universally satisfactory solution to the inequities of origin deprivation anytime soon; technological advances (such as social media, DNA testing) and the kindness of strangers (reunion 'search angels,' open records activists) now seem to be leading the way in highlighting to those who may take genetic connections for granted, the importance of 'knowing one's past.'

---

[78] On kinship care in the United Kingdom, see Nandy et al. (2011) available at http://www.buttleuk.org/data/files/Research_Documents/FULL_REPORT_finalkinship.pdf (accessed 23.04.13); see also Selwyn et al. (2013) available at http://www.buttleuk.org/data/files/Research_Documents/Ch_10__11_of_The_Poor_Relations.pdf (accessed 23.04.13).

[79] See further Nuffield Council on Bio-Ethics '*Donor Conception: Ethical Aspects of Information Sharing*' (April 2013) available http://www.nuffieldbioethics.org/sites/default/files/Donor_conception_report_2013.pdf (accessed 28.04.13). See also comments on the Nuffield Report by E Blyth available http://www.bionews.org.uk/page_290373.asp?dinfo=7qYLvnCXFa3JsxzobhayPv5J&PPID=290513 (accessed 29.04.13) and the differing perspective of C Smart available http://www.bionews.org.uk/page_290058.asp?dinfo=7qYLvnCXFa3JsxzobhayPv5J&PPID=290513 (accessed 29.04.13).

References

Ainsworth, M. D. S., et al. (1978). *Patterns of attachment: A psychological study of the strange situation*. Hillsdale: Erlbaum.

Alber, E. (2004). The real parents are the foster parents: Social parenthood among the Baatombu in Northern Benin. In F. Bowie (Ed.), *Cross-cultural approaches to adoption*. London: Routledge.

Alston, P. (1984). Conjuring up new human rights: A proposal for quality control? *American Journal of International Law, 78*, 607–615.

Alston, P. (2005). Ships passing in the night: The current state of the human rights and development debate seen through the lens of the millennium development goals. *Human Rights Quarterly, 27*(3), 755–829.

Alston, P., & Robinson, M. (Eds.). (2005). *Human rights and development: Towards mutual reinforcement*. Oxford: Oxford University Press.

Bainham, A. (2008). Removing babies at birth: A more than questionable practice. *Cambridge Law Journal, 67*(2), 260–262.

Baldassi, C. L. (2004–2005). The quest to access closed adoption files in Canada: Understanding social context and legal resistance to change. *Canadian Journal of Family Law, 21*, 212–265.

Baldassi, C. (2006). The legal status of aboriginal customary adoption across Canada: comparisons, contrasts and convergences. *University of British Columbia Law Review, 39*, 63–100.

Baran, A., & Pannor, J. (1989). *Lethal secrets*. New York: Warner.

Blaffer, S. (2003). Mothers and others. In A. Podolefsky, & P. J. Brown (Eds.), *Applying anthropology*.

Blyth, E. D. (2002). How it feels to be a child of donor insemination. *British Medical Journal, 324*, 797.

Blyth, E. D., & Farrand, A. (2004). Anonymity in donor-assisted conception and the UN Convention on the rights of the child. *The International Journal of Children's Rights, 12*, 89–104.

Bonney, H. (2002). The psychopathogenic power of secrecy: Child development and family dynamics after heterologous insemination. *Journal of Psychosomatic Obstetrics and Gynecology, 23*(3), 201–208.

Bowlby, J. (1980). *Attachment and loss (volume 3) loss, sadness and depression*. London: Hogarth.

Brandon, J., & Warner, J. (1977). AID and adoption: Some comparisons. *British Journal of Social Work, 7*(3), 335–341.

Brown, L. E. (2001, July). The role of perceived similarity to parents in adopted children's adjustment. *Dissertation Abstracts International: Section B: The Sciences and Engineering, 62*(1-B), 538.

Cahn, N., & Singer, J. (1999). Adoption, identity and the constitution: The case for opening closed records. *University of Pennsylvania Journal of Constitutional Law, 1*(50), 162–165.

Carp, E. W. (1998). *Family matters: Secrecy and disclosure in the history of adoption*. Cambridge: Harvard University Press.

Carsten, J. (2000). Knowing where you've come from: Ruptures and continuities of time and kinship in narratives of adoption reunions. *Journal of the Royal Anthropological Institute (NS), 6*, 687–703.

Carsten, J. (2004). *After kinship: New departures in anthropology*. Boston: Cambridge University Press.

Cotter-Busbee, P., & De Meyer, T. A. (Eds.). (2012). *Two worlds: Lost children of the Indian adoption projects*. Greenfield: Blue Hand Books.

Demian, M. (2004). Transactions in rights, transactions in children: A view of adoption from Papua new guinea. In F. Bowie (Ed.), *Cross-cultural approaches to adoption*. London: Routledge.

Eekelaar, J. (1994). The interests of the child and the child's wishes: The role of dynamic self determination. In P. Alston (Ed.), *The best interests of the child: Reconciling culture and human rights* (pp. 42–61). Oxford: Clarendon.

Eriksen, T. H. (2004). Place, kinship and the case for non-ethnic nations. *Nations and Nationalism, 10*(1/2), 49–62.

Feast, J., & Howe, D. (1997). Adopted adults who search for background information and contact with birth relatives. *Adoption and Fostering, 21*(2), 8–15.

Franco-British-Irish Colloque on Family Law. (2011, May). *The best interests of the child in the recent case-law of the European Court Of Human Rights*. Dublin: COE.

Franklin, S. (2003). Re-thinking nature-culture: Anthropology and the new genetics. *Anthropological Theory, 3*, 65–85.

Freeman, M. (1996). The new birth right? Identity and the child of the reproduction revolution. *The International Journal of Children's Rights, 4*, 273–297.

Freidlander, M. L. (2003). Adoption: Misunderstood, mythologized, marginalized. *The Counseling Psychologist, 31*, 745–752.

Grace, V. M. (2008). The psycho-social politics of paternity in the case of male-donated gametes. *Women's Studies International Forum, 31*, 257–262.

Haimes, E. (1988). Secrecy: What can artificial reproduction learn from adoption? *International Journal of Law, Policy and the Family, 2*, 46–61.

Halbmeyer, E. (2004). "The one who feeds has the rights": Adoption and fostering of kin, affines and enemies among the Yukpa and other Carib-speaking Indians of lowland South America. In F. Bowie (Ed.), *Cross-cultural approaches to adoption*. London: Routledge.

Hargreaves, K. (2006). Constructing families and kinship through donor insemination. *Sociology of Health and Illness, 28*(3), 261–283.

Hill, C., & Edwards, M. (2009). Birth family health history. *Adoption and Fostering, 33*(2), 45–53.

Johnson, D. E. (2002). Adoption and the effect on children's development. *Early Human Development, 68*(1), 39–54.

Kim, G. S., Suyemoto, K. L., & Turner, C. B. (2010, April). Sense of belonging, sense of exclusion, and racial and ethnic identities in Korean transracial adoptees. *Cultural Diversity and Ethnic Minority Psychology, 16*(2), 179–190.

King, M., & Trowell, J. (1992). *Children's welfare and the law: The limits of legal intervention*. London: Sage.

Kirton, D., Feast, J., & Howe, D. (2000). Searching, reunion and transracial adoption. *Adoption and Fostering, 24*(3), 6–18.

Levy-Schiff, R. (2001). Psychological adjustment of adoptees in adulthood: Family environment and adoption-related correlates. *International Journal of Behavioral Development, 25*, 97–104.

Lifton, B. J. (2010, January). Ghosts in the adopted family. *Psychoanalytic Inquiry, 30*(1), 71–79.

McGee, G., Brakman, S. V., & Gurmankin, A. (2001). Gamete donation and anonymity: Disclosure to children conceived with donor gametes should not be optional. *Human Reproduction, 16*(10), 2033–2038.

Miall, C. E. (1987, January). The stigma of adoptive parent status: perceptions of community attitudes towards adoption and the experience of informal social sanctioning. *Family Relations, 36*(1), 34–39.

Miller, G. H. (2002). The psychological best interest of the child is not the legal best interest. *Journal of the American Academy of Psychiatry and the Law, 30*, 196–200.

Mnookin, R. H., & Kornhauser, L. (1979). Bargaining in the shadow of the law: The case of divorce. *Yale Law Journal, 88*, 950.

Morgan, L. H. (1877). *Ancient society*. London: MacMillan.

Nandy, S., Selwyn, J., Farmer, E., & Vaisey, P. (2011). *Spotlight on kinship care*. Bristol: University of Bristol.

Neil, E. (2007). Post-adoption contact and openness in adoptive parents' minds: Consequences for children's development. *British Journal of Social Work*, 1–19.

Nicholas, B. (1962). *An introduction to Roman law*. Oxford: Oxford University Press.

Parkes, P. (2004). Fosterage kinship and legend: When milk was thicker than blood? *Comparative Studies in Society And History*, *46*(3), 587–615.

Powell, K., & Afifi, T. D. (2005). Uncertainty management and adoptees' ambiguous loss of their birth parents. *Journal of Social and Personal Relationships*, *22*, 129–151.

Raynor, L. (1980). *The adopted child comes of age*. George, Allen and Unwin: London.

Rushton, A. (2007). Outcomes of adoption from public care: Research and practice issues. *Advances in Psychiatric Treatment*, *13*, 305–311.

Ryburn, M. (1995). Adopted children's identity and information needs. *Children and Society*, *9*(3), 41–64.

Samson, C. (1997). Rights as the reward for simulated cultural sameness: The Innu in the Canadian colonial context. In J. K. Cowan, et al. (Eds.), *Culture and rights*: *anthropological perspectives*. Cambridge: Cambridge University Press.

Sandler, J., & Joffe, W. G. (1969). Towards a basic psychoanalytic model: The theory of parent-infant relationship. *International Journal of Psychoanalysis*, *41*, 585–595.

Schechter, M., & Bertocci, D. (1990). The meaning of the search. In: D. Brodzinsky, & M. Schechter (Eds.), *The psychology of adoption*. New York: Oxford University Press.

Schooler, J. E. (2008). Making the decision: To search or not to search. In J. E. Schooler, & B. L. Norris (Eds.), *Journeys after adoption*: *Understanding lifelong issues* (pp. 111–123). Connecticut: Greenwood.

Selwyn, J., Farmer, E., Meakings, S., & Vaisey, P. (2013, April). *The poor relations? Children and informal kinship carers speak out* (Chapters 10–11). Buttle Org, UK Summary Report.

Strathern, M. (2005). *Kinship, law and the unexpected*: *Relatives are always a surprise*. New York: Cambridge University Press.

Talbot, C., & Kidd, P. (2004, April). Special guardianship orders-issues in respect of family assessment. *Family Law Journal*, *34*, 273.

Triseliotis, J. (1973). *In search of origins: The experiences of adopted people*. London and Boston: Routledge & Kegan Paul.

Turkington, S., & Taylor, B. J. (2008). Post adoption face to face contact with birth parents: Prospective adopters' views. *Child Care in Practice*, *15*(1), 21–38.

Turner, A. J., & Coyle, A. (2000). What does it mean to be a donor offspring? The identity experience of adults conceived by donor insemination and the implication for counselling and therapy. *Human Reproduction*, *15*, 2041–2051.

Verrier, N. (1993). *The primal wound*: *understanding the adopted child*. Lafayette, CA: Verrier.

Verrier, N. (2003). *Coming home to self: The adopted child grows up*. Baltimore: Gateway.

Wegar, K. (1997). *Adoption, identity and kinship*: *The debate over sealed birth records*. New Haven: Yale University Press.

Weider, H. (1978). On and whether to disclose about adoption. *Journal of the American Psychoanalytic Association*, *26*, 793–811.

White, H. (1980). The value of narrativity in the representation of reality. In W. J. J. Mitchell (Ed.), *On narrative*. Chicago: University of Chicago Press.

Woodhouse, B. B. (2005–2006). Waiting for loving: The child's fundamental right to adoption. *Capital University Law Review*, *34*, 298–329.

Yngvesson, B., & Mahoney, M. A. (2000). As one should, ought and wants to be: Belonging and authenticity in identity narratives. *Theory, Culture and Society*, *17*, 77–110.

Ziller, R. C. (1969, September). The alienation syndrome: A triadic pattern of self-other orientation. *Sociometry*, *32*(3), 287–300.

Case Law

European Court of Human Rights

Anayo v Germany [2010] ECHR 2083 (21 December)
Harroudj v. France (October 2012) App No 43631/09
Kearns v France ECHR 10 January 2008 (Application no 35991/04)
Neulinger and Shuruk v. Switzerland (2010) [GC], no. 41615/07, ECHR
R and H v United Kingdom (2011) 54 EHRR 28
Yousef v Netherlands [2003] 1 FLR 210

United Kingdom

A & S (Children) v Lancashire County Council [2012] EWHC 1689 (Fam) (21 June 2012)
Re L (Care: Assessment: Fair Trial) [2002] 2 FLR 730
Re G [2013] EWHC 134 (Fam)
K (Children), Re [2011] EWCA Civ 1064 (20 July 2011)
S v L [2012] UKSC 30 (11 July 2012)

United States/Canada

Kelly v Superintendent of Child Welfare and Williams (1980) 23 BCLR 299 (SC)
Pratten v British Columbia (Attorney General) [2012] BCCA 480
Re Adoption of BA (1980) 17 RFL (2d) 140 (Man Co Ct)
Re Adoption of Baby Boy S (1997) 705 A.2d.822 (NJ) Super.Ct.Ch.Div
Re Baby Boy K (1999) 701 NY S 2d 600 NY Fam Ct
Re Bridget R et al (Minors) (1995) 25 USCA 1901 BO93520
Ross v PEI (Supreme Court, Family Division, Registrar) (1985) 56 Nfld & PEIR 248 [1985] PEIJ o 1 (PEISC) (QL)

The manufacturer's authorised representative in the EU is Springer Nature Customer Service Centre GmbH, Europaplatz 3, 69115 Heidelberg, Germany. If you have any concerns regarding our products, please contact ProductSafety@springernature.com

Printed and bound by CPI Group (UK) Ltd, Croydon, CR0 4YY
30/09/2026
02225216-0019